EXOTIC
SMOOTHNESS AND PHYSICS
DIFFERENTIAL TOPOLOGY AND SPACETIME MODELS

EXOTIC SMOOTHNESS AND PHYSICS

DIFFERENTIAL TOPOLOGY AND SPACETIME MODELS

TORSTEN ASSELMEYER-MALUGA
Humboldt University, Germany

CARL H BRANS
Loyola University, USA

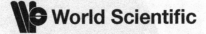

NEW JERSEY · LONDON · SINGAPORE · BEIJING · SHANGHAI · HONG KONG · TAIPEI · CHENNAI

Published by

World Scientific Publishing Co. Pte. Ltd.
5 Toh Tuck Link, Singapore 596224
USA office: 27 Warren Street, Suite 401-402, Hackensack, NJ 07601
UK office: 57 Shelton Street, Covent Garden, London WC2H 9HE

British Library Cataloguing-in-Publication Data
A catalogue record for this book is available from the British Library.

EXOTIC SMOOTHNESS AND PHYSICS
Differential Topology and Spacetime Models

Copyright © 2007 by World Scientific Publishing Co. Pte. Ltd.

All rights reserved. This book, or parts thereof, may not be reproduced in any form or by any means, electronic or mechanical, including photocopying, recording or any information storage and retrieval system now known or to be invented, without written permission from the Publisher.

For photocopying of material in this volume, please pay a copying fee through the Copyright Clearance Center, Inc., 222 Rosewood Drive, Danvers, MA 01923, USA. In this case permission to photocopy is not required from the publisher.

ISBN 978-981-02-4195-7

Printed in Singapore

For our wives
Andrea and Anna

Preface

This book is about the physics of spacetime at a deep and fundamental level, encoded in the mathematical assumptions of differentiability. We understand the phrase, "physics of spacetime" in the sense that general relativity has taught us, namely that the geometry, the topology, and now the smoothness of our spacetime mathematical models have physical significance.

Our aim is to introduce some of the exciting developments in the mathematics of differential topology over the last fifteen or twenty years to a wider audience than experts. In particular, we are concerned with the discoveries of "exotic" (sometimes called "fake" or "non-standard") smoothness (differentiable) structures on $\mathbb{R}^4$ and other topologically simple spaces. We hope to help physicists gain at least a superficial understanding of these results and their potential impact on physical theories involving spacetime models, i.e., *all* fundamental theories. Diffeomorphisms, the basic morphisms of differential topology, are the mathematical representations of the physical notion of transformations between reference frames. As we have learned from Einstein the investigation of these transformations can lead to deep insights into our physical world, as embodied for example in his General Relativistic theory of spacetime and gravity. What the mathematicians have discovered is that the global properties of diffeomorphisms are not at all trivial, even on topologically trivial spaces, such as $\mathbb{R}^4$. Yet in general relativity and other field theories physicists continue to assume that the global covering of such spacetime models with smooth reference frames is trivial. This is strongly reminiscent of the assumption of geometric triviality (flatness) of spacetime physics before Einstein. We hope that by presenting an overview of the mathematical discoveries, we may induce physicists to consider the possible physical significance of this newly discovered wealth of previously unexamined spacetime structures.

Within each section we attempt to accompany the mathematical presentation with parallel narratives of related physical topics as well as more informal "physical" descriptions if feasible. The abundant cross-fertilization of physics and mathematics in recent differential topology makes this endeavor quite natural.

Chapter 1 is an introduction, discussing the traditional interaction of physics and mathematics and speculations that this interaction extends to these differential topology results. We also review some "exotic" and unexpected or at least counter-intuitive facts in elementary topology and analysis to provide somewhat easier analogs to the more technically challenging "exotic" mathematics coming later. We then survey possible physical consequences of these unexpected structures. Chapter 2 begins with a review of some of the mathematical tools and techniques of algebraic topology. Chapter 3 concentrates on the notion of "smoothness" as defined by the introduction of differential structures on topological spaces. The field of differential topology is built on these constructions. In Chapter 4 we provide our first introductory look at some additional structures: bundles, geometry and gauge theory. These topics are considered in more detail later. Chapter 5 delves more deeply into gauge theory and introduces the concepts of moduli space especially those associated with solutions of the Yang-Mill theory. The important, but rather technical, tools associated with surgery and Morse theory are presented in Chapter 6. The remaining chapters then deal with the surprising and we believe physically intriguing discoveries of "exotic" smoothness in the unexpected context of topologically uncomplicated spaces. The first results (1957) along these lines were obtained by Milnor, his exotically smooth seven spheres. These are relatively easy to construct and analyze, and can be immediately associated with physics as having the underlying topology of the bundle of Yang-Mills $SU(2)$ gauge fields over compactified spacetime, S^4. These topics occupy us in Chapter 7. For the next 20 years or so, much work was done in the field, but some important outstanding questions remained. By various techniques, it was shown that Euclidean topology, $\mathbb{R}^n$ of dimension n can carry only the standard smoothness structures for all $n \neq 4$. The discovery that the missing dimension, 4, the dimension of classical spacetime, concealed a wealth of surprising results came as the results of the work of Freedman, Donaldson et al, having roots in the physically motivated work on moduli space of Yang-Mills connections over a four-manifold. We review the original techniques for studying exotic $\mathbb{R}^4$ in Chapter 8 and the more current ones based on Seiberg-Witten theory in Chapter 9. Chapter 10 discusses possible physical implications and Chapter 11 introduces some ideas still in the speculative stage.

We have been warned by many colleagues of the difficulty inherent in writing this book with its mix of recent deep mathematics probably unfamiliar to many physicists and physical topics similarly unfamiliar to mathematicians. Nevertheless, the long and rich history of the interplay of these two disciplines, and a wide variety of books aimed at non-experts on these subjects, makes us believe that this is not a hopeless task. However, we must remind the reader that since the presentation is aimed at "non-experts," we will necessarily be somewhat superficial and informal in our treatment of many topics, both from the physics and mathematics sides. Full rigor will sometimes be sacrificed, but not at the price of introducing egregious error we hope.

In all of our work we were patiently helped by many colleagues. Especial thanks should go to our mathematician colleagues who showed admirable patience in helping us through the mathematical intricacies leading to the beautiful exotica. Among these we must single out Duane Randall whose patient, expert, and always friendly help was absolutely essential to the completion of this book. We have also received help from others, including R. Gompf, T. Lawson, T. Mautsch, A. Nestke, D. Randall, and H. Rose. We repeat the standard phrase that this work incorporates the knowledge and insights of many people, but any errors are ours alone. This statement is especially true in a book such as this which includes topics from a very wide area. We also were aided by the financial support of a LASpace grant. The second author gratefully acknowledges the hospitality of the Institute for Advanced Studies, Princeton, a Humboldt Senior Research Prize, the hospitality of Friedrich Hehl and others at the Institut für Theoretische Physik, Köln, and the J. C. Carter Professorship at Loyola. Melissa Minneci and Stella VonMeer generously contributed excellent and much needed proof reading assistance. Most importantly, the second author is deeply indebted to Ron Fintushel, then at Tulane, for introducing him to the exciting possibilities of exotic smoothness during stimulating talks. Finally, the second author must affirm the indispensable help of Bill Wilson, Bob Smith and their students, without whom he could not have contributed to this work.

Contents

Preface		vii
1.	Introduction and Background	1
	1.1 Interaction of Physics and Mathematics	1
	1.2 Manifolds: Smoothness and Other Structures	4
	1.3 The Basic Questions	6
	1.4 Some Basic Topological Exotica	9
	1.4.1 Whitehead continua	9
	1.4.2 Weierstraß functions	10
	1.5 The Physics of Certain Mathematical Structures	10
	1.6 The Physics of Exotic Smoothness	13
	1.7 In Sum	13
2.	Algebraic Tools for Topology	15
	2.1 Introduction	15
	2.2 Prerequisites	16
	2.3 Concepts in Algebraic Topology	20
	2.3.1 Homotopy groups	21
	2.3.2 Singular homology	25
	2.4 Interplay between Homotopy and Homology	31
	2.5 Examples	32
	2.6 Axiomatic Homology Theory	33
	2.7 Conclusion	34
3.	Smooth Manifolds, Geometry	35
	3.1 Introduction	35

3.2	Smooth Manifolds	35
3.3	de Rham Cohomology	41
3.4	Geometry: A Physical/Historical Perspective	49
3.5	Geometry: Differential Forms	52

4. Bundles, Geometry, Gauge Theory   55

4.1	Introduction	55
4.2	Bundles	55
4.3	Geometry and Bundles	62
	4.3.1 Connections	64
4.4	Gauge Theory: Some Physics	69
4.5	Physical Generalizations, Yang-Mills, etc.	82
4.6	Yang-Mills Gauge Theory: Some Mathematics	83

5. Gauge Theory and Moduli Space   85

5.1	Introduction	85
5.2	Classification of Vector and Principal Fiber Bundles	86
5.3	Characteristic Classes	101
5.4	Introduction of Spin and $Spin_C$ Structures	114
5.5	More on Yang-Mills Theories	119
5.6	The Concept of a Moduli Space	124
5.7	Donaldson Theory	126
5.8	From Donaldson to Seiberg-Witten Theory	135

6. A Guide to the Classification of Manifolds   151

6.1	Preliminaries: From Morse Theory to Surgery	153
	6.1.1 Morse theory and handle bodies	153
	6.1.2 Cobordism and Morse theory	159
	6.1.3 Handle bodies and surgery	161
6.2	Application of Surgery to Low-dimensional Manifolds	169
	6.2.1 1- and 2-manifolds: algebraic topology	169
	6.2.2 3-manifolds: surgery along knots and Thurston's Geometrization Program	173
6.3	Higher-dimensional Manifolds	177
	6.3.1 The simply-connected h-cobordism theorem	177
	6.3.2 The non-simply-connected s-cobordism theorem*	180
6.4	Topological 4-manifolds: Casson Handles*	182
6.5	Smooth 4-manifolds: Kirby Calculus	187

	6.6	Why is Dimension 4 so Special?	190
	6.7	Constructing 4-manifolds from Intersection Forms	193
		6.7.1 The intersection form	193
		6.7.2 Classification of quadratic forms and 4-manifolds	197
		6.7.3 Some simple manifold constructs	200
	6.8	Freedman's Classification	203
7.	Early Exotic Manifolds	205
	7.1	Introduction	205
	7.2	Some Physical Background: Yang-Mills	206
	7.3	Mathematical Background: Sphere Bundles	207
	7.4	Milnor's Exotic Bundles	208
	7.5	Coordinate Patch Presentation	211
	7.6	Geometrical Consequences	213
	7.7	Eells-Kuiper Smoothness Invariant	215
	7.8	Higher-dimensional Exotic Manifolds(Spheres)	215
	7.9	Classification of Manifold Structures	221
8.	The First Results in Dimension Four	231
	8.1	The Smoothing of the Euclidean Space	231
	8.2	Freedman's Work on the Topology of 4-manifolds	234
	8.3	Applications of Donaldson Theory	237
	8.4	The First Constructions of Exotic $\mathbb{R}^4$	239
		8.4.1 The first exotic $\mathbb{R}^4$	241
	8.5	The Infinite Proliferation of Exotic $\mathbb{R}^4$	243
		8.5.1 The existence of two classes	245
	8.6	Explicit Descriptions of Exotic $\mathbb{R}^4$'s	248
	8.7	Other Non-compact 4-manifolds	250
9.	Seiberg-Witten Theory: The Modern Approach	253
	9.1	The Construction of the Moduli Space	254
	9.2	Seiberg-Witten Invariants	257
	9.3	Gluing Formulas	259
	9.4	Changing of Smooth Structures by Surgery along Knots and Links	260
	9.5	The Failure of the Complete Smooth Classification	263
	9.6	Beyond Seiberg-Witten: The Cohomotopy Approach	264
10.	Physical Implications	267

10.1	The Principle of Relativity	267
10.2	Extension of Metrics	270
10.3	Exotic Cosmology	272
10.4	Global Anomaly Cancellation of Witten	275

11. From Differential Structures to Operator Algebras and Geometric Structures — 281

- 11.1 Exotic Smooth Structures and General Relativity — 281
- 11.2 Differential Structures: From Operator Algebras to Geometric Structures on 3-manifolds — 291
  - 11.2.1 Differential structures and operator algebras — 292
  - 11.2.2 From Akbulut corks to operator algebras — 297
  - 11.2.3 Algebraic K-theory and exotic smooth structures — 301
  - 11.2.4 Geometric structures on 3-manifolds and exotic differential structures — 303

Bibliography — 307

Index — 317

Chapter 1

Introduction and Background

1.1 Interaction of Physics and Mathematics

The history of physics, especially over the last several hundred years, is replete with examples of the cross fertilization between Physics and (otherwise) Pure Mathematics. Examples and anecdotes abound. In the context of this book, the development of geometry (and its generalizations) as mathematics and as physics is especially noteworthy. In the 1700's mathematicians began in earnest the questioning of the minimal structure of Euclid's axioms, in particular, the necessity or not of including the postulate pertaining to parallel lines. Max Jammer [Jammer (1960)] summarizes this history very well. Underlying the mathematical discussion is the additional question of whether or not the axioms are physical or mathematical in nature. Of course, today, we are quite comfortable with the separation of pure mathematics from physics, but this has not always been so clear. Thus, for example, as described in [Jammer (1960)], Gauss actually performed a physical experiment with surveying equipment to determine if the sum of the angles in a triangle is indeed π, as it should be in flat, Euclidean, geometry. He bounced light off of three mirrors constituting the three vertices of a triangle. Of course, with the technology available to him at the time, such an experiment could be done with only crude accuracy, but it presaged a whole set of experiments on the behavior of light rays undertaken over the last thirty years or so within the solar system.

Such work by Gauss, Riemann, Lobachevski and others on the apparently very abstract and non-physical subject of "non-Euclidean" geometry was precisely what was needed to provide the foundation for Einstein's theory of General Relativity, in which gravity is described in terms of the geometric properties of spacetime. The path by which Einstein was led to consider what must have appeared to him to be very abstruse and abstract mathematics as a possible tool for physics has recently been reviewed in the various volumes celebrating the centennial of his birth. A very nice summary is provided by Norton[Norton (1992)].

Later investigations of Einstein's theory led to the natural introduction of non-trivial topology in addition to geometry. In the meantime, the parallel development of quantum theory and quantum field theory has led to the introduction into physics of branches of mathematics such as function theory, Hilbert spaces, bundle theory, moduli space structures, etc. In fact, the second half of the twentieth century has seen a virtual explosion of applications of various branches of mathematics, many of which were considered to be of only abstract interest, to physics. Conversely, in many cases the direction of "applicability" has been reversed, some of which will be touched on in this book. Questions of interest in theoretical physics have turned out to have value in the pursuit of "pure" mathematics.

In summary, the rich interplay between physics and mathematics is obvious to contemporary workers. Certainly, there is no theorem that says "Good mathematics makes good physics," but certainly there is strong anecdotal evidence that this has been true in many important situations. The purpose of this book is to introduce to physicists some recent exciting discoveries in pure mathematics that prove the existence of non-trivial structures on spacetime models which have always been assumed by physicists, and probably most mathematicians, to be trivial only. Can these new structures have physical significance?

The relevant mathematical arena is *"differential topology,"* a very descriptive name since it is concerned with global (topology) smoothness (differential) questions. Since almost all current physical theories make use of calculus at some level, the notion of differentiation on any spacetime model is certainly essential to physics. Furthermore, since the early days of relativity, the importance of global features of spacetime, i.e., topology, has been apparent. Recently, of course, topological features of various theoretical models have been important in a much wider class of theoretical constructs such as quantum field theory and attempts to quantize general relativity. The basic question motivating the studies in this book is whether or not there is any non-trivial relationship between the purely *local* nature of differentiation and the *global* nature of topology. The answer, since the pioneering work of Donaldson, Freedman, et al., is a resounding "yes," opening the door to obvious re-investigations of some fundamental assumptions of theoretical physics.

Before beginning the detailed study of the mathematics surrounding contemporary differential topology, let us recall briefly the roles that spacetime models play in physics, and what is required of them in this process. The history and philosophy of this subject is rich and much too involved and extensive for us to consider here. See the book of Jammer[Jammer (1960)] for one overview. Also, the history of general relativity is obviously intimately related to the development of spacetime structures, so the studies generated by the anniversary of Einstein's birth provide a more recent look at this subject. It is clear that spacetime models serve at least two roles:

(1) A spacetime model carries structures such as topology, smoothness, and geometry. To some extent or another, these features seem to have real, observable consequences[1].
(2) A spacetime model serves only as a computational "scratch pad," on which theories are expressed, calculations done, and experimental predictions made. Apart from this purpose, the model has no direct physical significance.

It is important to keep in mind these distinct roles. Someone trying to "understand" the physics of spacetime has an entirely different set of standards than one, perhaps as a worker in quantum field theory, who merely regards spacetime as a necessity for expressing field equations, perturbation theory, path integrals, etc. To the former, the innate difficulties in trying to give operational significance to spacetime in the light of quantum uncertainty principles for measurements are of profound importance. To the latter, these issues are peripheral at best and spacetime exists only as a platform for calculating integrals. Of course, most physicists at one time or another in their research probably find themselves taking each of the positions. In this book, of course, we clearly are prejudiced toward position (1).

---

[1]The observability of these structures and thus their operational significance continues to be a deep unresolved issue. See, e.g., [Brans (1980)],[Brans (1999)]

1.2 Manifolds: Smoothness and Other Structures

Consider the following sequence:

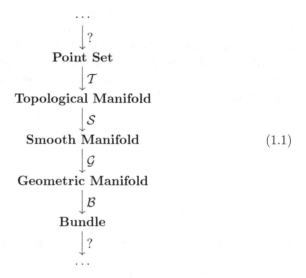

$$
\begin{array}{c}
\cdots \\
\downarrow ? \\
\textbf{Point Set} \\
\downarrow \mathcal{T} \\
\textbf{Topological Manifold} \\
\downarrow \mathcal{S} \\
\textbf{Smooth Manifold} \\
\downarrow \mathcal{G} \\
\textbf{Geometric Manifold} \\
\downarrow \mathcal{B} \\
\textbf{Bundle} \\
\downarrow ? \\
\cdots
\end{array}
\tag{1.1}
$$

representing what seems to be the minimal set of necessary component assumptions that go into the underlying spacetime models of almost all physical theories.

At the lowest level some sort of *point set* structure seems to be required. That is, except for a brave few who attempt to replace point-set structures by something "derived" from more basic quantum structures the idea that spacetime is a point set seems to be universally, if generally tacitly, assumed. Historically the quantification of geometry led to the identification of this point set with Euclidean space, $\mathbb{R}^n_{\text{PS}}$, whose points are the ordered sets of n real numbers. At its foundation, physics has not strayed too far from this point set model.

First, the transition $\mathcal{T}$: With the development of topology over the last 100 years, it has become obvious that more than the identity of points in a set is of significance. In fact, the notion of *limit* defining a *topology* is indispensable to any reasonable spacetime model. In modern usage, a *topology* can be defined in terms of *open sets*, or their generators, *neighborhoods*. Historically the "default" topology for numerical point sets such as $\mathbb{R}^n_{\text{PS}}$ has generally been that defined by the real numbers which have provided the point set structure, the *Euclidean metric topology*. Thus, in the transition $\mathcal{T}$ in 1.1 from $\mathbb{R}^n_{\text{PS}}$ to $\mathbb{R}^n_{\text{TOP}}$, the latter uses neighborhoods defined by the balls generated by the real numbers chosen to define the point set[2]. But current mathematics and physics make use of a generalization of pure global Euclidean topology while still maintaining this structure locally. This leads to the notion of a *topological manifold* as a point set with topology which is *locally* Euclidean. That is, each point has a neighborhood homeomorphic to $\mathbb{R}^n_{\text{TOP}}$. This latter, tautologically topological manifold, is the one most often used by physicists outside of general relativity, with little thought to alternatives.

Next, $\mathcal{S}$: Since physics uses fields and field equations, we need the machinery to perform *differentiation*. This is provided by a definition of how to do calculus locally, a *"smoothness (or differentiable) structure."* The definition and study of such structures, *in the large*, will be the main preoccupation of this book. At its most basic, we need to define which (real-valued) functions on $\mathbb{R}^n_{\text{TOP}}$ are to be regarded as "smooth," or equivalently "differentiable," and then decide how to construct the derivative. For simplicity, we restrict the notion of differentiability to "infinitely differentiable," or C^∞ in the usual notation. There are several routes to defining smoothness structures, the transition $\mathcal{S}$ in 1.1, and much of this book will be dedicated to this topic.

Current physics is based on the tacit assumption that this transition, $\mathcal{S}$, is trivial, that there is some *natural, standard smoothness* structure on $\mathbb{R}^n_{\text{TOP}}$ given by the *topological* coordinates, x^i. So, in what seems to be a natural, minimal, and harmless (but in reality not) process, let us define the class of smooth functions on this smooth structure, $\mathcal{F}_0(\mathbb{R}^n_{\text{TOP}})$, where the "0" means the standard structure, to be those real valued functions of these topological coordinates, $f(x^i)$, which are C^∞ differentiable in the usual real analysis sense. This defines $\mathbb{R}^n_{\text{DIFF},0}$ as a candidate for "the" standard smooth Euclidean manifold.

Of course, it soon becomes clear that these naive procedures may not be well defined because of built-in assumptions that might not be obvious to those concerned only with physical applications. In particular, within the topological category we are really only concerned with equivalence classes under homeomorphisms. Such topological changes to $\mathbb{R}^n_{\text{TOP}}$ and the corresponding $\{x^i\}$ can play havoc with the resulting notions of differentiability using the path above. That is, continuous functions need not be smooth! This fact is at the heart of the issues of central concern to this book.

Finally, assuming the smoothness issue has been adequately addressed,

---

[2] The open ball of radius ϵ centered on x_0^i is defined by $\{x^i | \sum (x^i - x_0^i)^2 < \epsilon^2\}$.

physics needs geometry and more: $\mathcal{G}$, the establishment of a geometry by appending a metric, connection, etc., and $\mathcal{B}$, some sort of bundle structure for defining "internal symmetry" spaces and their connections, etc.[3] These topics will be discussed in the next chapter.

Such questions as these are addressed on the mathematical side by appropriately named *category theory,* which provides an organized way of looking into the study of specific properties such as topology, smoothness, etc., on families of sets. The isomorphisms of a given category correspond to the physical notion of "equivalence" of the corresponding spacetime models. The book by Geroch[Geroch (1985)] provides a useful summary of this and other topics germane to our subject.

1.3 The Basic Questions

The fundamental issues with which we are concerned lie in defining the step $\mathcal{S}$ in 1.1 imposing a smoothness structure on a given topological manifold. However, something like this ambiguity even arises earlier in $\mathcal{T}$ where some point set isomorphisms can drastically alter the topology. As a specific example consider $\mathbb{R}^n_{\text{PS}}$. In the definition of the standard Euclidean topology giving $\mathbb{R}^n_{\text{TOP}}$, the standard metric,

$$d(x,x_0)^2 \equiv \sum_{i=1}^n (x^i - x_0^i)^2, \qquad (1.2)$$

is used. However, there are clearly infinitely many point set isomorphisms of $\mathbb{R}^n_{\text{PS}}$ with $\mathbb{R}^m_{\text{PS}}$ for any other m. For example, if I is the unit interval, consider $(x,y) \in I^2$. Map $(x,y) \to z \in I^1$ by interlacing the binary digits of x and y. This is clearly a one-to-one map, $I^2 \to I^1$, that is not a homeomorphism. That is, the *topological* notion of dimension is *not* determined by the point-set one apparently inherent in the definition of $\mathbb{R}^n_{\text{PS}}$. Thus, we could impose a topology on $\mathbb{R}^n_{\text{PS}}$ using one such isomorphism of it with $\mathbb{R}^m_{\text{PS}}$ and the Euclidean metric in 1.2 but with n replaced by m. This clearly results in a different topology, $\mathbb{R}^n_{\text{TOP}}$, for the *same* point set $\mathbb{R}^n_{\text{PS}}$. Although there seems little or no obvious motivation for studying such alternative topologies as physical models at the present time, this example does point out ambiguities in definitions that seem so natural as to almost be unique.

---

[3]The order in which these two steps are listed in 1.1 is somewhat arbitrary since it could be argued that the imposition of a metric only comes after an establishment of the principal bundle of frames. However, this point is not critical here.

Returning to the issue of defining $\mathcal{S}$ in 1.1, consider the simple example of $n = 1$, so that the topological space is simply the real line with points identified with real numbers and defining the usual Euclidean topology. For clarity, let us use p to denote one such point, so that p is a real number and a point in a topological space. Now, suppose we define a smoothness, $\mathcal{S}$, by defining a global coordinate patch, with coordinate x numerically equal to the real number p,

$$x(p) = p, \quad p(x) = x. \tag{1.3}$$

Here $p \in \mathbb{R}_{\text{TOP}}$ and $x \in \mathbb{R}$, where the latter is the set of real numbers with standard smoothness. Then a function $f : \mathbb{R}_{\text{TOP}} \to \mathbb{R}$ will be smooth if and only if it is C^∞ in the usual real variable sense when expressed in terms of the coordinate, x. If $f_c(x)$ is the coordinate expression, then

$$f_c(x) \equiv f(p(x)). \tag{1.4}$$

Let $\mathcal{F}$ denote this class of smooth functions. Trivially then, the variable x is itself smooth.

However, suppose we had first performed a homeomorphism of $\mathbb{R}_{\text{TOP}}$ onto itself, replacing p by p^3. The topology of the manifold clearly has been unchanged. Now let us define a new global patch and coordinate, y, in terms of this homeomorphic image,

$$y(p) = p^3, \tag{1.5}$$

defining a new smoothness, $\mathcal{S}'$, on the same topological space. Now the class of smooth functions, $\mathcal{F}'$, is defined to be those functions $f'(p)$ such that $f'_c(y) \equiv f'(y^{1/3})$ is C^∞ in the usual sense. Clearly, $\mathcal{F} \neq \mathcal{F}'$. The identity map is an element of $\mathcal{F}$, but, since $y^{1/3}$ is not differentiable at the origin, it is not an element of $\mathcal{F}'$. Thus, we have a simple example of one point set, with two *different* smoothness structures, $\mathcal{S} \neq \mathcal{S}'$.

From the viewpoint of physics, and many mathematical applications, the difference we have established in this example is not *essential*. In fact, mathematicians have found that the most fruitful object for study in differential topology is the equivalence class of smooth structures under diffeomorphisms, that is, homeomorphisms that are smooth when expressed in terms of local coordinates. This equivalence class is the proper object, "differential manifold," with the mathematical category being normally denoted as DIFF. Later chapters will contain more details on this category. For now, let us look at a simple 1-dimensional example provided by the homeomorphism,

$$h(p) = p^{1/3}, \tag{1.6}$$

of $\mathbb{R}_{TOP}$ onto itself which is surprisingly a diffeomorphism! Its coordinate expression is simply the identity map,

$$h_c(x) \equiv y(h(p(x))) = x. \tag{1.7}$$

Thus, from the viewpoint of differential topology, these two *different* smoothness structures on $\mathbb{R}_{TOP}$ are actually *diffeomorphic* and equivalent. Diagrammatically,

$$\begin{array}{ccc} \mathbb{R}^1_{TOP} & \xrightarrow{x} & \mathbb{R}^1 \\ h \downarrow & & \mathbb{I} \downarrow \\ \mathbb{R}^1_{TOP} & \xrightarrow{y} & \mathbb{R}^1 \end{array} \tag{1.8}$$

is commutative. The two horizontal maps are coordinate maps, defining $\mathcal{S}$ and $\mathcal{S}'$, respectively, the left downward map, h, in 1.6 is a homeomorphism, and the combined map expressed in the two coordinate systems, $\mathbb{I}$, is the identity diffeomorphism. The existence of such a diagram provides the fundamental definition of the equivalence, mathematical *and* physical, of two different smoothness structures.

In parallel with the mathematics, physics after Einstein is thoroughly imbued with the idea from General Relativity that all coordinations of spacetime should be physically equivalent. From the viewpoint of physics, diffeomorphisms are the mathematical embodiment of the notion of coordinate, reference frame, transformations. In other words, from the viewpoint of physics the two structures are equivalent, $\mathcal{S} \sim \mathcal{S}'$.

This discussion leads naturally to some basic questions:

Question 1 (existence): *Does there exist any smoothness structure on a given topological manifold?*

and, if so,

Question 2 (uniqueness): *Is the smoothness structure on a given topological manifold unique up to diffeomorphism?*

We will expend much of our effort on these questions. As stated, Question 2 is too general for present mathematical tools, but special forms of it for restricted classes of topological manifolds, has turned out to be of central importance in recent differential topology and forms the basis for this book. Our main concern will be

Question 3 (Euclidean): *How many smoothness structures (up to diffeomorphisms) can be put on Euclidean topological spaces, $\mathbb{R}^n_{\text{TOP}}$?*

To summarize the results discussed more thoroughly in the rest of this book, the answer is

Answer 3(exotic $\mathbb{R}^4_{\text{DIFF}}$): *Up to diffeomorphisms, there is one and only one smoothness on $\mathbb{R}^n_{\text{DIFF}}$ for $n \neq 4$, and uncountably many for the remarkable case of $n = 4$.*

For $\mathbb{R}^1_{\text{DIFF}}$, these assumptions are fairly easy to establish. A very nice and compact proof is provided in the Appendix of Milnor's book[Milnor (1965b)]. For $n = 2, 3$ the proof is much more technical and difficult. For $n > 4$, developments in cobordism theory settled the question. However, until the discoveries of Donaldson, Freedman, et al., the $n = 4$ case remained an open question. Their surprising result for this spacetime dimensional case is the main motivation of this book.

1.4 Some Basic Topological Exotica

1.4.1 *Whitehead continua*

An important thread in understanding exotic smoothness is associated to the assumptions of how properties are inherited in the process of forming mathematical products of spaces. That is, if

$$X = M \times N, \tag{1.9}$$

how are various mathematical properties of M and N related to their point set product, X? Thus, the product formation in 1.9 may be *topological* or *smooth*, etc. Our chief concerns will be in the smooth category, but it is instructive to look at a more basic class of non-intuitive results in low dimensions as provided by *Whitehead continua*[Whitehead (1935)],[Bing (1959)],[Glimm (1960)],[McMillan (1961)].

Whitehead constructed an open, contractible three-dimensional topological manifold, W, which has the following exotic properties:

- W is not homeomorphic to $\mathbb{R}^3$, but,
- $\mathbb{R}^1 \times W$ is homeomorphic to $\mathbb{R}^4$.

In other words, *it is not correct to assume that when an $\mathbb{R}^1$ is factored in*

$\mathbb{R}^4$ the result will necessarily be $\mathbb{R}^3$.

This too is a profoundly counter-intuitive result. The construction of Whitehead spaces can be visualized using an infinite sequence of twisting tori inside each other. For a discussion with diagrams see [McMillan (1961)]. The limit of the infinite iteration of this process produces a set whose complement in $\mathbb{R}^3$ is a Whitehead space. What the implications of this construction are for the smooth case are not now fully understood, but seem to be highly intriguing. In fact, these spaces are used in handlebody constructions of exotic manifolds.

1.4.2 Weierstraß functions

A naive conjecture from elementary calculus is that every function which is continuous over some interval must be at least piecewise smooth, i.e., its derivative exists except at isolated points. "Physical" intuition might well suggest that this conjecture is valid. However, it is not, as demonstrated by the "Weierstraß" functions, such as

$$W(t) = \sum_{0}^{\infty} a^k \cos(b^k t), \qquad (1.10)$$

where $|a| < 1$. Clearly, this series is absolutely convergent to a continuous function for all t. However, naive term by term differentiation under the summation results in

$$W'(t) \stackrel{??}{=} -\sum_{0}^{\infty} (ab)^k \sin(b^k t). \qquad (1.11)$$

If $|ab|$ is chosen to be greater than one, the convergence of this series is dubious at best. In fact, it can be shown rigorously that the derivative of $W(t)$ does not exist anywhere over certain intervals. For more details on such functions see a standard real analysis book such as [Stromberg (1981)].

1.5 The Physics of Certain Mathematical Structures

Finally, let us again come back to the issue of the importance of choice of mathematical structures for physical theories. Einstein's general relativity is nothing other than a theory of the physical importance of the choice of the mathematical structure *geometry*. Later developments indicate that the same may be true with *topology*. In this book we are suggesting that something similar could conceivably be true for the choice of the even more abstract idea of *smoothness*. Because this last concept is more difficult to "visualize" than geometry or even topology, let us conclude this introduction with a toy model of another mathematical structure, *complex structure*.

Our model begins with two-dimensional vacuum electrostatics. Using familiar vector notation, the electrostatic field is represented by a vector on a two-manifold, $\mathbf{E} \in T(M)$, which satisfies

$$\nabla \times \mathbf{E} = 0, \quad \nabla \cdot \mathbf{E} = 0. \tag{1.12}$$

For simplicity, let $M = \mathbb{R}^2$, and define the *standard* complex structure by

$$z = x + iy, \tag{1.13}$$

and the complex function

$$\mathcal{E} \equiv E_x - iE_y. \tag{1.14}$$

Then, it is well known that the physical equations, 1.12, are equivalent to the complex analysis statement that $\mathcal{E}$ is a holomorphic function, or

$$\frac{\partial}{\partial \bar{z}} \mathcal{E} = 0. \tag{1.15}$$

Now suppose that we decide to express the physical theory of vacuum electrostatics in terms of the statement 1.15. Recall that we are choosing this path to explore the possibility that some analytic structure might influence physics. So, what happens if we *change* the complex structure defined by 1.13. Is there a "complex relativity" in action here? Is the physics dependent on the choice of complex structure?

Before getting into this, let us recall the notion of relativity in the spacetime geometry of standard general relativity. There the basic field can be taken as the metric, which is generally expressed explicitly in terms of components relative to a particular coordinate patch. The basic principle of general relativity then asserts that the more physical expression, "change of reference frames," associated with a change of coordinates leaves the physics itself invariant. Thus, the physical field is represented by the equivalence class of local metric component representations mod coordinate changes. In other words, the physics is not contained in a particular functional form of local coordinate components, but rather in the equivalence class. The "practical" question of whether or not two coordinate presentations of metrics are equivalent is solved by extracting all possible invariant information from the metric. This is a problem that leads to the definition of curvature and its invariants.

So, returning to our toy complex model, suppose we decide that the corresponding complex relativity principle would be that the physics is defined by fields and equations expressed in the same form in different, but biholomorphic (the complex analogue of diffeomorphic) complex structures. So, for example, suppose we choose our complex structure to be defined by

$$z' = x - iy. \tag{1.16}$$

Then clearly z' is not an analytic function of z, but does the physics change? The answer is no, since the underlying recoordination: $(x,y) \rightarrow (x,-y)$ provides a biholomorphism between the z and the z' complex structures. Or, if $\mathbf{F}(x,y) = (x,-y)$, then the statement that the following diagram is commutative with $\mathbb{I}$ being the identity biholomorphism,

$$\begin{array}{ccc} \mathbb{R}^2 & \xrightarrow{z} & \mathbb{C} \\ \mathbf{F}\downarrow & & \mathbb{I}\downarrow \\ \mathbb{R}^2 & \xrightarrow{z'} & \mathbb{C} \end{array} \tag{1.17}$$

The problem of determining the equivalence class by evaluating complex invariants is a deep one which we will not go into here. Suffice it to say that the statement

No bound: *every non-constant holomorphic function is unbounded for standard analyticity on* $\mathbb{C}$,

is a well known fact. If we now take the complex structure form of vacuum electrostatics seriously, we would have

No bound(physics): *Every non-constant plane electrostatic vacuum field is unbounded.*

So, can we explicitly find another complex structure for which "No bound" is false? The answer is yes! Let

$$\mathbf{F} : (x,y) \rightarrow (x'',y''), \tag{1.18}$$

be a diffeomorphism of the entire plane onto the unit open ball at the origin. Then

$$z'' = x'' + iy'', \tag{1.19}$$

defines a different complex structure for which the self-defined holomorphic function z'' itself is non-constant and unbounded. Thus, there is no biholomorphism between z'' and the standard z. In terms of the physical interpretation, electrostatics with z'' is truly different from the standard electrostatics, since for this new structure, the "No bound(physics)" statement is false for this z'' complex structure.

1.6 The Physics of Exotic Smoothness

Can differential topology really have anything to do with physical theories? Clearly the answer to this question must be "Yes" because of the principle of general relativity. In light of this principle, the physical content of theories must be invariant under changes of local coordinate patches, *provided that the new smoothness structure is diffeomorphic to the original one*. This is in fact the prototype of "gauge" theory. However, the discovery of exotic smoothness structures shows that there are *many*, often an infinity, of non-diffeomorphic and thus physically inequivalent smoothness structures on many topological spaces of interest to physics. Because of these discoveries, we must face the fact that there is no *a priori* basis for preferring one such structure to another, or to the "standard" one just as we have no *a priori* reason to prefer flat to curved spacetime models. We note that these exotic structures are by definition all *locally* equivalent, so the local expression of physical laws is unchanged. This leads to the apparently paradoxical fact that the implications of exotic smoothness are global, but not in the topological sense!

Unfortunately, the technical difficulties encountered in applying these new results have resulted in only qualitative results for physical applications so far. In the last two chapters we review some of these results and speculate on new ones.

1.7 In Sum

The general question of *equivalence* for various mathematical structures is a fundamental one in mathematics, and, by extension, to physics. Of course defining what "equivalence" should mean is the indispensable first step. Much of the beauty of mathematics is in the discovery of counter examples to intuitively anticipated equivalences, such as the Weierstraß functions and the Whitehead continua examples discussed above.

In general relativity two structures are equivalent if one is obtained from the other by a coordinate transformation (diffeomorphism)[4]. We have long since passed the point in physics of wondering about the principle of general relativity, of saying that two different coordinate statements of the flat metric are physically equivalent while no coordinate transformation can

---

[4] Generalizations of this phenomenon lead to "gauge theory" in contemporary mathematical physics.

take a flat metric into a non-flat one. Our toy plane vacuum electrostatic model above displays a similar situation in which a complex structure has replaced geometry, and biholomorphisms have replaced diffeomorphisms: z is equivalent to z' but not to z''.

The motivating factor for this book is the exploration of possible extension of these ideas to the realm of differential or smooth structures. We thought all such were equivalent on $\mathbb{R}^4$. We now know this is not true. Does this present a situation parallel to the discovery of non-euclidean geometries as models for spacetime?

Chapter 2

Algebraic Tools for Topology

2.1 Introduction

In this chapter we review certain mathematical concepts specifically related to algebraic tools for investigating **topology**, assuming that the reader is familiar with the basic concepts of topology. The understanding of definitions is often enhanced by counterexamples, so we will try to present such with physical interpretations when possible.

Why are apparently abstract mathematical subjects such as topology of interest to physics? We tried to answer this question in the introductory chapter. Basically the answer is to be found in the fact that physical theories have evolved in such a way as to require a spacetime model consisting not only of a point-set (events), but one with a notion of limits (topology) and some mechanism for doing calculus (smoothness). The simplest, and historically first, way of accomplishing this is to assume that spacetime is topologically and smoothly Euclidean, that is, one in which events are identified with ordered sets of real numbers, and calculus is done in the usual way with these numbers. Of course, as it stands, this model is insufficient for any sort of relativity principle, since it assumes one global set of coordinates without any way to re-coordinatize, or, in a physical sense, to allow other reference frames, or observers. The generalizations from this primitive model required by extensions of the physical relativity principles to include *arbitrary* reference frames ultimately leads us into the realm of topology and differential topology of manifolds. In fact, one can discern a certain parallel in the developments of physics and mathematics with respect to the basic "relativity" question: Are two *presentations* of a given structure, e.g., topology, smoothness, geometry, or a physical theory, truly distinct or merely different representations of the same structure as evidenced by an equivalence map, homeomorphism, diffeomorphism, isometry between the two? Progress for studying this problem for topology and geometry as been made by finding readily computable constructs such as homology, homotopy, curvature, etc., widely used in both physics and mathematics. Recent progress with respect to the problem of equivalent smoothness structures is the main motivation of this book. The subject necessarily has both mathematical and physical implications.

So perhaps in this sense, the histories of physics and mathematics have seen parallel developments with respect to "relativity principles." In physics we have long been concerned with distilling the essence of physical theories moduli any specific representation of the model. This includes not only (external) spacetime coordinate transformations, but also internal, gauge transformations. Similarly, mathematics has sought to study equivalence classes of structures, such as point-set, topological, and smoothness, under corresponding "isomorphisms."

The reader can find full treatments of the foundations of these subjects in many books. There are many excellent textbooks and reference books on algebraic topology. We recommend especially those of Greenberg and Harper[Greenberg and Harper (1981)] and Vick[Vick (1994)].

2.2 Prerequisites

First, let us identify and describe the basic **functional categories** of interest. The standard notation, C^0, C^n, C^∞, C^ω, defines the sets of functions which are respectively continuous, continuous together with their first n derivatives, continuous with derivatives of all order, and finally, analytic. Unless otherwise specified we will be interested in only the real, $\mathbb{R}$, and complex, $\mathbb{C}$, fields for functions and coordinates. For the most part we will use the terms **smooth** and **differentiable** to mean C^∞.

These functional categories for real functions are naturally extended to spaces. In particular, one of the central themes of this book is the relationship between various structures (categories, **CAT**) on spaces which are locally topologically Euclidean, that is, topological manifolds. Here note three particular such structures, topological (**TOP**), piecewise linear (**PL**) and smooth or differentiable (**DIFF**). In decomposing a manifold into coordinate patches (an atlas), we have the following possibilities:

- the transition functions are homeomorphisms (TOP case)
- the transition functions are piecewise-linear homeomorphisms (PL case)

- and the transition functions are diffeomorphisms (DIFF case).

How are these three related?

In dimension 1, 2 [Rado (1925)] and 3 [Moise (1952); Cerf (1968)] the three cases TOP, PL and DIFF coincide, i.e., every topological manifold admits an essentially unique PL and a DIFF structure. In the higher dimensions ≥ 5 *obstruction theory* plays an important role. The locally trivial property of manifolds means that locally all three CAT's exist. But the problem of continuing these local structures is in general non-trivial. Obstruction theory studies this problem and defines "obstructions" (generally cohomology elements) for such a continuation. For example, there is a single obstruction to the extension of the TOP structure to PL defined by Kirby and Siebenmann [Kirby and Siebenmann (1977)]. This obstruction lies in $H^4(M, \mathbb{Z}_2)$. The extension of the TOP structure to DIFF is basic to our interests in this book and will be studied in some detail in 6.3. In all dimensions < 7 the two cases PL and DIFF coincide, i.e., it is always possible to "round the corners."

Of course our main interest is in the gap between TOP and DIFF. As mentioned above, tools exist for dimensions less than four and greater than four, but four itself remained inaccessible until the 1980's.

Remark 2.1.

There are some interesting counter examples of the difference between *continuous*, C^0, and *smooth*, C^∞ functions. Intuitively it might seem reasonable to conjecture that a continuous function would be differentiable except perhaps at discrete jumps of slope. However, a class of functions introduced by Weierstraß in the 1800's of the form $\sum_{k=1}^{\infty} b^k \cos(a^k x)$, with $|b| < 1$, but $|ab| > 1$ is everywhere continuous, but nowhere smooth. See Stromberg ([Stromberg (1981)] pp 562-563) for more details. This is an interesting example for this book since we are intimately concerned with the difference between *topology*, e.g., C^0, and *smoothness*, e.g., C^∞. Of course our problems are inexorably tied to global questions.

We assume familiarity with the definition of a **topological space** in terms of **neighborhoods, open/closed sets**, and the notions of **continuous, open,** and **homeomorphic** maps between two such spaces.

Some terms of special interest follow:

- **Metric**: This is a positive definite symmetric function which satisfies the triangle inequality, $d(x,y) + (d(y,z) \leq d(x,z)$. These properties are distilled as the essential elements of what a metric should be from the Pythagorean metric of flat Euclidean geometry. We must point out, however, that the term "metric" is used in spacetime physics to refer to an indefinite (but non-singular) form, the so-called Minkowski or Lorentz signature metrics. We will try to be careful to distinguish these cases when appropriate.

- **Metric space**: This is a topological space for which metric disks, $\{x|d(x,x_0) < a\}$, for fixed x and positive real number a are a basis. To be a metric space is a very strong restriction and such spaces have to fulfill the so-called fourth separation axiom.
- **Euclidean space**: This is the prototype for all spacetime models of physics, and indeed, for spaces used in much of mathematics. The Euclidean n-space, $\mathbb{R}^n$ is a set whose points are the ordered set of real numbers. In addition, the topology is given by the metric

$$d(x,y) = \sqrt{\sum_{i=1}^{n}(x^i - y^i)^2} \quad .$$

Also, note that the question of whether or not the further structure of *smoothness* should be induced by these global coordinates is a central issue in this book.

A couple of important constructs in Euclidean space are:

- **Disk** or synonymously **Ball**: This is a closed subset of $\mathbb{R}^n$ defined by $\{x|d(x,x_0) \leq a\}$, for some fixed x_0 and real, positive a.
- **Sphere**: This is the boundary of a **disk**, a closed subset of $\mathbb{R}^n$ defined by $\{x|d(x,x_0) = a\}$, for fixed x_0 and positive real a.

Finally, we mention a few important terms:

- **Compactness**: A topological space X is compact if for every open covering of X one can choose a finite sub-covering. In particular, every closed subset of a compact metric space is also compact and vice versa. Also, in $\mathbb{R}^n$ every disk and sphere are necessarily compact and every compact set can be contained in some disk.
- **Product topology**: The point-set product $X \times Y$ of two topological spaces is defined as the set of tuples $(x,y) \in X \times Y$ with $x \in X$ and $y \in Y$. This space is then given the product topology. If there are two open subsets $U \subset X$ and $V \subset Y$ containing x and y, respectively, then $U \times V$ is an open subset of the product topology.
- **Quotient topology**: Now we define the process of "dividing out" some relationship. Consider an equivalence relation R, i.e., a subset R of the

product $X \times X$ of a topological space with itself, where $(x,x) \in R$, $(x,y) \in R$ then $(y,x) \in R$ and if $(x,y) \in R$ and $(y,z) \in R$ then $(x,z) \in R$. The space
$$X/R = \{[x] \mid x \in X\}$$
of equivalence classes $[x] = \{y \in X \mid (x,y) \in R\}$ is a topological space with the finest topology such that the inclusion map of a point into its equivalence class is continuous.

Remark 2.2.

The reciprocal relationship of cancellation between products and quotients familiar in ordinary arithmetic does not hold in the topological category. A very interesting counter example is provided by Whitehead manifolds, first defined in 1935, [Whitehead (1935)]. These are open, contractible 3-manifolds, W, that are *not* homeomorphic to standard $\mathbb{R}^3$, but, as shown by Glimm, [Glimm (1960)], the topological product $\mathbb{R}^1 \times W$ *is* homeomorphic to $\mathbb{R}^4$. Thus, in this product, "factoring out" $\mathbb{R}^1$ from $\mathbb{R}^4$ does not necessarily result in $\mathbb{R}^3$.

Next, unless otherwise stated, we assume that our topological models are

- **Hausdorff separable,** that is, disjoint points are contained in disjoint open sets and the topology is generated by a countable basis of neighborhoods,[1]

and

- **Topological manifolds**, that is, each point is contained in a (topological) coordinate patch, U, with a coordination, $\phi_U : U \to \mathbb{R}^n$ which is a homeomorphism.

Counter example: A standard example of a non-Hausdorff manifold consists of three intervals, $X = I_1 \cup I_2 \cup I_3$, where $I_1 = \{x | 0 < x < 1\}, I_2 : \{y| -1 < y \leq 0\}, I_3 : \{z| -1 < z \leq 0\}$, with topology generated by the standard Euclidean disks about every point except $y = 0$ and $z = 0$. The neighborhoods of these points are sets of the form $\{-a < y \leq 0\} \cup \{-b < z \leq 0\} \cup \{0 < x < c\}$ for all of a, b, c positive. Thus, the disjoint points $y = 0$ and $z = 0$ are not contained in disjoint open sets. Furthermore, a single convergent series, $x = 1/n$ converges to both of these disjoint points. There are many obvious examples on topological spaces which are not locally homeomorphic to $\mathbb{R}^n$.

---
[1] Every metric space is Hausdorff separable.

2.3 Concepts in Algebraic Topology

In this section we continue with the theme of looking for tools or "metrics" for determining whether or not two structures are "equivalent," in this case, topologically **homeomorphic**. Recall that in manifold theory, or its physical correlative, general relativity, a central question is whether or not two spaces or spacetime models are diffeomorphic, that is, essentially the same, just presented, or represented, in different coordinate systems. So too here, if we describe two different topological spaces from different routes, how can we tell whether or not the resulting spaces are topologically equivalent?

Examples abound:

- Consider planes and spheres, $\mathbb{R}^n$, and S^n. How do we know that these two are not homeomorphic? In this case, there is an easily applied topological tool to distinguish these two, namely, the second is **compact** while the first is not.
- What about $\mathbb{R}^1$ and $\mathbb{R}^2$? We can easily provide explicit maps between the two as point sets, but can we find a homeomorphism? That the answer to this question is "no" is most easily derived from the tools of homology theory discussed below. The result is that the homology groups in the dimension of the space are different. In particular,

$$H_m(\mathbb{R}^n, \mathbb{R}^n - 0) = \delta_{mn}\mathbb{Z}, \; m > 1.$$

- Consider two compact spaces of the same dimension, S^2 and the torus, $T^2 = S^1 \times S^1$. Are these two spaces homeomorphic? The study of this question led to the notion of topological **genus**, which is a measure of the number of "holes" in a space.

The study of such problems has led to the development of various mathematical techniques. Here we will concentrate on **algebraic topology**. As its name implies this subject is concerned with algebraic structures built from topological ones. Appropriate choices for these structures, will provide tools to detect important changes of topology by doing algebra. Since these structures depend only on the topology, we can say that two spaces with different algebraic structures are **not** homeomorphic. However, in general the converse will not be true. One of the most important long standing problems in mathematics revolves around this issue for the apparently elementary space, S^3. The Poincaré conjecture, described more carefully below, is that any topological manifold with the same **homotopy** groups as standard S^3 must necessarily be homeomorphic to it. Later, we will visit the Poincaré sphere, a topological manifold with the same **homology** groups as standard S^3, but known to be not simply connected, and thus not homeomorphic to standard S^3.

We review and summarize the important algebraic topological structures, **homotopy** and **homology** in the following.

(1) **Homotopy.** Consider equivalence classes of maps defined by continuous one-dimensional deformations. In particular, define the homotopy groups of a space by considering such deformation-equivalent classes of maps of spheres into the space, subject to certain restrictions. This results in a family of groups, $\pi_n(M, x_0), n = 0, 1, \ldots$. Since these groups are invariant under homeomorphisms, we know that if $\pi_n(M, x_0) \neq \pi_n(M', x_0')$ for some n then the two spaces are **not** homeomorphic.

(2) **Singular Homology.** First define **simplices**, disks presented in a convenient form. Next, define the notion of **boundary**, ∂, for such objects. We then investigate a space, M, by looking at the maps of simplices into it, carrying into M the notions of boundary, etc., leading ultimately to **singular homology groups, H_***. Again these groups are invariant under homeomorphisms and thus are sufficient metrics for determining that two spaces are **not** homeomorphic.

(3) **deRham cohomology.** This technique, available only for smooth manifolds, constructs algebraic structures, cohomology, dual to homology, using exterior forms and the exterior differential as the boundary operator. The pairing of this with homology is through integration over singular simplices and their boundary.

(4) **Axiomatic approach.** This extracts the essentials of what homology and cohomology can say about the topology of a space into a set of axioms. We only mention this briefly to point out that tools are available for determining when different approaches, simplicial, singular, deRham, etc., all produce the same information.

2.3.1 *Homotopy groups*

Define groups, $\pi_n(X, x_0)$, for each manifold X relative to a point x_0 in X and each $n > 0$ by considering homotopy-equivalent classes of maps of n-spheres into X.[2] These groups are invariant under homeomorphisms and thus contain sufficient information to determine that two spaces are not topologically equivalent. However, as with homology, these groups are not sufficient for determining the converse. Thus, for example, we will find that $\pi_m(\mathbb{R}^n, 0)$ are all trivial, but as homology theory shows these spaces for different n are not topologically equivalent.

First consider the definition of **homotopy** as applied to a pair of maps, $f, g : X \to Y$.

Definition 2.1. Let $f, g : X \to Y$ be continuous functions. f and g are **homotopic** to each other, denoted by $f \simeq g$ if there is a continuous function $F : X \times [0, 1] \to Y$ with $F(x, 0) = f(x)$ and $F(x, 1) = g(x)$ for all $x \in X$.

The function F provides a *deformation* of one map into the other. Clearly, this relation is an equivalence relation. The equivalence class of homotopic maps between X and Y will be denoted by

$$[X, Y] = \{f : X \to Y \text{ continuous }\}/\simeq \ .$$

This relation leads to the notion of homotopy-equivalence of spaces.

Definition 2.2. Two topological spaces X and Y are **homotopy-equivalent**, if there are two smooth maps $f : X \to Y$ and $g : Y \to X$ so that

$$f \circ g \simeq Id_Y \qquad g \circ f \simeq Id_X$$

where Id_X and Id_Y are the identity maps on X and Y, respectively.

In general define

$$\pi_n(X, x_0) = [(S^n, s_0), (X, x_0)] \ ,$$

the homotopy equivalence class of maps of the pointed sphere into the pointed space. Since we have used the word "group" to refer to them we must define a combining operation.

Start with the $n = 1$ case, π_1, the **fundamental group**. This is the loop space, modulo smooth deformations or contractions. There are several ways to define the group combining operation. We choose one which is easily extendible from $n = 1$ to the general case. Let $S^1 \vee S^1$ be the one-point union defined by

$$S^1 \vee S^1 = S^1 \times \{s_1\} \cup_{s_0 \equiv s_1} \{s_0\} \times S^1 \subset S^1 \times S^1 \ni (s_0, s_1).$$

---

[2] Since we will be interested in pathwise connected spaces only, the choice of x_0 will be irrelevant.

Now define the product $\gamma_1 \star \gamma_2 : S^1 \to X$ of the maps $\gamma_1, \gamma_2 : S^1 \to X$ to $\gamma_{12} : S^1 \to X$ as defined geometrically by the process described in the following sketch in which we deform S^1 into the one-point union of two other circles, with naturally defined map. We then combine this map with the maps γ_1, γ_2 to define the product, $\gamma_1 \star \gamma_2$. This provides a group

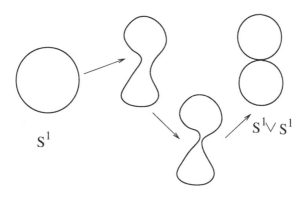

Fig. 2.1 Definition of the map $S : S^1 \to S^1 \vee S^1$

product structure which will in general not be abelian, since there is no map homotopic to the identity which switches the upper and lower circles in this diagram. The proof of the associativity of this product can be found for instance in [Bredon (1993)] Proposition 14.16.

A product structure for $n > 1$ can be defined by the same sort of diagram, replacing S^1's with S^n's. For the higher dimensional spheres, however, it is easy to see that a rotation about an axis passing through the pinched equator can exchange the upper and lower spheres, and still be continuously deformed to the identity. Thus, the product structure for $n > 1$ is abelian. So,

Fact: *The sets, $\pi_n(X, x_0)$, are endowed with natural product structures. For $n > 1$ this structure is abelian, but for $n = 1$, that is, the fundamental group, this may not be true.*

In general, we will only be concerned with pathwise connected spaces. For this class, a change in the reference point, $x_0 \to y_0 \in X$, only changes homotopy groups by isomorphism, so we will drop the identification of the fixed point,

$$\pi_n(X, x_0) \sim \pi_n(X, y_0) \equiv \pi_n(X).$$

Consider some examples. First, if X is contractible to x_0 then $\pi_n(X) = \pi_n(x_0)$. This latter is easily seen to consist of a single element for each n. This single element is generally called 0, but recall that for $n = 1$ it is the unit multiplicative element. So, in particular, $\pi_n(\mathbb{R}^n) = 0$.

The first non-trivial space is the circle itself. Consider the map $n : \mathbb{R} \to S^1$ given by $n(r) = \exp(i2\pi r)$. A closed loop around the circle is given by $n|_{[0,1]}$ and a closed loop going twice around the circle is $n|_{[0,2]}$. Intuitively it seems obvious that **neither** is homotopic to the other and so they represent two different elements in $\pi_1(S^1)$. It is an interesting exercise to construct a rigorous proof of this fact, however. See for example the book of Massey, [Massey (1967)]. Continuing with this argument we obtain an isomorphism $\pi_1(S^1) = \mathbb{Z}$.

What about $\pi_q(S^n)$ for $q < n$? It turns out that each such group is zero but a rigorous proof of this fact is not easy. We need to show that every continuous map $f : S^q \to S^n$ is contractible. An easy way to do this would be to show that the image of every such map does not contain all points of S^n. If that were the case, the image of f would be contractible so that the map would be homotopic to a map to a single point, thus establishing the statement. It would seem to be "intuitively obvious" that the continuous image of a lower dimension sphere in a higher dimensional one is not onto. However, this is not true, in fact there are continuous maps S^q onto S^n. Examples can be constructed from generalizations of "space filling" curves. Perhaps the most famous of these is that due to Peano, which Munkres describes in great detail in his book[Munkres (1975)]. This function maps the closed unit interval continuously onto the closed unit square in the plane. As a "dimension changing" map, it is in fact a fractal curve, and is pictorially presented by an iterative scaling sequence of replacements of straight line segments by triangular segments. While continuous, this map cannot be a homeomorphism since the homological tools developed below will establish that Euclidean spaces of different dimensions cannot be homeomorphic.

However, the existence of such maps does raise a problem for the computation of $\pi_q(S^n)$. The resolution is provided by the cellular approximation theorem quoted in Spanier [Spanier (1966)], p 404. This implies that even though $f: S^q \to S^n$ may be onto, it is homotopic to one whose image is *not* the entire S^n space, as required. This argument again illustrates the importance of carefully questioning those spacetime model assumptions based only on notions of being "intuitively obvious."

What about the other groups $\pi_m(S^n)$ for $m > n$? The calculation of homotopy groups is notoriously non-trivial and attacks on the problem have led to the development of many ancillary tools. One of these is the Hurewicz theorem relating homotopy to homology (often easier to compute). Other tools are provided by bundle theory, as evidenced by the fact that Steenrod spends about one third of his classic work, *The Topology of Fibre Bundles*[Steenrod (1999)] to the homotopy theory of bundles.

The importance of the fundamental group, $\pi_1(X)$, the set of equivalence classes of closed loops in X, leads to the definition that X is **simply connected** if and only if $\pi_1(X) = 0$, that is, if every closed loop is continuously contractible to a point. As the example above shows S^1 is not contractible. An important tool for studying non-simply connected spaces is the notion of **covering space**. $\tilde{X}$ is a covering space for X if $\tilde{X}$ is simply connected and there is continuous epimorphism,

$$f: \tilde{X} \to X,$$

which is a local homeomorphism. Thus, if $x \in X$, there is a neighborhood, U of x such that f is a homeomorphism when restricted to each component, U_a, of $f^{-1}(U)$. The inverse map, f^{-1} is called a **lifting**, and the lift of every closed path in X is a path in $\tilde{X}$ which may or may not be closed. For the example S^1 above the map $n(r)$ is a covering map from $\mathbb{R} \to S^1$.

Remark 2.3.

Physicists will of course be familiar with the important role that closed loops play in some approaches to gauge theories in general and quantum gravity in particular. The action resulting from parallel translation around a closed loop generated by a gauge connection is an important tool for studying such theories. We will say more on this in our later discussion of **gauge theories** and **bundle formalisms**.

2.3.2 *Singular homology*

The original attempts to "quantify" topology with algebraic structures were based on what is now known as **simplicial homology**. The spaces considered are called **triangulable** which means that they can be topologically decomposed into "simple" disk-like pieces of dimensions $n = 0, 1, 2, ...$, called **simplices**. If we denote the "triangulated" space by K formal linear combinations of these simplices constitute the simplicial **chain** groups, $C_n(K)$, of various dimensions. The next step is to define the notion of **boundary** and look for chains which have null boundaries (cycles) but which are themselves not boundaries of a higher dimensional element. Denote the n-cycles by $Z_n(K)$ and chains which are themselves boundaries by $B_n(K)$. The occurrence of "cycles which are not boundaries" is then measured by the nth **simplicial homology group**, of K defined by $Z_n(K)$ mod $B_n(K)$, or

$$H_n(K) = Z_n(K)/B_n(K).$$

An alternative, perhaps more widely used, tool is **singular homology**. This is based on probing the target space, say X, by continuous images of simplices and constructing the algebraic structure, the chains, boundaries, etc., on these maps, rather than completely triangulating K. First, choose a sequence of naturally ordered simplices, one in each dimension, defined by some natural basis, $e_0, e_1, ...$ in $\mathbb{R}^\infty$. The standard **n-simplex** is defined by

$$\sigma_n \equiv \{x = \sum_{i=0}^{n} t_i e_i | \sum_{i=0}^{n} t_i = 1, \ 0 \leq t_i \leq 1\}. \tag{2.1}$$

Thus σ_0 is a point, e_0, σ_1 is the line from e_0 to e_1, σ_2 is a triangle formed by the tips of e_0, e_1, e_2, etc. Note that each such simplex contains $n + 1$ **faces**, σ_n^j defined by

$$\sigma_n^j \equiv \{x \in \sigma_n | t_j = 0\}. \tag{2.2}$$

Clearly the union of faces defines qualitatively what is meant by the boundary of the simplex, but to construct an **algebraic** tool we need to quantify and, in particular, **orient** these constructions.

First, define a **singular n-simplex**, f as a continuous map from σ_n into X. This map can be explicitly stated as a function with real arguments, the coordinates used in (2.1). Thus f is explicitly a function of $n + 1$ real variables, t_i, restricted to $\sum_{i=0}^{n} t_i = 1$. We earlier talked informally of forming the abelian chain group by taking formal sums of simplices. To make this more precise, choose the **coefficient group**, G, an abelian group, to be one of the three: the reals, $R($ or $\mathbb{R}^1)$, the integers, $\mathbb{Z}$, or the binary set, the integers mod 2, $\mathbb{Z}_2$. The **n-singular chain group with coefficients** G is technically defined as the "free abelian group with G

coefficients generated by singular n-simplices." This can most easily be understood as the set of formal sums,

$$C_n(X;G) = \{c_n = \sum_f x_f f | x_f \in G, \; f : \sigma_n \to X\}, \tag{2.3}$$

and where only a finite number of x_f are not zero. From this construction it is easy to see that C_n has a natural abelian structure.

Next comes the definition of boundary operators, $\partial_n : C_n \to C_{n-1}$. First define the face maps,

$$F_n^j : \sigma_{n-1} \to \sigma_n^j \subset \sigma_n,$$

by

$$F_n^j(t_0, ..., t_{n-1}) = (t_0, ..., 0, ...t_{n-1}) \in \sigma_n, \tag{2.4}$$

where the 0 is in the jth place. Finally, the **boundary** operator, ∂_n, taking $C_n(X;G) \to C_{n-1}(X;G)$ is defined by linearity starting from

$$\partial_n f_n = \sum_{j=0}^{n}(-1)^j f_n \circ F_n^j. \tag{2.5}$$

Clearly $-1 \in G$, so this is defined. What we are defining here is a formal sum of oriented boundaries in a sense familiar vector integration in physical applications in three-space, such as Stokes and divergence theorems. In fact, shortly this integration aspect will be used in the definition of **deRham** cohomology in the next chapter.

A straightforward computation using the alternating signs in (2.5), leads to the important result, a defining feature for the topological notion of boundaries,

$$\partial_{n-1} \circ \partial_n = 0. \tag{2.6}$$

At this point define the notion of **cochains** as the abelian groups dual to chains. That is, $C^n(X) = \text{Hom}(C_n(X) \to G)$. This then leads to a natural **coboundary** operator, $\delta^n : C^n \to C^{n+1}$, with $\delta^{n+1} \circ \delta^n = 0$. Finally, C_*, C^* denote the complex of chains, cochains of all dimensions. Similar notation will be applied to derived structures, such as H_*, H^*.

We now proceed to define **homology groups** to detect generalized "holes" in a space, by looking for chains which themselves have null boundary, but are not themselves boundaries. For example consider two loops, 1-chains, in a punctured plane, figure 2.2. Both chains clearly have no boundary, but the one on the left cannot itself be realized as the boundary of any 2-chain, whereas the one on the right can be. Homology provides a tool to quantify this distinction in an algebraic manner.

So, begin by defining two subsets of $C_q(X)$, the **cycles** and the **bounds**,

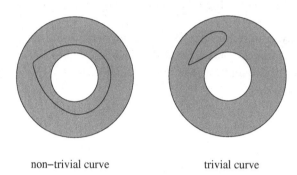

non–trivial curve   trivial curve

Fig. 2.2 Non-trivial and trivial curves on the punctured disk

- Chains with no boundary,
$$Z_q(X) = \{\, c \in C_q(x) \mid \partial_q(c) = 0 \,\} = ker(\partial_q)$$
These are called q-**cycles**.
- The set of chains which are themselves boundaries,
$$B_q(X) = \{\partial_{q+1}(c')|c' \in C_{q+1}\} = im(\partial_{q+1})$$
This is the set of q-**bounds**.

According to the remarks above, we look for q-cycles which are not $q+1$-boundaries using the equivalence relation:

- Two q-chains $c_1, c_2 \in C_q(X)$ are **homologous** to each other, if and only if $c_1 - c_2 \in B_q(X)$.

The equivalence classes define the (singular) **homology groups**,
$$H_q(x) = \frac{Z_q(X)}{B_q(X)}.$$
Perhaps the most important property of homology groups is summarized by the fact that homology is a topological invariant,

Theorem 2.1. *If X and Y are homeomorphic then*
$$H_q(X) \cong H_q(Y) \quad \forall q \in \mathbb{N} \quad .$$

In other words, equality of homology groups is a **necessary**, but, unfortunately not **sufficient**, condition for two spaces to be topologically equivalent. There are many examples of non-homeomorphic spaces which have identical homology groups. In terms of our central relativity task of determining the equivalence of structures, homology provides only a partial answer.

Actually, there is the stronger statement that H_q is a **functor** from the category of topological spaces to the category of abelian groups, or, if

$$f : X \to Y, \tag{2.7}$$

then there is an induced homomorphism of abelian groups,

$$f_q : H_q(X) \to H_q(Y). \tag{2.8}$$

An important extra tool is provided by **relative homology**, which we will briefly review now. Suppose A is a subset of X, so $C_*(A)$ is naturally defined as a sub-complex of $C_*(X)$. Then generalize the notion of cycle to include those singular chains whose boundaries lie entirely in A, labeled $Z_*(X, A)$. Thus

$$Z_q(X, A) \equiv \{c \in C_q(X) | \partial_q c \subset A\}. \tag{2.9}$$

Similarly, $B_*(X, A)$ are chains which are homologous to chains in A,

$$B_q \equiv \{c \in C_q(x) | c = \partial_{q+1} c' + a, \ c' \in C_{q+1}(X), \ a \in C_q(A)\}. \tag{2.10}$$

The relative homology complex, $H_*(X, A)$ is then defined

$$H_q(X, A) = Z_q(X, A) / B_q(X, A). \tag{2.11}$$

The usefulness of relative homology is based on the so-called "exact homology sequence,"

$$\cdots H_q(A) \xrightarrow{i_q} H_q(X) \xrightarrow{\pi_q} H_q(X, A) \xrightarrow{\Delta_q} H_{q-1}(A) \cdots \tag{2.12}$$

In this equation, the map i_q is a natural chain injection map, π_q is a projection one, and the map Δ_q is generated by the boundary operator, projected to relative homology. Finally an important fact about (2.12) is that it is an example of an **exact sequence,** that is, it is a sequence of homomorphisms of abelian groups having the property that the **image** of one map is exactly the **kernel**[3] of the following one. Another important feature is an extension of the homology **functorial** property of maps. We write $f : (X, A) \to (Y, B)$ for a map from X to Y which also maps the subset

---
[3] The kernel of a map is the set of elements mapped into the zero of an abelian group.

A into $B \subset Y$. Then this map induces homology maps, f_* such that the following diagram,

$$\begin{array}{ccccccccc}
\cdots H_q(A) & \xrightarrow{i_q} & H_q(X) & \xrightarrow{\pi_q} & H_q(X,A) & \xrightarrow{\Delta_q} & H_{q-1}(A) \cdots \\
& \downarrow f_* & & \downarrow f_* & & \downarrow f_* & & & (2.13) \\
\cdots H_q(B) & \xrightarrow{i_q} & H_q(Y) & \xrightarrow{\pi_q} & H_q(Y,B) & \xrightarrow{\Delta_q} & H_{q-1}(B) \cdots
\end{array}$$

is **commutative**, that is $f_* \cdot \Delta_q = \Delta_q \cdot f_*$, etc.

Another very useful exact sequence is provided by the **Mayer-Vietoris** theorem. Let X be covered by the union of the interior of two sets, $X = \text{Int}U \cup \text{Int}V$. Then it is possible to define chain homomorphisms, g_*, H_*, Δ_* such that following sequence is exact,

$$\cdots H_q(U \cap V) \xrightarrow{g_q} H_q(U) \oplus H_q(V) \xrightarrow{h_q} H_q(X) \xrightarrow{\Delta_q} H_{q-1}(U \cap V) \cdots$$

This theorem is especially useful for investigating the homology of a space which can be covered by a pair of open sets with known homology properties as we will see in our brief presentation of some examples below.

Finally, we mention the useful **excision** theorem which essentially says that the excision of certain open sets from an homology pair does not change the homology. Specifically, let U be an open set with closure, $\bar{U} \subset A \subset X$. Then the homology sequence $H_*(X, A)$ is isomorphic to that of $H_*(X - U, A - U)$.

We should point out that we have been assuming one fixed coefficient group, G. In general, it can occur that two spaces have the same homology with respect to one G but different homologies relative to another. It turns out that the integers provide a "universal coefficient" group as described in the **universal coefficient theorem,** discussed in many textbooks, for example, Vick[Vick (1994)], page 73. This essentially means that if homology groups agree when $G = \mathbb{Z}$ then they will for any other G.

Recall that the H_p are formed by quotients of abelian groups. As a result the elements of H_p fall into two classes, the *free* elements, those σ_p for which $g\sigma_p = 0$ implies $g = 0$, and *torsion* elements for which $g\sigma_p = 0$ for some $g \neq 0$. For example, if $Z_p = \mathbb{Z}$ and $B_p = 2\mathbb{Z}$, that is, the even integers, then $H_p = \mathbb{Z}_2$, and multiplication of any element by 2 gives zero. We summarize, using $G = \mathbb{Z}$,

$$H_q(K) = Free(H_q(K)) \oplus Tor(H_q(K)) = \left(\bigoplus_{i=1}^{b_q} \mathbb{Z}\right) \oplus \left(\bigoplus_{i=1}^{N} \mathbb{Z}_{n_i}\right)$$

Define b_q (= rank of the homology group H_q) as the q-th **Betti number** of K. Then define the weighted sum $\chi(K)$

$$\chi(K) = \sum_{i=0}^{n}(-1)^i b_i$$

as the **Euler characteristic** $\chi(K)$ of K, dim $K = n$. Of course, the Euler characteristic is also a topological invariant.

2.4 Interplay between Homotopy and Homology

So far we have introduced two different ways to construct algebraic tools which are topological invariants of a space, namely their homotopy and homology groups. The study of the interaction between these two has a long and fruitful history in topology. In fact, one of the most widely studied conjectures in topology over the last 100 years or so is the famous **Poincaré conjecture**,

Poincaré: *If a compact topological manifold without boundary has the same homotopy groups as a three-sphere, S^3, must it be homeomorphic to S^3?*

While we certainly can't answer this question, we can point out some useful known facts. First, if a pair of maps, $f, g | X \to Y$, are homotopic, then they produce identical homomorphisms $H_*(X) \to H_*(Y)$ as in (2.8). In particular, if X is a **deformation retract**[4] of Y then $H_*(X)$ is isomorphic to $H_*(Y)$.

Next are the famous **Hurewicz results**. Recall that the homology groups are all abelian, as are all homotopy groups after the fundamental group. Also recall that we have restricted ourselves to path-connected spaces.

Now, since closed loops and spheres are simplices, with null boundaries, that is, cycles, there is a natural map $\pi_* \to H_*$. Hurewicz showed that in fact, if $\pi_q(X) = 0$ for $1 \leq q < n$ then this map is an isomorphism in dimensions in all $q \leq n$. That is, the first non-trivial homotopy group is isomorphic to the homology group in this dimension, and all $H_q(X) = 0, 1 \leq q < n$.

Clearly dimension $n = 1$ has to be exceptional since the fundamental

---
[4] X is a deformation retract of Y if $X \subset Y$, with inclusion map, i, and there is a map $g : Y \to X$ with $i \cdot g$ homotopic to the identity.

group can be non-abelian. However, in this case Hurewicz showed

$$\pi_1/[\pi_1,\pi_1] \equiv H_1,$$

that is, the abelianized fundamental group is isomorphic to the first homology group. Recall that we are assuming path-connectedness. Chapter 12 of [Greenberg and Harper (1981)] provides an extensive look at applications of this relationship between π_1 and H_1 for many simple but important examples.

2.5 Examples

Of course, the entire purpose of homotopy and homology theory is to provide a tool for answering the fundamental topology equivalence questions, so actually calculating these groups is an important task. At the end of the section above on homotopy we computed a few homotopy groups. However, unfortunately and perhaps surprisingly, the task of computing these groups even for relatively simple spaces such as spheres is difficult in general. In fact, there is no known formula for $\pi_n(S^m)$ for general n, m.

However, the computational task for homology groups is often easier and we will review a few examples. The simplest non-empty space is a single point, pt. Because every map to a point is identical to every other one, it is easy to see that the alternating signs in the boundary operator result in the fact that if f_{2n-1} and f_{2n} are singular simplices of odd/even dimension, then

$$\partial f_{2n-1} = 0, \quad \partial f_{2n} = f_{2n-1}, n > 0,$$

so that $Z_n(\text{pt}) = B_n(\text{pt})$, $n > 0$, while $Z_0(\text{pt}) = K$, $B_0(\text{pt}) = 0$, and

$$H_n(\text{pt}) = \begin{cases} K & n = 0, \\ 0 & n > 0. \end{cases} \quad (2.14)$$

Next, we use the homotopy invariance property to note that any **contractible** space, that is, one for which the identity map is homotopic to a map to a single point, has the same homology as this point, (2.14). In particular, the Euclidean spaces and disks are homologically trivial,

$$H_n(\mathbb{R}^m) = H_n(D^m) = \begin{cases} K & n = 0, \\ 0 & n > 0, \end{cases} \quad (2.15)$$

for any m.

An application of the Mayer-Vietoris sequence enables us to investigate the homology of spheres. In fact, S^n can be written as the union of two disks, with overlap including the equator,
$$S^n = D_1^n \cup D_2^n, \quad D_1^n \cap D_2^n = I^1 \times S^{n-1}.$$
the last term on the right side is clearly homotopic to S^{n-1} itself. This, together with the homological triviality of the disks themselves leads to the following part of Mayer-Vietoris,
$$\cdots 0 \xrightarrow{h_q} H_q(S^n) \xrightarrow{\Delta_q} H_{q-1}(S^{n-1}) \xrightarrow{g_{q-1}} 0 \cdots$$
the exactness of this sequence then implies that $H_q(S^n)$ is isomorphic to $H_{q-1}(S^{n-1})$ for $q > 1, n > 1$. These are but a few illustrative examples, and may more may be found in the standard texts such as [Vick (1994)],[Greenberg and Harper (1981)].

But we cannot leave the discussion of spheres without mentioning the famous counterexample to the *homology* version of the Poincaré conjecture. In fact, Poincaré was able to construct a class of spaces called **homology three-spheres**, spaces Q with $H_*(Q) = H_*(S^3)$, but with different fundamental groups, so that Q cannot be homeomorphic to S^3. See the discussion pages 150ff in Greenberg and Harper[Greenberg and Harper (1981)] for a construction of such spaces using **knots**, which of course have been playing increasingly important roles in theoretical physics recently. Thus, these spaces provide powerful counter examples showing that while homology equality is *necessary,* it is not *sufficient* for topological equivalence, i.e., homeomorphism.

Finally, we briefly mention the important technique of **attaching spaces with maps.** Let $A \subset X$, and $f : A \to Y$, then define $Y_f = X \cup_f Y$ as the space obtained from $X \cup Y$ by identifying $x \in A \subset X$ with $f(x) \in Y$. An important special case is the attachment of cells, $X = D^n$, $A = \partial D^n = S^{n-1}$.

This leads to the definition of an important special class of spaces, the **CW complexes** as spaces defined by a decomposition,
$$X^0 \subset X^1 \cdots \subset X^n = X,$$
in which X^0 is a finite set of points and X^k is obtained from X^{k-1} by attaching a finite number of k-cells.

2.6 Axiomatic Homology Theory

In addition to the singular approach to homology there is also the simplicial one we mentioned briefly in the introduction, and at least two other fairly well-known approaches for cohomology, those of Čech and deRham. The latter will be discussed in the next chapter.

However, this multiplicity of techniques leads to the question of how unique the results are. This problem was studied extensively by Eilenberg and Steenrod, who were able to distill certain essential aspects of what a homology/cohomology should be in what is now known as the set of **Eilenberg-Steenrod** axioms. For more detail on these see Vick[Vick (1994)], chapter 3. Informally, we can summarize these results by saying that an algebraic chain complex constructed from topological spaces satisfying the **functorial** and **boundary** properties expressed in (2.13), the **excision** theorem, together with the point property expressed by (2.14), is uniquely determined by the coefficient group, K on the right side of this equation.

2.7 Conclusion

We have reviewed, all too briefly, some of the tools available for constructing algebraic tools, "metrics" in a certain sense, to investigate a particular topology that has been applied to a given point set. It is important to recall that all of this is *prior* to having imposed any smoothness or definition of differentiability which will be discussed in the following chapter. Nevertheless we will find that these algebraic topological tools are necessary for our investigation of the various notions of smoothness that can be imposed on certain spaces. Furthermore, in the physical sense of the relativity principle, it is crucial to know whether or not our construction of various spacetime models has led to physically equivalent results at the topological level, i.e., whether there is a homeomorphism of one onto the other. Algebraic topology provides us with powerful tools for studying these questions.

Chapter 3

Smooth Manifolds, Geometry

3.1 Introduction

In this chapter we consider the basic subject of our considerations: **smooth manifolds**. We start by reviewing the essential definitions and tools for studying these objects. In particular, we return to the algebraic topological considerations of the previous chapter using the **de Rham cohomology** which smoothness allows us to introduce. Next we review some basic **differential geometry** first using an older, coordinate explicit formalism more familiar to most physicists, and then the differential form techniques of Cartan.

3.2 Smooth Manifolds

Starting from a topological manifold as defined in the previous chapter, we now get to the heart of our problem: "smooth manifolds." This subject is sometimes referred to as **differential topology**, and the reader can find excellent and complete treatments of the subject in many places. We recommend especially books by Bredon[Bredon (1993)], Bröcker and Jänich[Bröcker and Jänich (1987)], or Hirsch[Hirsch (1976)]. Also, Milnor[Milnor (1965a)] provides an informal review of many important basic topics in a compact book. Treatments of the same subject slanted toward physicists can be found in Trautman[Trautman (1984)] or Nakahara[Nakahara (1989)].

Recall that "smooth" is synonymous with "infinitely differentiable" for our purposes. These spaces form the basis for almost all spacetime models in mainstream theoretical physics, since the latter requires at least local coordinates with which to do calculus on fields. For this to be consistent, the notion of smooth in one coordinate patch must be consistent with that in an overlapping one, or, equivalently, one set of coordinates must be smooth functions of any other where both are defined.

To put this more precisely, we use a standard definition:

- **Smooth Manifold:** Let M be a Hausdorff topological space covered by a (countable) family of open sets, $\mathcal{U}$, together with homeomorphisms, $\phi_U : U \in \mathcal{U} \to U_R$, where U_R is an open set of $\mathbb{R}^n$. This defines M as a topological manifold. For smoothness we require that, where defined, $\phi_U \cdot \phi_V^{-1}$ is smooth in $\mathbb{R}^n$, in the standard multivariable calculus sense. The family $\mathcal{A} = \{\mathcal{U}, \phi_U\}$ is called an **atlas** or a **differentiable structure**. Obviously, $\mathcal{A}$ is not unique. Two atlases are said to be compatible if their union is also an atlas. From this comes the notion of a **maximal** atlas. Finally, the pair $(M, \mathcal{A})$, with $\mathcal{A}$ maximal, defines a **smooth manifold**.

An important extension of this construction yields the notion of **smooth manifold with boundary**, M, defined as above, but with the atlas such that the range of the coordinate maps, U_R, may be open in the half space, $\mathbb{R}^n_+$, that is, the subspace of $\mathbb{R}^n$ for which one of the coordinates is non-positive, say $x^n \leq 0$. As a subspace of $\mathbb{R}^n$, $\mathbb{R}^n_+$ has a **topologically** defined **boundary**, namely the set of points for which $x^n = 0$. Use this to define the (smooth) **boundary** of M, ∂M, as the inverse image of these coordinate boundary points.

Remark 3.1.

For notational brevity, the symbol representing the atlas is generally omitted and M is referred to as a smooth manifold. An open set, U, of an atlas will be called a **coordinate patch** and the list of n real functions, $\phi_U^i, i = 1, \ldots, n$ describing the corresponding map, ϕ_U, will be local **coordinates**. Note that we follow the standard physical notation of representing the list of n coordinates with superscripts. This provides the mathematical formalism for the physical notion of a **reference frame**.

From this definition naturally flows the definition of **smooth** maps between manifolds: $f : M \to N$ is smooth if its expression in terms of local coordinates is C^∞ in the usual real variable sense. If the map is also a homeomorphism and its inverse is smooth, then it is a **diffeomorphism**. Explicitly, if U, V are elements of the atlases of M and N respectively, then $\phi_V \cdot f \cdot \phi_U^{-1}$ is a map from an open subset of $\mathbb{R}^n$ onto another which must be a homeomorphism and smooth together with its inverse in the usual sense of n-variable real calculus. From this it is easy to build the category, DIFF, of smooth manifolds and maps.

- **"Relativity" of Manifolds:** The category, DIFF, is central to our interests in this book. From the mathematical viewpoint the natural equivalence question is whether or not two smooth manifolds are **equivalent** in the sense of being **diffeomorphic** to each other. From the physical viewpoint, a diffeomorphism represents a global change of coordinates. The foundational principle of General Relativity is that the laws of physics should be invariant under such changes. However, if two manifolds are *not* diffeomorphic, then there is some underlying physical difference between them.

  Misunderstandings between physicists and mathematicians can easily arise on this issue. Thus diffeomorphic manifolds with different smoothness structures are distinguished as "different" manifolds in most mathematical literature, whereas physicists are apt to tacitly assume a relativity principle and identify them as the "same" manifold. Perhaps this issue can be clarified by a simple analogous matter in differential geometry. Consider "two" metrics on $\mathbb{R}^2$,

$$ds_1^2 = dx^2 + dy^2, \tag{3.1}$$

and

$$ds_2^2 = \cosh^2 x\, dx^2 + dy^2. \tag{3.2}$$

  Taking these two presentations literally, (3.1) and (3.2) are clearly different, and might be called "different metrics" by mathematicians. However, physicists are likely to immediately recognize that (3.2) can be obtained from (3.1) simply by the replacement $x \to \sinh x$, a diffeomorphism (mathematically) or a change of reference frame (physically).

- **An important example:** The following example may help to clarify some critical points. Let M be the topological manifold $\mathbb{R}^1$. Thus a point of M is simply a real number, say p. This identification leads to a natural smoothness structure on M generated by an atlas consisting of a single set, $U = M$, and coordinate $\phi_U(p) = x$ where x is numerically equal to p. The maximal atlas generated by this makes M into a smooth manifold, called **standard** $\mathbb{R}^1$. Now consider a second maximal atlas generated by the single chart $U' = M$, but with coordinate $\phi'_{U'}(p) = y$ where y is $p^{1/3}$. Clearly the the two atlases are incompatible since $p^{1/3} = \phi'_{U'}(p)$ is not a smooth function of $p = \phi_U(p)$. However, the two manifolds are diffeomorphic since the **homeomorphism**, $h : p \to p^3$, is the identity diffeomorphism when expressed in the local coordinates, $\phi'_{U'}(h(\phi_U^{-1}(x))) = (x^3)^{1/3} = x$. We would not get any new physics from M' as compared to M since the two differ simply by changes of reference frames. Actually, the uniqueness of the smooth structure on $\mathbb{R}^1$ is extremely important to the rest of smooth manifold theory since much of the remaining structure of manifold theory such as vectors, forms, etc., depends on the definition of *smooth* real valued functions. If the smoothness on $\mathbb{R}^1$ were not unique up to diffeomorphism, the rest of differential topology would be drastically different!

- **Triviality of Structures:** A central, driving, question in much of differential topology is whether or not a given structure is "trivial." Here triviality means that a structure, such as an atlas, can be reduced to one in simplest form, one coordinate patch for an atlas. In this sense, as we will learn later, every smoothness structure on $\mathbb{R}^n$ is trivial for all $n \neq 4$, but an infinite number are not trivial for $n = 4$. When we expand to bundle and other structures defined locally relative to some atlas, the structure will be trivial if the atlas can be reduced to one coordinate patch. Another way to look at the question of triviality is to pose it in terms of whether a local construction can be extended to a global one. Much of the developments of differential topology can be traced to exploration of such questions.

- **Tangent vector**: Roughly, the modern definition of a tangent vector is as "differentiation in a direction." More precisely, let f be a smooth function on M, then v_p is a tangent vector at $p \in M$, if it provides a real number, $v_p(f)$, in such a way that

    (1) v_p depends only on the germ of the function. That is, $v_p(f) = v_p(g)$ if $f = g$ over some neighborhood of p,

    (2) v_p is linear. That is, $v_p(c_1 f + c_2 g) = c_1 v_p(f) + c_2 v_p(g)$ for real constants c_1, c_2, and,

    (3) v_p satisfies the Leibnitz rule: $v_p(fg) = f(p)v_p(g) + g(p)v_p(f)$.

    This is naturally extended to **smooth tangent vector fields**, defined as those for which $v_p(f)$ is smooth in p for each smooth f. Within a local patch, the coordinate representation of points, $x^i = \phi_U^i(p)$, provides a natural basis for vector fields over U, so we can write[1]

    $$v_p = v^\mu(p)\frac{\partial}{\partial x^\mu},$$

---

[1] We remind readers (especially mathematicians) of our summation convention commonly used in relativity that indices repeated in products are to be summed.

providing the component representation of vectors, with the standard "contravariant" transformation properties. Notation: the vector space of tangent vectors at p is denoted by $T_p(M)$. Also, a smooth map, $f : M \to N$, naturally defines $f_* : T_p(M) \to T_{f(p)}(N)$. Note the important definition of **bracket** of two vector fields:

$$[v, u](f) \equiv v(u(f)) - u(v(f)).$$

- **One-Form**: These are algebraic duals to tangent vectors, with naturally inherited notions of smoothness. A natural way to define a one-form is as the **exterior derivative**, d, of a function, df. This form is defined by

$$df(v_p) = v_p(f).$$

We will use Greek letters to denote forms. As with vectors, local coordinates provide a basis for one-forms,

$$\alpha_p = \alpha_\mu(p) dx^\mu,$$

with local components transforming "covariantly." Notation: the vector space of one-forms at p can be denoted as T_p^*, signifying the duality, or, by Ω_p^1, a special subset of the exterior algebra defined immediately below. Also, smooth $f : M \to N$, leads to $f^* : T_{f(p)}^*(N) \to T_p^*(M)$.

- **Exterior Algebra and Calculus**: The totally anti-symmetric part of the tensor algebra generated by one-forms is of especial interest. Restricting to smooth fields, the family of q-forms is denoted by $\Omega^q(M)$, with algebraic product, the **wedge**,

$$\wedge : \Omega^q(M) \otimes \Omega^r(M) \to \Omega^{q+r}(M),$$

linearly and smoothly. An element of $\Omega^q(M)$ provides a totally anti-symmetric linear map on the tensor product of q tangent vectors. The **exterior derivative**, d, defined above maps $\Omega^p(M)$ linearly and differentially into $\Omega^{p+1}(M)$. Note especially that $dd = 0$. This structure will be the basis for **de Rham cohomology** discussed more thoroughly below.

- **Orbits, Exponential Maps**: The idea of "flux lines" associated with force fields is familiar to all physicists. The mathematical formulation of this idea is embodied in the notion of **orbits**. Given $v(p) \in T_p(M)|_U$ with U a coordinate patch, define the local orbit through p_0 to be the locally defined path, $p(t)$, expressed in local coordinates by the unique solution to

$$\frac{dx^\mu(p(t))}{dt} = v^\mu(p(t)), \quad \text{with } p(0) = p_0. \tag{3.3}$$

From basic studies of ordinary differential equations, we know that if v_p is smooth (stronger results can be stated, of course) a unique solution to (3.3) exists locally. This leads to the definition of the **exponential map**, [Kobayashi and Nomizu (1963)], taking an open set in the tangent vector bundle (defined below) onto an open set of M diffeomorphically. This is a very useful tool as will be seen in the section on Morse theory below. We should note that if a Riemannian geometry is defined on M the term **exponential map** can have a different meaning, using geodesics generated at least locally by the vectors at one fixed point of the manifold. See, for example, [Kobayashi and Nomizu (1963)], volume 1, page 140.

Finally, we close this section with a necessarily too brief reference to two important special manifolds.

First, **complex manifolds** are smooth manifolds as above, but with the added condition that the local coordinates can be represented as complex numbers (so the dimension is necessarily even), and, in some atlas, the transition functions are **biholomorphic**. As we have learned from the simple theory of functions of one variable the difference between smooth and complex analytic is profound and has vast implications. However, because of space limitations, we restrict ourselves essentially to this definition that will be used in certain important examples, such as complex projective spaces, $\mathbb{CP}^n$.

Next, a **Lie group,** G, is a topological group which is also a smooth manifold with local expressions of group products also smooth. Actually, a deep result is that the atlas can be chosen so that the transition functions are not just smooth but real analytic. Furthermore, group multiplication provides a natural translation of a neighborhood of any point to that of any other. This, together with analyticity, implies that the full group topological and manifold structure in the connected component of the identity is determined by the behavior of tangent vectors in any neighborhood of the identity. The group action of left multiplication leads to the generation of a canonical (left-invariant) vector field for each element of the tangent space at the identity. The set of left-invariant vector fields is called the **Lie algebra,** $\mathfrak{g}$, of G. Given a basis for $\mathfrak{g}$, say $\{v_i\}$,

$$[v_i, v_j] = c^k{}_{ij} v_k.$$

It turns out that the quantities $c^k{}_{ij}$ must be constants and are called the **structure constants** of $\mathfrak{g}$. The pioneering work of E. Cartan has shown how these constants completely determine the group structure, topology, and geometry of a wide class of Lie groups, the semi-simple ones.

The smoothness structure is only the first step in preparing a spacetime model for physics. Two major superstructures are **Geometry** and **Bundles**. In later sections in this chapter we review several formalisms for studying differential geometry in more or less historical order. In a following chapter on bundles we return to differential geometry but in a more powerful modern formulation.

First, let us review an important **topological** tool that arises from the smoothness of a manifold, namely, **de Rham cohomology**.

3.3 de Rham Cohomology

de Rham cohomology is built on the notion of differential forms and the exterior algebra generated by them. Of course, we need smoothness to make sense of this, so for the rest of this section, we tacitly assume that all relevant objects are smooth. Furthermore, for simplicity, we assume that M is orientable, with fixed orientation.

Tangent vectors and their duals, one-forms, were briefly introduced in a previous section. Recall that a tangent vector has been defined as a differentiation operator on functions. Conversely, every function $f : M \to \mathbb{R}$ induces a linear map $(df)_x : T_xM \to \mathbb{R}$, the differential of f at $x \in M$. In local coordinates this differential can be written

$$(df)_x = \frac{\partial f}{\partial x^\mu} dx^\mu \ .$$

This set of the differentials spans the linear space, T_x^*M, the cotangent space (dual of T_xM). Clearly the basis $\partial_\nu = \frac{\partial}{\partial x^\nu}$ of T_xM and the basis dx^μ of T_x^*M are dual to each other

$$(dx^\nu)\left(\frac{\partial}{\partial x^\mu}\right) = \left(\frac{\partial}{\partial x^\mu}\right)(x^\nu) = \delta^\nu_\mu,$$

where δ^ν_μ denotes the Kronecker delta function. With the help of linear algebra we can form multilinear functions $T_xM \times \ldots \times T_xM \to \mathbb{R}$. Among these multilinear functions there is the class of alternating linear n-forms defined by the relation

$$\omega(X_1, \ldots, X_i, \ldots, X_j, \ldots X_n) = -\omega(X_1, \ldots, X_j, \ldots, X_i, \ldots, X_n) \quad (3.4)$$

for the exchange of one pair of indices i, j and where the X_i's are elements of T_xM. Denote by $A^n(T_xM)$ the set of alternating functions satisfying (3.4).

A differentiable function which assigns to each point $x \in M$ a member $\omega_x \in A^p(T_xM)$ is a **differential p-form** ω on M. The set of such forms is denoted by $\Omega^p(M)$. Now construct such functions from the one-forms, elements of T_x* by algebraic means. The following definition is useful in defining p-forms,

Definition 3.1. Let ω and η be two alternating p and q-forms, respectively. The form $\omega \wedge \eta$ is an alternating $p+q$-form and thus the wedge product $\wedge$ is defined by

$$\omega \wedge \eta = (-1)^{pq} \eta \wedge \omega \ .$$

If x^i are local coordinates, then $\{dx^i\}$ is a basis for one-forms and $\{dx^{i_1} \wedge \ldots \wedge dx^{i_p} | i_1 < \ldots < i_p\}$, provide a basis for p-forms. Finally, over the coordinate patch an arbitrary $\omega \in \Omega^p$ can be represented

$$\omega = \sum_{\mu_1 < \ldots < \mu_p} f_{\mu_1,\ldots,\mu_p}(x) dx^{\mu_1} \wedge \ldots \wedge dx^{\mu_p}. \qquad (3.5)$$

The $\binom{n}{p}$ independent functions $f_{\mu_1,\ldots,\mu_p}$ are the local components of ω.

Given a map $g: M \to N$ and, as above, $g_*: T_xM \to T_{g(x)}N$ denotes the push-forward of a tangent vector, then the pull-back $g^*: \Omega^p(N) \to \Omega^p(M)$ of a differential p-form is defined by:

$$g^*(\omega)(X_{1_x},\ldots,X_{p_x}) = \omega(g_*(X_{1_x}),\ldots,g_*(X_{p_x})),$$

where $X_{1_x},\ldots,X_{p_x}$ are tangent vectors to M at x and ω is an element of $\Omega^p(N)$. Thus a dual of a map changes the direction, i.e., if g and h are smooth mappings of manifolds then $(g \circ h)^* = h^* \circ g^*$ but $(g \circ h)_* = g_* \circ h_*$.

Note that $\Omega^0(M)$ is the space of all real smooth functions $M \to \mathbb{R}$, so the differential is a linear map $d: \Omega^0(M) \to \Omega^1(M)$. The extension of this map to all p-forms ω also called the **exterior derivation** $d\omega$ is simply given by the local coordinate description:

$$d\omega = \sum_{\mu_1 < \ldots < \mu_p} df_{\mu_1,\ldots,\mu_p}(x) \wedge dx^{\mu_1} \wedge \ldots \wedge dx^{\mu_p},$$

because $f_{\mu_1,\ldots,\mu_p}(x)$ is an element of $\Omega^0(M)$. This gives a linear map $d: \Omega^p(M) \to \Omega^{p+1}(M)$.

The main properties of the exterior derivation are collected in the following theorem:

Theorem 3.1. Let ω and η be two differential forms of rank p and q respectively. Then

$$d(\omega \wedge \eta) = d\omega \wedge \eta + (-1)^p \omega \wedge d\eta \ . \qquad (3.6)$$

Furthermore, for any differential form, ω,

$$d \circ d\omega = 0 \ . \tag{3.7}$$

The first part of the theorem is easily confirmed. For the second part note that an arbitrary p-form can be written as a sum of forms of the type $\omega_f = f dx^1 \wedge \ldots \wedge dx^p$. Then

$$d\omega_f = df \wedge dx^1 \wedge \ldots \wedge dx^p = \sum_{\mu=1}^{n} \frac{\partial f}{\partial x^\mu} dx^\mu \wedge dx^1 \wedge \ldots \wedge dx^p,$$

so that

$$dd\omega_f = \sum_{\nu=1}^{n}\sum_{\mu=1}^{n} \frac{\partial^2 f}{\partial x^\nu \partial x^\mu} dx^\nu \wedge dx^\mu \wedge dx^1 \wedge \ldots \wedge dx^p.$$

From the lemma of Schwarz the second derivative is symmetric while the expression $dx^\nu \wedge dx^\mu$ is anti-symmetric so the sum is zero. From linearity, this argument leads to $dd\omega = 0$ for all forms ω.

(3.7) has a deep and important relationship to $\partial \circ \partial c = 0$ in homology theory. In fact, these two equations are dual to each other. Thus d is a coboundary operator, and

$$Z^p(M) = ker\{d : \Omega^p(M) \to \Omega^{p+1}(M)\},$$

is the set of closed forms, or p-**cocyles**. The vector space,

$$B^p(M) = im\{d : \Omega^{p-1}(M) \to \Omega^p(M)\},$$

is the set of p-**coboundaries**, the vector space of all exact p-forms. Finally, define the pth **de Rham cohomology group** as the quotient,

$$H^p_{dR}(M) = \frac{Z^p(M)}{B^p(M)}. \tag{3.8}$$

Before proceeding to justify this structure as a cohomological one in the sense used in the chapter on algebraic topology, we should point out that the coefficient group here must be the reals, $\mathbb{R}$, not the integers $\mathbb{Z}$. The reason for this, of course, is that the basic "chain" groups, Ω^p, are described in terms of real smooth functions. To be non-trivial, the range of such a function cannot be merely $\mathbb{Z}$, and it is not useful to regard f and $\sqrt{2}f$ as functionally independent. As a result of this, the topological information encoded in de Rham groups is less complete than that of singular homology/cohomology based on $\mathbb{Z}$ as a coefficient group. Of course, the de Rham construction satisfies the Eilenberg-Steenrod uniqueness axioms, so the de

Rham groups in (3.8) are isomorphic to the singular cohomology ones with real coefficients.

The elementary Euclidean metric vector calculus in three dimensions uses the familiar terms **gradient, curl,** and **divergence**. These can be interpreted in terms of de Rham structures. First, we must realize that as usually presented three dimensional vector calculus does not distinguish contravariant vectors, or simply vectors, from covariant ones, one-forms because the Euclidean metric provides a natural isomorphism between the two which can be interpreted as an identification. Let $M = \mathbb{R}^3$, with coordinates $\{x^n | n = 1, 2, 3\}$ with the standard Euclidean base vectors, $\mathbf{e_n}$, but now regarded as "covariant" vectors, identified with the one-form basis, dx^n, in current terminology. The "vector cross product," $\mathbf{e_n} \times \mathbf{e_m}$ becomes the exterior $dx^n \wedge dx^m$, after using the natural (Poincaré duality) identification of one- and two-forms in dimension three. The de Rham complex for this space is

$$0 \to \Omega^0(M) \xrightarrow{d} \Omega^1(M) \xrightarrow{d} \Omega^2(M) \xrightarrow{d} \Omega^3(M) \xrightarrow{d} 0$$

with $d^2 = 0$. The map $\Omega^0(M) \xrightarrow{d} \Omega^1(M)$ is nothing other than the gradient $grad(f)$ leading to a "covariant" vector field. $\Omega^1(M) \xrightarrow{d} \Omega^2(M)$ becomes the $curl(V)$ of a vector field V identified with a one-form. Finally, the map $\Omega^2(M) \xrightarrow{d} \Omega^3(M)$ is the divergence $div(V)$ of a vector field V with respect to the previous identification j and the natural duality identification $dx^1 \wedge dx^2 \wedge dx^3 \to 1$. Then we obtain the complex:

$$0 \to \Omega^0(M) \xrightarrow{grad} \Omega^1(M) \xrightarrow{curl} \Omega^2(M) \xrightarrow{div} \Omega^3(M) \to 0$$

and the familiar relations: $curl \circ grad = 0$ and $div \circ curl = 0$. If we generalize from $\mathbb{R}^3$ to an arbitrary three manifold and go from *local* to *global*, the information in the de Rham cohomology determines whether the equations $curl(V) = 0$ and $div(V) = 0$ have global solutions $V = grad(f)$ and $V = curl(A)$, respectively. In algebraic terms, can we identify $ker(curl) = im(grad)$ and $ker(div) = im(curl)$? The de Rham cohomology measures the deviation of the exactness of the above sequence, i.e., $H^p_{dR}(M) = 0$ for $p > 0$ guarantees the existence of the solutions f so that $V = grad(f)$ for $curl(V) = 0$ and A so that $V = curl(A)$ for $div(V) = 0$.

The circle, S^1, provides an easily understood non-trivial global example. Let θ be the usual angle on S^1, which is of course *not* a globally defined smooth function since $\theta = 0$ is the same point as $\theta = 2\pi$. Nevertheless, $d\theta$ is a well defined smooth element of $\Omega^1(S^1)$. Clearly $d\theta \in Z^1(S^1)$, so this set is not empty, but this one-form is not the differential of any smooth function, and in fact

$$H^1_{dR}(S^1) = Z^1(S^1)/B^1(S^1) = \mathbb{R}, \tag{3.9}$$

agreeing with the singular cohomology group of this space. This is a simple illustration of the interpretation of the de Rham complex as a **co-** homology, that is as the dual to some homology group.

Generalizing from the circle example, we describe some of the tools needed to justify the use of the term "co" for de Rham cohomology by using integration of forms over singular simplices as the means for defining de Rham cohomology as dual to singular homology (with real coefficients). The classic book of de Rham himself[de Rham (1984)] is still an excellent reference for the following discussion, and much more. Another text on the subject is by Bott and Tu[Bott and Tu (1982)].

First, define the integral of a differential form over the image of a singular simplex in a manifold. Basically, we use local coordinate charts to define integration in terms of ordinary Riemannian integration over subsets of $\mathbb{R}^n$. For simplicity start with an n-form with compact support $U = supp(\omega)$ in some coordinate patch. In terms of these coordinates, $\omega = f(x^1, \ldots, x^n)dx^1 \wedge \ldots \wedge dx^n$ on $\phi_U(U) = U_R \subset \mathbb{R}^n$. Define

$$\int_U \omega = \int_{U_R} f(x^1, \ldots, x^n)dx^1 dx^2 \cdots dx^n,$$

where the right-hand integral is an ordinary Riemann integral. By straightforward use of local patches this definition can be extended to subsets of M which are images of singular simplices. For other degrees, let $\eta \in \Omega^p(M)$ be a p-form on M and let $\sigma : \sigma_p \to M$ be a singular p-simplex. The simplex σ_p is a subset of $\mathbb{R}^p$ and we know how to integrate a p-form on this set, defining

$$\int_\sigma \eta = \int_{\sigma_p} (\sigma^*)\eta,$$

as the integral of p-form on a p-dimensional submanifold of M defined by the image of σ. For an arbitrary p-chain $c = \sum_k n_k(\sigma)_k$ with different singular simplices $(\sigma)_k$ and $n_k \in \mathbb{R}$ linearity leads to

$$\int_c \eta = \sum_k n_k \int_{(\sigma)_k} \eta.$$

This leads to

$$\Psi(\omega) : S_p(M) \to \mathbb{R},$$

which is a homomorphism $\Psi : \Omega^p(M) \to Hom(S_p(X), \mathbb{R})$ where S_p are the p-simplicial chains and $Hom(A, B)$ is the set of homomorphism between A and B. To formulate the relation between homology and cohomology we need the following important theorem:

Theorem 3.2. Stokes theorem
Let ω be a p-form and c a $p+1$-chain then

$$\int_c d\omega = \int_{\partial c} \omega.$$

If we use the notation $\langle c, \omega \rangle$ for $\Psi(\omega)(c) = \int_c \omega$ then we can write Stokes theorem in the form

$$\langle c, d\omega \rangle = \langle \partial c, \omega \rangle,$$

showing that ∂ and d are adjoint to each other with respect to the pairing $\langle \, , \, \rangle : S_p(M) \times \Omega^p(M) \to \mathbb{R}$. Let $\omega \in Z^p(M)$ be a cocycle and $c \in Z_p(M)$ be a cycle. For any $\eta \in \Omega^{p-1}(M)$ and chain $b \in S_{p+1}(M)$,

$$\langle (c + \partial b, \omega + d\eta) \rangle = \langle c, \omega \rangle,$$

extending the pairing to $\langle \, , \, \rangle : H_p(M, \mathbb{R}) \times H^p_{dR}(M) \to \mathbb{R}$, ultimately leading to the following theorem:

Theorem 3.3. *(de Rham theorem)*
The de Rham cohomology $H^*_{dR}(M)$ is the dual of the (real) singular homology $H_*(M, \mathbb{R})$.

A simple linearity argument leads to the fact that

Theorem 3.4. *The wedge product $\wedge$ extends to the de Rham cohomology class:*

$$\wedge : H^p_{dR}(M) \times H^q_{dR}(M) \to H^{p+q}_{dR}(M),$$

making the de Rham cohomology into a ring.

Thus de Rham cohomology has a (natural) product inducing a ring structure. In fact, this is true of any cohomology theory where such a product is called the cup product, $\cup$.

Recall we have assumed a fixed orientation for M. Two important objects will be needed: a volume form and the Hodge duality operator. For these we must choose some arbitrary, but fixed, (pseudo-) Riemannian metric g defining the line element locally in physics notation as $ds^2 = g_{\mu\nu}dx^\mu dx^\nu$. If x^i are local coordinates in some coordinate patch U, then the volume element is defined by patching together forms such as

$$\nu_{M_U} = \sqrt{|\det g_{\mu\nu}|}dx^1 \wedge \ldots \wedge dx^n.$$

The result is a globally defined volume form, dependent only on M and the choice of metric, $\nu_{M,g}$. invariant with respect to coordinate transformations. Next, define a linear map $\star : \Omega^p(M) \to \Omega^{n-p}(M)$ called the **Hodge star** locally by

$$\star(dx^{\mu_1} \wedge dx^{\mu_2} \wedge \ldots \wedge dx^{\mu_p}) = \frac{\sqrt{|\det g_{\mu\nu}|}}{(n-p)!}\epsilon^{\mu_1\mu_2\ldots\mu_p}_{\rho_{p+1}\rho_{p+2}\ldots\rho_n} dx^{\rho_{p+1}} \wedge dx^{\rho_{p+2}} \wedge \ldots \wedge dx^{\rho_n},$$

where $\epsilon^{\mu_1\mu_2\ldots\mu_p}_{\rho_{p+1}\rho_{p+2}\ldots\rho_n}$ is the totally antisymmetric tensor. The proof of the global validity of this definition is straightforward.

Remark 3.2.

In the simple Euclidean three-vector notation we reviewed earlier, the dual of the exterior product provides the basis for identifying the cross product of two vectors as a vector. Physics refer to this as a "pseudovector" because of its obvious dependence on choice of orientation. In fact, because of this the magnetic field three-vector is a pseudo-vector, and a famous Nobel prize winning experiment in the 1960's verified that the fundamental atomic force, the weak force, is also orientation dependent.

Note especially that $\star 1 = \nu_M$, naturally and uniquely patched from coordinate segments. This operation is idempotent up to ± 1,

Theorem 3.5. *If (M, g) is a Riemannian manifold then*
$$\star \star \omega = (-1)^{p(n-p)}\omega,$$
and if (M, g) is a Lorentz manifold then
$$\star \star \omega = (-1)^{p(n-p)+1}\omega \quad .$$

Of course the operation extends to the de Rham cohomology as well and we obtain a map $\star : H^p_{dR}(M) \to H^{n-p}_{dR}(M)$ mapping each generator of $H^p_{dR}(M)$ to each generator of $H^{n-p}_{dR}(M)$, leading to

Theorem 3.6. *(Duality)*
Let (M, g) be a smooth compact manifold with an arbitrary, fixed metric g then
$$\star : H^p_{dR}(M) \to H^{n-p}_{dR}(M)$$
is an isomorphism.

In particular, the group $H^0_{dR}(M)$ is (Hodge-)dual to $H^n_{dR}(M)$. The generator of $H^n_{dR}(M)$ is the volume form ν_M generating $H^0_{dR}(M)$ by duality.

Now combine the Hodge-dual with the induced dual from the pairing $\langle .,. \rangle : H_p(M, \mathbb{R}) \times H^p_{dR}(M) \to \mathbb{R}$ to obtain

Theorem 3.7. *(Poincaré duality)*
Let M be a compact manifold without boundary of dimension $\dim M = n$ then there is a isomorphism
$$H_p(M, \mathbb{R}) \to H^{n-p}_{dR}(M),$$
mapping one generator of the free part of the homology to one generator of the de Rham theory. The torsion group of H_p is mapped to zero.

Note that the compactness and the nonexistence of a boundary is crucial to the validity of this result. Historically, the dual of H_p for simplicial homology was constructed from a dual cell decomposition of the space where every p cell is mapped to a $n-p$ cell.

Now define the pairing:
$$<\omega, \eta> = \int_M \omega \wedge \star \eta,$$
for two p forms $\omega, \eta \in \Omega^p(M)$. Analogously, define the dual δ of the derivative d by:
$$<d\zeta, \eta> = <\zeta, \delta\eta> \quad \to \quad \delta = (-1)^{p(n-p)+1} \star d\star,$$
for a $p-1$ form ζ. This derivative $\delta : \Omega^p(M) \to \Omega^{p-1}(M)$ with $\delta^2 = 0$ lowers cohomology degree, and in fact, because of the (Hodge-)duality, the cohomology induced by δ is the same as that from d. In fact, a cocycle $[\omega]$ is also coclosed $\delta[\omega] = 0$.

Finally we obtain the manifold generalization of the Laplacian,
$$\triangle_p = (d+\delta)^2 = d\delta + \delta d : \Omega^p(M) \to \Omega^p(M).$$
Now, assume that the chosen metric is definite. Then an important result is
$$<\triangle_p \omega, \omega> = <(d\delta + \delta d)\omega, \omega> = <d\omega, d\omega> + <\delta\omega, \delta\omega> = 0.$$
This leads to the famous theorem of Hodge,

Theorem 3.8. *(Hodge theorem)*
The kernel of the Laplacian $\triangle_p$, i.e., the vector space generated by all solutions $\triangle_p \omega = 0$ is isomorphic to the pth de Rham cohomology group $H^p_{dR}(M)$.

This is a very important interpretation of cohomology in terms of the solution space of a differential equation. Again, note, however, that this applies only to closed manifolds with definite metric. For example, for Lorentz signature metrics, the pairing $< \triangle_p \omega, \omega >$ may be zero even though neither factor is.

3.4 Geometry: A Physical/Historical Perspective

Einstein's profound contribution was to suggest that spacetime has a physical structure, **geometry,** described by a **metric** of Lorentzian signature, and that this describes the physical field **gravity**. From the metric there is a naturally induced torsion free **connection**. Geodesics defined by this metric/connection then describe the action of the "force" of gravity on particles. See for example, some of the earlier relativity texts, such as Bergman [Bergmann (1942)], Adler et al. [Adler *et al.* (1965)] or Weinberg [Weinberg (1972)]. The older notation emphasizes local coordinate representations, with vectors and tensors introduced in terms of matrices representing their components in a particular coordinate patch. Distinctions are made between **contravariant** vectors represented by component matrices with superscripts and **covariant** ones with subscript components. These components "transform" as dx^μ and $\partial/\partial x^\mu$ respectively. From this beginning the full "tensor" formulation can be generated.

Einstein's relativity started with the **special principle of relativity**, stating that all laws of physics should be the same, i.e., invariant in form, in all **inertial** reference frames. The latter frames are those in which Newton's mechanics are valid, or in which a free particle is unaccelerated. One of the laws of physics is Maxwell's electrodynamics, which predicts a definite constant value of the speed of electromagnetic waves in a vacuum, which happens to be the same as the observed speed of light, usually denoted by c. So Einstein's principle requires the extremely non-intuitive assumption that all relatively moving observers should measure the same speed for a light wave. From this, Einstein developed the physics of length and time comparisons in terms of a mathematical formulation, the Lorentz transformations. Ultimately, this theory of special relativity can be summarized mathematically by asserting that spacetime has a flat metric, of Lorentz signature, $(-, +, +, +)$ and that the inertial reference frames are those with coordinates, x^μ, for which the metric expression is

$$ds^2 = \eta_{\mu\nu} dx^\mu dx^\nu, \quad \eta = diag(-1, 1, 1, 1). \tag{3.10}$$

After this major step, two important issues arose for Einstein:

(1) Why restrict physics to a particular class of "inertial" reference frames?
(2) How can gravity fit into this spacetime formalism?

Point (1) led to a "generalization" of special relativity. Physicists often learn of this in explicit local coordinate formulation, which we review next. Point (2) led to **Einstein's general relativity** as a theory of gravity. We will discuss that in the next section.

First, in local coordinates we can replace (3.10) by

$$ds^2 = g_{\mu\nu}dx^\mu dx^\nu, \tag{3.11}$$

with variable metric components, $g_{\mu\nu}(p)$, often described as the components of a covariant (symmetric) tensor field of rank 2. This **metric tensor** provides an "infinitesimal distance/time" interval, a scalar derived from coordinate displacement, dx^μ. In addition to providing a "physical" metric for infinitesimal displacements in (3.11), the metric and its inverse provide isomorphisms between the spaces of contravariant and covariant vectors.

Of course, a basic mathematical question is whether or not a given smooth manifold can support a metric. In the case of Riemannian (that is, definite) signature, the answer is yes. Since every coordinate patch is diffeomorphic to $\mathbb{R}^n$ a metric exists locally. The question then is whether or not this local construction can be "patched" together to give a global one. The proof that this can be done for definite metrics is best defined in terms of bundle cross sections, as discussed below. For indefinite signatures, such as Lorentzian of physical four-spacetime, the answer is yes only if a certain topological condition is met. Again, this is a question of whether or not a given construction can be extended from local to global.

So now, in keeping with the idea of generalizing special relativity, we must allow arbitrary, not necessarily linear, smooth transformations. In this case, the ordinary coordinate derivatives,

$$A^\mu_{,\nu} \equiv \frac{\partial A^\mu}{\partial x^\nu},$$

of the components of a vector do not result in components of a tensor. The same obviously applies to the components of the metric tensor. However, the peculiar combination

$$A^\mu_{;\nu} \equiv A^\mu_{,\nu} + \begin{Bmatrix} \mu \\ \nu\rho \end{Bmatrix} A^\rho,$$

transform linearly and thus define the components of a tensor where the **Christoffel** symbols are defined as

$$\left\{ \begin{array}{c} \mu \\ \nu\rho \end{array} \right\} \equiv g^{\mu\eta}(g_{\nu\eta,\rho} + g_{\rho\eta,\nu} - g_{\nu\rho,\eta})/2. \qquad (3.12)$$

The components, $A^{\mu}_{;\nu}$, are called the **covariant derivative** components. Recall that a vector is a differentiation of functions. Use the preceding formulation to generalize this scalar differentiation to a **covariant derivative** of the vector field A along B,

$$\nabla_B A = B^\nu A^\mu_{;\nu} \frac{\partial}{\partial x^\mu}. \qquad (3.13)$$

The physical motivation provided to physicists for studying this formalism as described in early books, such as that of Bergmann[Bergmann (1942)], is to find a formalism in which the equations of physics are "invariant in form." To achieve this the equations must be of the form

$$\text{tensor} = 0,$$

such as

$$A^\mu_{;\nu} = 0,$$

for example. On the other hand

$$\left\{ \begin{array}{c} \mu \\ \nu\rho \end{array} \right\} = 0,$$

is not invariant under diffeomorphisms and is thus *not* a candidate for a physical law under the principle of general relativity. Actually, Einstein and Grossman briefly considered non-invariant field equations involving the Christoffel symbols before finally settling on the now standard (3.21) below. This interesting history is reviewed in a very readable fashion by Norton[Norton (1984)], reprinted in the Einstein studies, volume 1[Howard and Stachel (1986)].

The formation of tensors from vectors, etc., by covariant differentiation is a necessity for expressing physical theories invariantly. In fact, one (strong) form of the **Principle of Equivalence** can be expressed by saying that special relativistic theories are "generalized" by replacing ordinary with covariant differentiation.

We will not trouble the reader with further explicit description of this approach to general relativity, but proceed to the "differential form" one, which uses form-frames, or "anholonomic coordinates" adapted to the metric leading to the statement of Einstein's **General relativistic gravity** equations, (3.21).

3.5 Geometry: Differential Forms

An important improvement on the earliest differential geometric formalisms was provided by E. Cartan's idea of a "repère mobile," a moving frame which may or may not be "holonomic," that is, always along coordinate lines. Another expression to describe this formalism is **vielbein**, or for four-spacetime, **vierbein**. The "legs" are a basis for the tangent vectors chosen conveniently at each point. Here, we use the dual, form, basis originally suggested by Cartan, and explained extensively and graphically for the physics audience in the book of Misner, Thorne, and Wheeler[Misner et al. (1973)].

The basic idea is to choose a one-form basis at each point adapted to the metric. That is, one in which the metric takes some convenient constant form. In spacetime models we might write,

$$ds^2 = \eta_{\mu\nu}\theta^\mu\theta^\nu, \tag{3.14}$$

with the $\{\theta^\mu\}$ constituting a basis for one-forms (with dual $\{e_\mu\}$ for vectors) so defined that the metric components $\eta_{\mu\nu}$ are some constant, standard values. For example, the matrix η is diag(-1,+1,...) in physics (Lorentzian geometry), and the unit matrix in most mathematical applications (Riemannian geometry). The "gauge" freedom inherent in (3.14) keeping the form of η fixed is thus some classical Lie group, $SO(1, n-1)$, or $SO(n)$. The old Christoffel symbol description of the connection can then be summarized by one-forms, $\omega^\mu{}_\nu$, the **connection forms** uniquely defined by

$$d\theta^\mu = \theta^\nu \wedge \omega^\mu{}_\nu, \tag{3.15}$$

$$\omega_{\mu\nu} + \omega_{\nu\mu} = d\eta_{\mu\nu} = 0. \tag{3.16}$$

Here the indices are "lowered" with the metric components, $\eta_{\mu\nu}$, as usual.

As one-forms, the connection forms have components relative to the θ basis,

$$\omega^\mu{}_\nu = \gamma^\mu{}_{\nu\rho}\theta^\rho.$$

In writing (3.15) and (3.16) we have made certain assumptions. (3.15) is the assumption that the connection is **torsion free**, and (3.16) is that the connection is **metric**. For a very complete exposition of generalizations of these equations and their possible physical significance see the review paper of Hehl et al. [Hehl et al. (1995)].

We now generalize the notion of covariant differentiation defined in terms of coordinate bases in (3.13) to the θ^μ, e_ν bases. Note that the e_ν components of a vector A are $\theta^\nu(A)$,

$$A = \theta^\nu(A)e_\nu.$$

The covariant derivative of A along B is then

$$\nabla_B A = (B(\theta^\mu(A)) + \theta^\nu(B)\omega^\mu{}_\nu(A))e_\mu. \tag{3.17}$$

Finally, the connection forms lead to the **Curvature two-form**, $\Omega^\mu{}_\nu$,

$$d\omega^\mu{}_\nu + \omega^\mu{}_\rho \wedge \omega^\rho{}_\nu = \Omega^\mu{}_\nu. \tag{3.18}$$

The components of the curvature form can then be expressed in a generalization of the coordinate form of the curvature tensor,

$$\Omega^\mu{}_\nu = \frac{1}{2}R^\mu{}_{\nu\rho\eta}\theta^\rho \wedge \theta^\eta. \tag{3.19}$$

From this the Ricci tensor is defined as usual,

$$R_{\mu\nu} \equiv R^\rho{}_{\mu\rho\nu}. \tag{3.20}$$

Einstein's theory of **General Relativity** was based on his remarkable observation that the physics of **gravity** could be understood in terms of the curvature of the spacetime manifold. Thus, geometric structures, metric, connection, and curvature describe the physical **gravitational field**. Explicitly Einstein proposed the gravitational field equations relating geometry to "matter" as

$$R_{\mu\nu} - \frac{1}{2}\eta_{\mu\nu}R = 8\pi G T_{\mu\nu}, \tag{3.21}$$

where $R = \eta^{\mu\nu}R_{\mu\nu}$ is the scalar curvature, G is Newtons gravitation constant and all matter is represented by the stress-energy tensor, with components $T_{\mu\nu}$ relative to the θ basis.

While an improvement over the older coordinate basis formalism, these equations are still replete with indices, a bane to most mathematicians and many physicists. Fortunately, the modern approach to differential geometry using bundles as described in the following chapter obviates many of these complications.

Chapter 4

Bundles, Geometry, Gauge Theory

4.1 Introduction

We continue with our study of structures on smooth manifolds, starting with the central topic of **bundles**, including the classical examples of tangent and frame bundles. We discuss the critical issue of **triviality** of a bundle. Next, we discuss **geometry** and **connections** in the bundle formalism. We close the chapter with a brief review of two important physical fields, the **electromagnetic** field and the **Dirac** spin 1/2 field. The $U(1)$ invariance of the Dirac field leads in a classical way to understanding its electromagnetic interactions in terms of a $U(1)$ bundle connection, which is then identified with the electromagnetic potential one-form. This procedure turns out to be the prototype for the physics of **gauge theory**. We provide a summary of the generalizations of importance to recent elementary particle physical models as well as mathematics. We conclude with mention of some mathematical applications. The reader may find this chapter to be an introductory survey to topics covered in much more detail in the next chapter.

4.2 Bundles

The notion of a **bundle**, E, is built from the intuitive idea of tying or attaching copies of one space, the **fiber**, F, to each point of another, the **base space**, M, in some standard way. Thus, the bundle construction can be thought of as a generalization of the cartesian product formation,

$$E \approx M \text{``}\times\text{''} F. \tag{4.1}$$

The notation "×" is meant to indicate that this is not literally a product but only a local one, "patched" together globally in some well-defined way. If (4.1) is actually a product, the bundle is said to be **trivial**. An example of a trivial bundle is the tangent vector bundle over euclidean $\mathbb{R}^n$ discussed below. This difference between local and global, non-trivial and trivial, product structures for bundles is closely related to the **gauge** formalism in physics, corresponding to requirements of local vs global symmetries. This distinction is at the heart of the various $SU(n)$ models of contemporary particle or "high energy" physics, as we will discuss in more detail below.

In general the patching is done in the TOP category but in this chapter, unless otherwise noted, we will restrict the discussion to DIFF, so that the patchings will be required to be smooth. This "patching" will involve a **bundle group**, G, which acts on E through its action on the fiber, F. Unless otherwise noted, G is also smooth, and thus a Lie group. The **projection**, π, maps the fiber space, E, onto the base space, M. A map, inverse to π, from the base space to a point in the bundle above the point is called a **section** of the bundle. It is clear that this construction epitomizes the fundamental physical operation of constructing a **physical field** as a section in which the fiber defines the possible values of the field, e.g., scalar, real or complex vector, spinor, etc. The bundle itself is the set of all values at all points for some physical field, represented at each point by an element of the fiber. The group action on fibers corresponds to some **symmetry group** of the field theory.

By way of introduction, we begin with the **tangent vector bundle** of a smooth manifold. In a sense this is the prototype of a bundle. Let M be a smooth manifold of dimension n. In section 3.2 we introduced the tangent space $T_p(M)$ at $p \in M$. Now put all of these spaces together to get $T(M) = \bigcup \{T_p(M) | p \in M\}$. This is the set of all ordered pairs (p, ξ) where $\xi \in T_p(M)$ together with the natural projection $\pi : T(M) \to M, \pi(p, \xi) = p$.

At this point, all we have defined is $T(M)$ as a point set. To provide the additional DIFF structure (smoothness) it is most convenient to make use of the localization already inherent in the smoothness of M itself. So, let $x : U \to \mathbb{R}^n$ be a coordinate chart of M with local coordinates $x^1, \ldots, x^n$ near p. Thus any tangent vector at a point of U can be written $a_U^\mu(\xi, x^\nu) \frac{\partial}{\partial x^\mu}$, and we require that these functions be smooth. Use this coordination of the vectors to provide the diffeomorphisms required to make $T(M)$ into a $2n$ dimensional smooth manifold, the **tangent vector bundle** to M. Thus, if $\pi^{-1}(U) = \hat{U} \subset T(M)$, then define $\phi_U : U \times \mathbb{R}^n \to \hat{U}$ by

$$\phi_U(p, a_U^\mu) = a_U^\mu \frac{\partial}{\partial x^\mu} \in T_p(M). \tag{4.2}$$

Consider another chart on V, with $y : V \to \mathbb{R}^n$, and $V \cap U \neq \emptyset$. Then over this intersection

$$\phi_U(p, a_U^\mu) = a_U^\mu \frac{\partial}{\partial x^\mu} = \frac{\partial y^\nu}{\partial x^\mu} a_U^\mu \frac{\partial}{\partial y^\nu} = \phi_V(p, g_{VU}{}^\nu{}_\mu a_U^\mu) = \phi_V(p, a_V^\nu). \tag{4.3}$$

These local coordinations, ϕ_U, ϕ_V provide matrix representations, a_U, a_V, elements of the **fiber**, $\mathbb{R}^n$, in this case, for vectors ξ over each point. The **transition** between the representations ϕ_U and ϕ_V is given by an element of $Gl(n, R)$, whose matrix representation is

$$g_{VU}{}^\nu{}_\mu \equiv \frac{\partial y^\nu}{\partial x^\mu} \in Gl(n, R), \tag{4.4}$$

$$a_V^\nu = \frac{\partial y^\nu}{\partial x^\mu} a_U^\mu, \tag{4.5}$$

the usual transformation role for "contravariant vector components."

This collection of maps from the intersection of coordinate patches to **bundle group**, $Gl(n, R)$, in this case, satisfying the cocycle condition,

$$g_{VU} g_{UW} = g_{VW}.$$

These transition group elements contain the essence of the bundle, tangent vector in this case, but also for the generalizations discussed below. In this case, we have defined the bundle coordinate patch transformations in terms of the smoothness patches, (4.4), of the base space, but this will not necessarily be true for the more general bundles defined below.

A natural question then is whether the bundle group can be reduced, that is, if the bundle is equivalent to one with a smaller group. Of course, reduction to the identity means triviality, that is, the bundle is equivalent to a global product.

As in Chapter 3, $T_p^*(M)$ is the dual space to the tangent space $T_p(M)$, leading to the cotangent bundle $T^*(M) = \bigcup \{T_p^*(M) | p \in M\}$. Recall that the space of one-forms is the dual to vectors, so the cotangent bundle is the one-form bundle. If ω is a one-form at p, then a local coordination leads to

$$\phi_U^*(p, a_{U\mu}) = a_{U\mu} dx^\mu \in T_p^*(M).$$

A discussion similar to that leading to (4.5) for vectors then results in the "covariant vector" transformation rules for the local one-form components, a_μ.

Example: a non-trivial vector bundle

Clearly, if $M = \mathbb{R}^n$, there is a global coordination of both M and $T(M)$ and the factoring in (4.1) is global. That is $T(\mathbb{R}^n)$ is a **trivial bundle**. This is a special case. In general a manifold admitting such a trivial tangent bundle is called **parallelizable**. Such a manifold will necessarily have a global family of n smooth linearly independent vector fields.

But not every manifold is parallelizable, so it is helpful to display an easily understood counter example of a non-trivial bundle $T(M)$ which is not diffeomorphic to $M \times \mathbb{R}^n$. Such an example is provided by the two-sphere S^2. Several methods lead to the conclusion that this space cannot support a globally non-zero smooth vector field. One of these is topological and involves the algebraic topological characteristic of the space called the "Euler number." However, here we look at the matter directly in terms of the coordinate bundle representation. Let us realize S^2 as the subspace of $\mathbb{R}^3 = \{(x,y,z)\}$ for which $x^2+y^2+z^2 = 1$. Consider the (extended) northern and southern hemispheres H_+ and H_-, defined by the extra conditions $z > -1$ and $z < +1$ respectively. Now choose coordinates on these two subspaces by stereographic projections and denote them by $(X^i_\pm), i = 1,2$ respectively. The range of these coordinates is the full space $\mathbb{R}^2$ and their transformation equations are clearly

$$X^1_+ = \frac{X^1_-}{(X^1_-)^2 + (X^2_-)^2} \qquad X^2_+ = -\frac{X^2_-}{(X^1_-)^2 + (X^2_-)^2}.$$

Letting ϕ be the azimuthal coordinate, the transformation matrix described in (4.4) here becomes at the equator

$$g_{-+}(\phi) = \left\{\frac{\partial X^i_+}{\partial X^j_-}\right\} = \begin{pmatrix} -\cos 2\phi & -\sin 2\phi \\ \sin 2\phi & -\cos 2\phi \end{pmatrix}. \tag{4.6}$$

Now, suppose the bundle were trivial, so that there exists a global basis for the tangent vectors, say (e_1, e_2). Then there would exists elements, $\mu_\pm$, of $Gl(2, R)$, defined over $H_\pm$ respectively by

$$\frac{\partial}{\partial X^i_\pm} = \mu^{\;j}_{\pm i} e_j, \tag{4.7}$$

and on the equator

$$g_{-+}(\phi) = \mu_-(\phi)(\mu_+(\phi))^{-1}. \tag{4.8}$$

It is a well known fact from group theory that every non-singular matrix can be written as a product of a symmetric matrix times an orthogonal one. So, we can "reduce" the maps $\mu_\pm$ to their orthogonal components,

$\mu_{0\pm}$. But each of $\mu_{0\pm}$, is then a map of the contractible disks, $H_\pm$, into $SO(2, R) = S^1$ and as such is homotopic to a constant map. But, from (4.8) we would then have that the map of S^1 on itself defined by $g_{-+} : \phi \to 2\phi$ in (4.7) is homotopic to a constant map. Since this is clearly impossible (a double winding of the circle on itself is not homotopic to the identity[1]), there is no such global basis, (e_1, e_2), and S^2 is not parallelizable.

Remark 4.1.
What we have just looked at in some detail for the special case of S^2 is the prototype of the more general bundle theoretic studies of the trivialization, or **obstruction** to trivialization of bundles that has been so important to recent mathematics. The classic work on the subject continues to be the book of Steenrod,[Steenrod (1999)].

Generalizing this argument to S^n is a natural and productive enterprise, as discussed in Steenrod, [Steenrod (1999)]. In dividing a sphere into two overlapping disks, with intersection $I \times S^{n-1}$, homotopic to S^{n-1} the coordinate patch overlaps then involve maps of S^{n-1} into $SO(n, R)$. Thus, the questions of triviality reduce to questions concerning the homotopy group, $\pi_{n-1}(SO(n, R))$.

For $n = 3$, it is known that $\pi_2(SO(3, R)) = 0$, so the argument against a global trivialization used in the S^2 case does not hold. While this is not a sufficient condition to guarantee triviality, it turns out that other arguments insure that $T(S^3)$ is trivial. For example, embedding S^3 in $\mathcal{R}^4 = \{t, x^\mu\}$, $\mu = 1, 2, 3$, we find that the set of of three vectors in $T(S^3)$ defined by

$$(t\partial/\partial x^\mu - x^\mu \partial/\partial t) + \epsilon_{\mu\nu\rho} x^\nu \partial/\partial x^\rho,$$

where $\epsilon_{\mu\nu\rho}$ is the standard alternating symbol, defines a global basis for $T(S^3)$.

Using the tangent vector bundle as a model, we can proceed to define a general **fiber bundle**. Again the basic idea is to associate to every point of a manifold (the base space) a copy of another space (the fiber), locally, in bundle charts. On the overlap of two charts, we have to choose a function mapping a fiber copy of the one chart to a corresponding copy of the other chart. These maps are assumed to be group actions of the bundle group on the fiber.

Definition 4.1. The structure (E, M, F, G, π) defines a **coordinate fiber bundle**, B, over M if

---
[1]We are making use here of some interesting elementary topology, namely that $\pi_1(S^1, x_0) = \mathbb{Z}$.

(1) E, M, and F are smooth manifolds and $\pi : E \to M$ is a smooth map.
(2) G is a Lie group of left actions on F.
(3) There is an open covering of the base space, $M = \bigcup U$ and diffeomorphisms, $\phi_U : U \times F \to \pi^{-1}(U)$ with $\pi \cdot \phi_U$ the identity on U.
(4) Suppose $x \in U \cap V \neq \emptyset$. Then if $f \in F$,

$$\phi_U(x, f) = \phi_V(x, g_{VU}(x)f), \tag{4.9}$$

with $g_{VU} : U \cap V \to G$ smoothly.

E is the **bundle space**, M the **base space**, and π is the projection map. F is the **fiber** and G the **structure group**. The maps ϕ_U are the local **bundle coordinates** and g_{VU} the **transition functions**. It is easy to see that these functions must satisfy the important relationships, the cocycle conditions,

$$g_{WU} = g_{WV} \cdot g_{VU},$$

with the combination taken as group multiplication.

What we have presented in this definition is a **coordinate bundle**. This is reminiscent of defining a smooth manifold by giving a particular local coordinate patch representation of it. So, in a similar fashion, we say that two coordinate bundles over the same base space are **bundle equivalent** if the union of their coordinate patches provides a self-consistent coordinate bundle. This results in a definition of a bundle in a coordinate independent way as an equivalence class. However, in physical, and many mathematical, applications, it is often most convenient to work in a particular coordinate representation.

Of course, bundles can be generated in ways other than a coordination, and another notion of bundle equivalence can be expressed in terms of bundle maps, $\text{Hom}(E, E')$. These are smooth maps preserving fibers, that is, commuting with projections in a natural way,

$$\begin{array}{ccc} E & \xrightarrow{f} & E' \\ \pi \downarrow & & \downarrow \pi' \\ M & \xrightarrow{f^*} & M' \end{array}$$

The notion of bundle equivalence can then be defined in a natural way in terms of such maps.

An additional important bundle construct is that of **section of the bundle**. A map $s : M \to E$ with $\pi \circ s = id_M$ is called a section of the vector bundle. The set of all sections of the bundle E over M is denoted by $\Gamma(M, E)$, or sometimes as $\Omega^0(M; E)$. The latter notation describes sections as 0-forms, functions, over M with values in E. In physical applications, a section is just a physical field. Geometrically, an important question is whether or not there is a global cross section to the principal (tangent) bundle of frames. If such a cross section exists, the tangent frame bundle is trivial, i.e., a product. We will return to these points in more detail in the next chapter.

General fiber bundles can be characterized by special features of the fiber. For example, if the fiber is a vector space of a field K and $G \subset Gl(n, K)$ then the bundle is a **vector bundle**. One-dimensional (either real or complex) vector bundles are called **line bundles**. Perhaps the most important type is a **principal bundle** for which $F = G$ and the action is left G multiplication. However, in this case, the group can also act on the fiber from the right, an important point in defining connections in the following chapter. If $e \in P$, a principal G bundle, and $g \in G$, then we can define eg as $\phi_U(\pi(e), g_U g)$, where $e = \phi_U(\pi(e), g_U)$. Since left and right multiplication commute, this result is independent of the choice of ϕ_U. An important class of problems is associated with the notion of the **reduction of the bundle group**. The bundle group is said to be reduced from G to $G_0 \subset G$ if a local trivialization can be found in which the transition functions, (4.9), are all in G_0. As a special case, the bundle is **trivial**, or a **product**, if and only if the group can be reduced to its identity element.

From a principal bundle other bundles with group G can be defined for fibers on which G acts using the same transition functions. These are called **associated bundles**. Suppose we have a principal bundle, P and a fiber, F, on which G acts from the left. Then define an equivalence relation on $P \times F$ by $(e, f) \approx (eg, g^{-1}f)$, for $g \in G$. The resulting equivalence set, $P \times_G F$, is then easily seen to be a fiber bundle, B, with group G and having the same transition functions as P. In most of our applications, F will be a vector space, generally denoted by V over the group G. Finally, note that there is an important natural right action of G on the associated bundle B defined in the natural equivalence class invariant way.

Remark 4.2.
The associated bundle transition functions can be explicitly obtained as follows. Suppose $m \in U \cap V \neq \emptyset$ and $e = \phi_U(m, \mathbf{1}) = \phi_V(m, g_{VU}) \in P$. Then an element, $b \in B$, in the associated bundle can be defined as an equivalence class, $[e, f]$, $f \in F$. Then define the local coordination of B by

$$\bar{\phi}_U(m, f) \equiv [\phi_U(m, \mathbf{1}), f] = [\phi_V(m, g_{VU}), f] = [\phi_V(m, \mathbf{1}), g_{VU} f] = \bar{\phi}_V(m, g_{VU} f).$$

4.3 Geometry and Bundles

An important application of bundle theory is to differential geometry and its generalizations. As discussed in 3.4 and 3.5, differential geometry is concerned with norms of vectors (metric) and methods for parallel displacement and covariant differentiation of tangent vector fields (connections). With tangent vectors replaced by elements of a more general fiber it is natural to look for generalizations of these differential geometric tools to bundles.

A connection solves the important problem of lifting a path in the base space to one in the fiber in such a way that the lifted path represents **parallel displacement** in some sense. Recall that the fiber elements are values of a field at the projected point. So, given a path in the base space, use the connection to lift it to the bundle, obtaining the parallel displacement of the field element along the path. From here, it is an easy step to define the notion of **covariant derivative** which is perhaps more usually associated to the idea of a connection by physicists. Nevertheless, the definition of connections and covariant differentiation finds its most natural home at the principal bundle level. This will be especially clear when we discuss gauge theories and their physical applications below.

We begin by considering a principal bundle, since the discussion of associated bundles in the preceding section allows us to produce connections in associated bundles from ones defined in the principal one. Or, we can think in terms of lifting vectors, by reducing a base space path to its tangent vector at a point, lifting the path and considering the resulting bundle tangent vector.

Note that the action of the group induces a natural subspace of bundle tangent space, $T_b(E)$, defined by paths along fibers, or equivalently, vectors whose projection under π is zero. This is called the **vertical vector** space. For a principal bundle, the dimension of this subspace is clearly the dimension of G. Let $b \in E$ be the value of $\phi_U(m, g)$ for some U, m, g. Then the vertical vector space at b, **Ver**$_b$ is the image of $T_g(G)$ under the map ϕ_U for fixed m. Because of the homogeneity of Lie groups there is a natural identification $T_g(G) \approx \mathfrak{g} \equiv T_1(G)$, where **1** is the group identity and $\mathfrak{g}$ is the **Lie algebra** of G. Using right translation, we can map $\mathfrak{g}$ into the bundle vector space naturally. Let $A \in \mathfrak{g}$ be tangent to the curve a_t in G, with $a_0 = \mathbf{1}$. Then A_b^* is defined to be the tangent to the curve ba_t at $t = 0$. The tangent to the curve at b is called the **fundamental vector** defined by $A \in \mathfrak{g}$. The set of fundamental vectors generates the vector subspace at b, vertical vectors, **Ver**$_b$. It is easy to see that the vertical vectors are precisely the set of vectors with zero projection to the base space.

Remark 4.3.
We pause to make explicit some of the abstract group notation used in the discussion of the notion of vertical vectors in a principal bundle. For notational simplicity, stay in one bundle coordinate patch, and locally identify an element of the bundle with a pair $(m, g) \in U \times G$. Left multiplication in G lifts to tangent vectors, as

$$L_h A_g(f) = A_g(f(hg)).$$

The family of left invariant vectors then constitutes the **Lie Algebra**, $\mathfrak{g}$ of G. Each element of $\mathfrak{g}$ is thus uniquely defined by an element of $A \in T_1(G)$. It is easy to show that if A is defined as differentiation along a path at the identity, a_t, $a_0 = 1$, then A_g is generated by the path ga_t. This notation carries over immediately to the bundle definition of the fundamental, **vertical**, vector A^*, as acting on $f(m,g)$ by

$$A^*_{(m,g)}(f) = \frac{df(m, ga_t)}{dt}\Big|_{t=0}.$$

Now consider right translation, R_h,

$$R_h(A^*_{(m,g)})(f) \equiv A^*_{(m,g)}(f \cdot R_h) = \frac{df(ga_t h)}{dt}\Big|_{t=0} = (ad(h^{-1})A)^*(f), \qquad (4.10)$$

where $ad(h^{-1})A$ is generated by the path $h^{-1}a_t h$.
Left invariant vector fields satisfy

$$L_h X_g = X_{hg}, \text{ that is, } X_g(f(hg)) = X_{hg}(f(hg)).$$

Clearly such a field is uniquely determined by its value at the identity, X_1. A similar definition applies to forms. Now the general definition of $\mathfrak{g}$ as the set of left invariant vector fields means that it can be identified

$$\mathfrak{g} = T_1(G). \qquad (4.11)$$

Let $\{E_{1a}, a = 1, ..., N\}$, where N is the dimension of the group, span $T_1(G)$. Then each element of $\mathfrak{g}$ as a left invariant field is defined by N constant components, X^a,

$$X_g = X^a L_g E_{1a} \qquad (4.12)$$

Lie structure. It is easy to see that for any $X, Y \in \mathfrak{g}$,

$$[X, Y]_g = L_g[X, Y]_1 = (C^a_{bc} X^b Y^c) L_g E_{ga},$$

giving the famous **structure constants**.
Let θ^a_g be the dual basis for forms at g,

$$\theta^a_g(E_{gb}) = \delta^a_b.$$

From this, the famous Maurer-Cartan equations,

$$d\theta^a_g = \frac{1}{2} C^a_{bc} \theta^b_g \wedge \theta^c_g,$$

can be derived. Then the **canonical form**, $\theta \in \Omega(G; \mathfrak{g})$ is defined by left translation of

$$\theta_1 = E_{1a} \otimes \theta^a_1. \qquad (4.13)$$

Let the Lie algebra generator indices, a, be identified with the components of the matrix representation of G, so $a \to {}^i{}_j$, and

$$E_{1a} \to \partial^i_{1j} \equiv \Big(\frac{\partial}{\partial g^j{}_i}\Big)_{|g=1}$$

So

$$E_{ga}(f(g)) \to \Big(\frac{\partial}{\partial h^i{}_j}\Big)_{|h=1} f(gh) = \Big(\frac{\partial f(u)}{\partial u^k_l}\Big)\Big(\frac{\partial (gh)^k_l}{\partial h^i{}_j}\Big)_{|u=g} = \frac{\partial f}{\partial g^k_l} g^k{}_r \delta^r_i \delta^j_l,$$

or
$$E_{ga} \to g^k{}_i \frac{\partial}{\partial g^k{}_j}.$$

In this case, a left-invariant field is, from (4.12),
$$X^*_g = g^k{}_i X_1{}^i{}_j \frac{\partial}{\partial g^k{}_j},$$

so, in terms of components,
$$X^{*i}_g{}_j = (g \cdot X_1)^i{}_j, \qquad (4.14)$$

and left translation merely multiplies the components.

Finally, we will attempt to make transparent the often used (e.g., see [Atiyah (1979)], p. 8, or [Moore (2001)], p. 12) notation $g^{-1}dg$ as an element of $\Omega^1(G; \mathfrak{g})$ by making use of explicit components. Suppose $X(g) = X^i{}_j(g)\frac{\partial}{\partial g^i{}_j}$. Then define

$$g^{-1}dg(X(g)) \equiv (g^{-1})^r{}_s X^s{}_i(g) \partial^i_{1r}.$$

So,
$$g^{-1}dg(X^*_g) = X_1 \in \mathfrak{g}. \qquad (4.15)$$

Making membership in $\Omega^1(G; \mathfrak{g})$ more evident, write
$$g^{-1}dg = (g^{-1})^i{}_j dg^j{}_k \otimes \partial^k_{1i}. \qquad (4.16)$$

Clearly (4.15) implies that
$$g^{-1}dg = \theta,$$

the fundamental form on the Lie group, G.

4.3.1 Connections

Having defined the vertical subspace, we still find arbitrariness in the choice of a complementary space to fill out $T_b(E)$. Resolving this ambiguity is precisely what a **connection** does. This complementary space is called the **horizontal space, Hor$_b$**, and is required to satisfy

$$T_b = \mathbf{Ver}_b \oplus \mathbf{Hor}_b, \qquad (4.17)$$

and

$$\mathbf{Hor}_{bg} = R^*_g \mathbf{Hor}_b. \qquad (4.18)$$

The condition (4.18) is essentially the "naturality" of the horizontal plus vertical decomposition as we move along fibers under the group action. The splitting in (4.17) and (4.18) is used to define the lift of vectors in $T(M)$ to horizontal vectors in $T(E)$ in a "natural" way. Given $V \in T_m(M)$ and some $b \in E$, $\pi(b) = m$. Choose any $X \in T_b(E)$ such that $\pi(X) = V$. Then

$\hat{V}_b$ is the horizontal component of X defined by (4.17) and defines the lift of V. Condition (4.18) is then the "naturality" condition,

$$\hat{V}_{bg} = R_g^* \hat{V}_b.$$

Given some (smooth) path in the base space, $u : I \to M$, we can find many smooth " lifts" of this path to one taking $I \to E$. Any local bundle coordination provides at least a part of this lift. What the connection provides is a uniquely defined **horizontal lift**, $\hat{u} : I \to E$, defined so that the tangent vector to $\hat{u}$ is the horizontal lift of the tangent vector to u.

A more easily handled alternative to (4.17) and (4.18), is provided by the **connection form**, $\omega \in \Omega^1(E; \mathfrak{g})$ taking a vector $X \in T(E)$ into its vertical component generated by $\omega(X) \in \mathfrak{g}$. Specifically,

$$\omega_b(X) = A \in \mathfrak{g}, \quad \mathbf{Ver}_b(X) = A_b^*(A), \tag{4.19}$$

so that

$$\omega_{bg} = adg^{-1}\omega_b. \tag{4.20}$$

Thus, (4.17) and (4.18) lead to the definition of ω satisfying (4.19) and (4.20). Conversely, these two equations lead to the definitions of **Hor, Ver** satisfying (4.17) and (4.18) by choosing

$$\mathbf{Hor}_b(X) = X - (\omega_b(X))^*.$$

Note that

$$\omega_b(A_b^*) = A. \tag{4.21}$$

This bundle definition of **connection** applies to any bundle and can be represented by generalizations of the Christoffel symbols of elementary coordinate geometry. Consider a local trivialization of the bundle given by $\phi_U : U \times G \to \pi^{-1}(U) \subset P$ for some principal bundle P. The pullback of this applied to forms provides

$$\phi_{U*} : \Omega^1(\pi^{-1}(U); \mathfrak{g}) \to \Omega^1(U; \mathfrak{g}) \oplus \Omega^1(G; \mathfrak{g}), \tag{4.22}$$

giving

$$\phi_{U*}(\omega_b) = \gamma_U + \chi_U. \tag{4.23}$$

Using (4.21), we see that χ must be the canonical form, θ. The behavior under right translations implies that the first term is of the form

$$\gamma_U = g^{-1} \Gamma_U g, \tag{4.24}$$

where Γ_U depends only on $m = \pi(b)$, and thus is an element of $\Omega^1(U; \mathfrak{g})$, which can be expanded in terms of local coordinates,

$$\Gamma_U(m) = \Gamma^i_{U\ j\mu} dm^\mu \otimes \partial^j_{1i}, \qquad (4.25)$$

where we are assuming a general vector bundle, with group G a subset of $Gl(n, \mathbb{K})$, thus represented by matrices, $(g^i{}_j)$. These Γ-symbols are the bundle generalization of the **Christoffel symbols** of elementary coordinate geometry. Thus (4.23) expands to

$$\phi_{U*}(\omega_b) = g^{-1}\Gamma_U(\pi(b))g + g^{-1}dg. \qquad (4.26)$$

Clearly the local representation of the connection form depends on the local trivialization, ϕ_U. From the bundle equation, (4.9), $\phi_U(m, g) = \phi_V(m, h)$ where $h = g_{VU}(m)g$. Note that the group elements are, in general, functions of the base space point, $m = \pi(b)$. Replacing $U \to V$, $g \to h$ in (4.26)

$$g^{-1}\Gamma_U(m)g + g^{-1}dg = g^{-1}g_{VU}^{-1}(m)\Gamma_V(m)g_{VU}(m)g + g^{-1}g_{VU}(m)^{-1}d(g_{VU}(m)g). \qquad (4.27)$$

The dependence of g_{UV} on the base space point, m, then implies that

$$\Gamma_U(m) = g_{VU}(m)^{-1}\Gamma_V g_{VU}(m) + g_{VU}^{-1}dg_{VU} \in \Omega^1(U; \mathfrak{g}). \qquad (4.28)$$

This equation is the bundle generalization of the non-linear Christoffel transformations under coordinate changes familiar in elementary coordinate differential geometry.

Consider the **lift** of $T_x(M)$ for the tangent bundle of a smooth manifold. Start by computing the lifts, $\hat{\partial}/\partial x^\mu$, of the bases, $\partial/\partial x^\mu$, using the local coordinate trivialization, $\phi_U(x, g)$. Since $\pi \hat{X} = X$,

$$\hat{\partial}/\partial x^\mu = \partial/\partial x^\mu + B^i{}_{j\mu} \partial/\partial g^i_j. \qquad (4.29)$$

From (4.26) and (4.25), the horizontal condition, $\omega(\hat{\partial}/\partial x^\mu) = 0$ gives

$$B^i{}_{j\mu} = -\Gamma^i_{U\ k\mu} g^k_j, \qquad (4.30)$$

so

$$\hat{X} = \phi^*_U(X^\mu \partial/\partial x^\mu - \Gamma^i_{U\ k\mu} X^\mu g^k_j \partial/\partial g^i_j). \qquad (4.31)$$

Now consider an explicit construction of the **bundle of frames** or **repere bundle**, BF, as a principle $Gl(n, R)$ bundle. Start with the point set $\{(m, \Xi)\}$, where Ξ is a frame at $m \in M$, that is a set of n linearly independent tangent vectors, $\Xi = \{\xi_\mu\}$, $\xi_\mu \in T_m(M)$. Since not every manifold is parallelizable, this cannot in general be a trivial bundle. So, local coordination is required to define this as a principal $Gl(n, R)$ bundle using smooth coordinate patches as follows. If $g = (g^\mu{}_\nu) \in G$, and U is a local coordinate patch over M, with coordinates x^μ, define

$$\phi_U(m, g) = (m, \Xi),$$

where

$$\xi_\mu = g^\nu{}_\mu \frac{\partial}{\partial x^\nu},$$

and the x^ν are local coordinates in U. The usual tangent vector bundle can then be reproduced as the associated vector bundle. Thus if $F = \mathbb{R}^n$ is the vector space of column matrices, then an element of the associated bundle, $BF \times_G F$ is an equivalence class, $[(m, \Xi), f]$, that can be identified with a tangent vector,

$$[(m, \Xi), f] \to f^\nu \xi_\nu \in T_m(M).$$

The equivalence

$$[(m, \Xi h), f] = [(m, \Xi), hf],$$

describes the usual "contravariant transformation law" for the vector components under a change in basis.

Finally, the last and perhaps most familiar term associated to a connection is that of **covariant derivative**. This is closely related to the **horizontal** lift idea above. Given a tangent vector to the base space, $X \in T_x(M)$, defined by differentiation along the path u through x. The horizontal lift, $\hat{u}$, then provides a path in any associated vector bundle, and differentiation along $\hat{u}$ provides ∇_X.

As usual, let P be a principal G bundle over M, with local coordination, ϕ_U. Now assume G is a linear group of matrices, $\{g^i{}_j\}$, acting on a vector space, $V = \{v^j\}$, and $E = P \times_G V$ be the associated bundle. A cross section, $\sigma \in \Omega^0(M; E)$ is a V **vector field**,

$$\sigma(x) = [p(x)g, g^{-1}v(x)] \in E, \tag{4.32}$$

where $[*, *]$ represents the G-equivalence class. The **covariant derivative**, ∇, can then be defined as a map

$$\nabla : \Omega^0(M; E) \to \Omega^1(M; E), \tag{4.33}$$

as follows. Let $X \in T(M)$, represented locally $X = X^\mu \partial/\partial x^\mu$, with lift, $\hat{X}$, defined in (4.31). Then

$$\nabla(\sigma)(X) = [p(x)g, \hat{X}(g^{-1}v(x))]. \tag{4.34}$$

At $g = 1$, in terms of the local ϕ_U,
$$\hat{X}(g^{-1}v(x))^i|_{g=1} = X(v^i(x)) + \Gamma^i_{U\ j\mu}v^j(x)x^\mu. \tag{4.35}$$
In more familiar notation, the covariant derivative expressed in (4.34), locally in (4.35), is written
$$\nabla_X \sigma.$$
The generalization to associated "tensor" bundles corresponding to representations, ρ, of G (so elements of V may have multiple indices, superscripts for vectors, subscripts for forms) is
$$\nabla_X \sigma_\rho = [p(x)g, \hat{X}\rho(g^{-1})(v(x))]. \tag{4.36}$$
For example, for one-forms (co-vectors in older terminology), $v^i \to v_i$, and $\rho(v)_j = v_i g^i{}_j$, so (4.35) becomes
$$\hat{X}(g^{-1}v(x))^i|_{g=e} = X(v_i(x)) - \Gamma^j_{U\ i\mu}v_j(x)x^\mu. \tag{4.37}$$
Alternatively, the connection can be considered as a differentiation
$$\nabla : \Omega^p(M; B) \to \Omega^{p+1}(M; B),$$
as follows. Given a connection on a principle bundle with group G and an associated vector bundle, $B = E \otimes_G V$, recall that $\Omega^p(M; B)$ are the B-valued p-forms over the base space, M. First consider the case $p = 0$. An element of $\Omega^0(M; B)$ is a smooth cross section, in this case a smooth vector field. The covariant derivative defines from this a B-valued (that is, vector field-valued) one-form. So, if $X \in \Omega^0(M; B)$ let $X(x) = [p, v(x)]$ be the associated bundle equivalence class definition of the section. Now consider some $Y \in T_x(X)$ defined by differentiation along u with lift $\hat{u}$. The vector field, X, along the path is represented by $[\hat{u}(t), \hat{v}(u(t))]$, and
$$\nabla_Y X = [\hat{u}(t), \frac{d}{dt}\hat{v}(u(t))|_{t=0}. \tag{4.38}$$
The extension to higher values of p is easy to define, using the Leibniz rules for a derivative operator.

Finally, we present these abstract definitions in perhaps more familiar terms, using local coordinate representations for the principal bundle of frames, P, and the associated tangent vector bundle. Let the bundle coordinate patches be the local manifold patches for the base space, U, with manifold coordinates $\{x^\mu\} : U \to \mathbb{R}^n$. The group, G, is the general linear group. Define the bundle maps from x and $g \in G$ by
$$\phi_U(x, g) = \{g^\mu{}_\nu \frac{\partial}{\partial x^\mu}\}. \tag{4.39}$$

Then, the most general connection form will have the local representation,

$$\omega = (g^{-1})^\nu{}_\rho [g^\eta{}_\mu \Gamma^\rho{}_{\eta\iota} dx^\iota + dg^\rho{}_\mu] \otimes \partial^\mu_1{}_\nu. \tag{4.40}$$

It is a straightforward exercise to show that this definition is independent of the manifold and bundle coordinations and does indeed satisfy the connection conditions (4.19) and (4.20).

The last topic of interest to us here is the **curvature form** defined by the connection. Starting from the connection as the $\mathfrak{g}$-valued one form on P, define Ω as the $\mathfrak{g}$-valued two form,

$$\Omega = d\omega + \omega \wedge \omega. \tag{4.41}$$

Here the wedge product in the second term on the right side includes the Lie bracket. The components of Ω can then be related to the curvature tensor components for the tangent vector bundle case, (3.19), by

$$\Omega = \frac{1}{2} R^\nu{}_{\mu\rho\sigma} \theta^\rho \wedge \theta^\sigma \otimes \partial^\mu_1{}_\nu. \tag{4.42}$$

4.4 Gauge Theory: Some Physics

In this section we will introduce ideas and constructs from physics which are necessary for understanding the broad area of **gauge theory** that has crept from physics into mathematics and been of significant value in studying the mathematics of differential topology. We will try to provide some background and physical motivation that will help to justify the introduction of these equations for the mathematicians.

For various reasons, it is easier to start off using the mathematically old-fashioned formalism of local coordinates and components. Later, we will summarize matters in the more compact and elegant formalism of bundle connections and differential forms.

Begin with the formalism appropriate to **Special Relativity** which assumes (global) coordinates and flat metric. So $M = \mathbb{R}^4 = \{x^\mu | \mu = 0, 1, 2, 3\}$ is the spacetime model, with standard smoothness and Minkowski metric, η. Recall also the summation and raising/lowering of indices conventions.

In classical (that is, pre-quantum) physics, the basic elements are **point particles**, whose histories are time-like curves $x^\mu(\tau)$ in M, restricted by some dynamical law, such as Newton's, modified for special relativity to the form

$$m \frac{d^2 x^\mu}{d\tau^2} = \mathcal{F}^\mu, \tag{4.43}$$

where m is a constant, the particle's mass, and $\mathcal{F}^\mu$ are the components of the **four-force**, representing external influences. $d\tau$ is defined in (4.45) below.

At this point, we pause to review some important and relevant issues at the basis of relativity. First, note that the Minkowski metric form, η, is central to this theory. The homogeneous group leaving this metric invariant in form (with positive determinant) is $SO(1,3)$, the proper **Lorentz group**. It was tacitly assumed in the original form of special relativity that the spacetime model, M, is simply standard $\mathbb{R}^4$. Each smooth atlas on M describes one physical **Reference Frame**. The x^μ can be thought of as results of physical measurements, perhaps only idealized, of locations in space and time provided by some physical reference frame. Those reference frames in which the metric components are the standard Minkowski values and the laws of mechanics as expressed in (4.43) are valid are called **Inertial Reference Frames, IRF's**. Of course, we must first assume that at least one IRF exists. Historically, Einstein was led to special relativity from study of **Maxwell's electromagnetic field theory**, summarized in a set of four partial differential equations for two 3-vector fields over pre-relativistic space and time. These equations are presented below in relativistic form as (4.49) and (4.50). These equations predict a specific speed, now called the **speed of light**, c, for all electromagnetic radiation. Einstein's first formulation of his **Principle of Special Relativity** was that this speed, c, should be the same for all observers. If we choose units in which this value is one,[2] the intervals which are zeros of

$$ds^2 = \eta_{\mu\nu}dx^\mu dx^\nu = -dt^2 + dx^2 + dy^2 + dz^2, \tag{4.44}$$

are invariant under change of IRF, mathematically a diffeomorphism. Such intervals are called **light-like**. If $ds^2 > 0$, the interval is **space-like**. Finally, if $ds^2 < 0$ the interval is **time-like** and ds^2 is replaced by the square of **proper time** interval,

$$d\tau^2 = -ds^2. \tag{4.45}$$

One further assumption of spacetime homogeneity (linearity for the diffeomorphisms) and overall scale invariance leads to the definition of $SO(1,3)$ as the group of matrix transformations, L, for which

$$L\eta L^T = \eta, \quad \det L = 1. \tag{4.46}$$

---

[2] Of course, since the speed of something depends on choice of units, which are arbitrary, we must be careful in formulating this principle. However, a full discussion of this would take us too far afield.

A final expression of the Principle of Special Relativity might now be

Principle of Special Relativity: *All laws of physics, including Maxwell's, must be invariant in form in all Inertial Reference Frames.*

This was of course later generalized by Einstein to the **Principle of General Relativity**, which requires all laws to be invariant in form under *all* (smooth) changes of reference frames, or, mathematically, all **diffeomorphisms** of M. The extension of these relativity principles to include **internal space transformations**, i.e., bundle group operations on fibers, gives rise to **gauge theory** as understood in contemporary physics.

Returning to mechanics, note that from the definition of $d\tau$ the velocity vector must have constant norm (square= -1), so in particular, (4.43) implies

$$\eta_{\mu\nu}\mathcal{F}^\mu \frac{dx^\nu}{d\tau} = 0. \tag{4.47}$$

The problem of understanding these influences, forces, led to classical **field theory**. Roughly, a physical F-**field** is an F valued function over M, later replaced by **section** of a bundle with fiber F. We will now focus on the **electromagnetic** field. One easy way to satisfy (4.47) is to have the force be of the form

$$\mathcal{F}^\mu = qF^{\mu\nu}\frac{dx_\nu}{d\tau}, \quad \text{with } F^{\mu\nu} + F^{\nu\mu} = 0, \tag{4.48}$$

a form actually provided by Maxwell's electromagnetic theory formulated in its complete form only some 40 years before Einstein's work. So, we can satisfy (4.48) by choosing q to be the electric **charge**, an intrinsic property of the particle and $F^{\mu\nu}$ as the components of an antisymmetric two-tensor representing **electromagnetic field**.

(4.48) provides the influence of the field on particles. But physics must also provide a theory of how the charges produce the field. This is provided by **Maxwell's equations,** expressed here as

$$F^{\mu\nu}_{,\nu} = J^\mu, \tag{4.49}$$

$$F_{\mu\nu,\rho} + F_{\nu\rho,\mu} + F_{\rho\mu,\nu} = 0, \tag{4.50}$$

where J^μ are the components of the **charge current density**, which acts as the "source" of the electromagnetic field and describes how matter produces electromagnetic fields. This form of particle-field interaction was the basis for fundamental physics before quantum theory. Finally, we note that these equations, (4.49) and (4.50), as well as their later forms, (4.53) and

(4.52) are invariant in form under Lorentz transformations, as required by special relativity.

Since Cartan we have learned that a more productive way to handle antisymmetric objects is by use of exterior forms and calculus. Thus define the electromagnetic field as a two-form

$$\mathbf{F} = \frac{1}{2} F_{\mu\nu} dx^\mu \wedge dx^\nu, \tag{4.51}$$

and (4.50) becomes simply

$$d\mathbf{F} = 0, \tag{4.52}$$

and (4.49) becomes

$$d * \mathbf{F} = *\mathbf{J}. \tag{4.53}$$

Note that we are using the Hodge star operator defined using the Minkowski metric of signature $-,+,+,+$ so

$$*1 = dx^0 \wedge dx^1 \wedge dx^2 \wedge dx^3, \quad **1 = -1,$$

and $\mathbf{J}$ is the current density one form, with

$$\mathbf{J} = -\rho dx^0 + J_x dx^1 \ldots, \tag{4.54}$$

ρ is charge density, J_x is the usual x-component of spatial charge current, etc. To relate this formalism to the pre-relativistic Maxwell theory, note

$$F_{01} = -E_x, \quad F_{23} = B_x, \text{etc.},$$

$$(*F)_{01} = B_x, \quad (*F)_{23} = E_x, \text{etc.}$$

and the two three-vectors, (E_x, E_y, E_z) and (B_x, B_y, B_z) are the pre-relativistic descriptions of the electric and magnetic fields, now unified in the single field, electromagnetic form, $\mathbf{F}$. From (4.54) the total **charge** in a spatial volume, V is

$$Q(V) = \int_V J^0 dx dy dz,$$

or considering V as a singular chain, and integration as the de Rham cohomology pairing, this equation can be written

$$Q(V) = V(*\mathbf{J}) = \int_V *\mathbf{J}. \tag{4.55}$$

From the Maxwell equation (4.53),

$$Q(V) = \int_V d * \mathbf{F} = \int_{S=\partial V} *\mathbf{F} = \Phi(S), \tag{4.56}$$

where $\Phi(S)$ is called the electric **flux** through the surface S.

Finally, the physical requirement of "conservation of charge" can be expressed locally as
$$J^\mu_{,\mu} = 0,$$
or
$$d * \mathbf{J} = 0. \tag{4.57}$$
So, conservation of charge is built into the Maxwell equations in the form (4.53).

Of course, since we are dealing with a topologically trivial space, (4.52) can be replaced globally by
$$\mathbf{F} = d\mathbf{A}, \tag{4.58}$$
for a one-form, $\mathbf{A}$, called the electromagnetic **gauge potential**. In pre-relativistic terms,
$$\mathbf{A} = -V dx^0 + A_x dx^1 \ldots,$$
with V the scalar potential of electrostatics. Then (4.53) becomes
$$d * d\mathbf{A} = *\mathbf{J}. \tag{4.59}$$
In terms of local spacetime coordinates in vacuum, $\mathbf{J} = 0$,
$$*d * d\mathbf{A} = \Box^2 \mathbf{A} = (-\frac{\partial^2}{\partial t^2} + \frac{\partial^2}{\partial x^2} + \frac{\partial^2}{\partial y^2} + \frac{\partial^2}{\partial z^2})\mathbf{A} = 0, \tag{4.60}$$
with solutions which are "moving shapes," that is, waves, moving with speed of light, 1 in standard units. So, again, Einstein's principle of special relativity by requiring invariance in form of Maxwell's equations, and thus (4.60), predicts the invariance of the speed of light.

These equations lead naturally to the physical idea of a **gauge transformation**. Physical observations are only of forces, and thus of $\mathbf{F}$. However, this physical field does not uniquely determine the gauge field, $\mathbf{A}$, which at this stage may only seem to be of formal importance. With the tools of exterior calculus at hand, it is of course quite easy to express this ambiguity:
$$\mathbf{F} = d\mathbf{A} = d(\mathbf{A} + d\lambda), \tag{4.61}$$
for any smooth scalar field λ. Physics expressed $\mathbf{A} \to \mathbf{A} + d\lambda$ as a gauge transformation, while mathematically it is obvious that this entire structure is ripe for expression in terms of de Rham cohomology. Recall that this discussion has been initially constrained to the topologically trivial case,

but if we relax this condition, then de Rham cohomology will give rise to **wormholes** replacing charges. For example, if $H_2(M) \neq 0$, there is a closed 2-cycle, S which is not a bound. So we could have $d * \mathbf{F} = *\mathbf{J} = 0$ but

$$\int_S *\mathbf{F} = \Phi(S) = Q \neq 0. \tag{4.62}$$

If this is not zero, the space containing a closed surface, S, not a bound, can support a $\mathbf{F}$ corresponding to non-zero charge, even though there is no charge in classical sense of (4.55). In fact, such exploration has been on going in physics since the pioneering work of Wheeler, Misner, et al., in the 1950's, under the general heading of **geometrodynamics**[Wheeler (1962)]. While very interesting in its own right, illustrating the potential applications of topology even in classical physics, this topic is not an integral part of our aim in this book, so we will leave wormholes with this brief mention.

Classical and contemporary physics have found a much more fruitful way to express the field equations and conservation laws of a theory from **actions** using **Lagrangians**. For simplicity begin again with local, coordinate representation, and assume that the physical field can be represented as an N-vector with components, $\phi^a, a = 1, \ldots, N$ in some vector space with properties discussed later. A Lagrangian density is then a function of the fields and their derivatives, $\mathcal{L}(\phi^a, \phi^a_{,\mu})$. If V is some 4-volume, the action, $\mathcal{S}$ is then defined by

$$\mathcal{S} = \int_V \mathcal{L} * 1.$$

The physical field equations are then to be derived from the "least-action" (more properly "extremal-action") principle

$$\delta_\phi \mathcal{S} = 0, \tag{4.63}$$

subject to certain conditions. An explicit evaluation of this is

$$\delta_\phi \mathcal{S} = \int_V [(\frac{\partial \mathcal{L}}{\partial \phi^a_{,\mu}} \delta \phi^a)_{,\mu} + (-\frac{\partial}{\partial x^\mu} \frac{\partial \mathcal{L}}{\partial \phi^a_{,\mu}} + \frac{\partial \mathcal{L}}{\partial \phi^a}) \delta \phi^a] * 1. \tag{4.64}$$

The first term, the "divergence," can be converted to a surface integral (we will do this using forms later),

$$B(V, \delta\phi) = \int_V (\frac{\partial \mathcal{L}}{\partial \phi^a_{,\mu}} \delta \phi^a)_{,i} * 1 = \int_{\partial V} \frac{\partial \mathcal{L}}{\partial \phi^a_{,\mu}} \delta \phi^a dS^\mu, \tag{4.65}$$

where dS^μ are the components of the three-area element of ∂V.

The standard approach to deriving field equations is then to require $B(V, \delta\phi) = 0$ either because ∂V is empty, or by requiring that $\delta\phi^a|_{\partial V} = 0$.

Given this condition then, we have that the variational principle, (4.63), produces the **Euler-Lagrange** equations,

$$-\frac{\partial}{\partial x^\mu}\frac{\partial \mathcal{L}}{\partial \phi^a_{,\mu}} + \frac{\partial \mathcal{L}}{\partial \phi^a} = 0. \tag{4.66}$$

This variational principle formalism for a field theory lends itself directly to a theorem of central importance to our gauge theory studies, **Noether's theorem** relating symmetries to conservation laws. Suppose the theory has a **symmetry group,** G, a Lie group. Using the explicit coordinate representation in (4.64), suppose a variation is the result of an "infinitesimal"[3] symmetry, so that δS is zero for any ϕ^a. If the fields also satisfy the field equations, then the quantity in the second set of parenthesis vanishes and

$$(\frac{\partial \mathcal{L}}{\partial \phi^a_{,\mu}}\delta\phi^a)_{,\mu} = 0,$$

or, in form notation,

$$d * \mathbf{J}_\delta = 0, \quad \mathbf{J}_\delta = C(\frac{\partial \mathcal{L}}{\partial \phi^a_{,\mu}}\delta\phi^a)\eta_{\mu\nu}dx^\nu. \tag{4.67}$$

for a symmetry variation, δ, and C any constant. $\mathbf{J}_\delta$ is then the conserved current associated to this symmetry.

Using this formalism, the Maxwell field equations (4.49), (4.50) can be derived from a Lagrangian,

$$\mathcal{L} = -\frac{1}{4}F_{\mu\nu}F^{\mu\nu} + J^\nu A_\nu, \tag{4.68}$$

In this case, the index a is the coordinate index μ, $\phi^a \to A^\mu$.

Before proceeding further, let us re-express the action formulation with the more satisfying index-free formalism of exterior forms. Note that

$$\mathbf{F} \wedge *\mathbf{F} = \frac{1}{4}F_{\mu\nu}(*F)_{\rho\eta}\epsilon^{\mu\nu\rho\eta} * 1 = \frac{1}{4}\frac{1}{2}F_{\mu\nu}\epsilon^{\iota\kappa}{}_{\rho\eta}F_{\iota\kappa}\epsilon^{\mu\nu\rho\eta} * 1.$$

This reduces to

$$\mathbf{F} \wedge *\mathbf{F} = \frac{1}{4}F_{\mu\nu}F_{\iota\kappa}(g^{\mu\kappa}g^{\nu\iota} - g^{\mu\iota}g^{\nu\kappa}) * 1 = \frac{1}{2}F_{\mu\nu}F^{\mu\nu} * 1 = (-E^2 + B^2) * 1.$$

The Lagrangian,(4.68), can then be expressed in terms of forms

$$\mathcal{S} = \int_V (-\frac{1}{2}d\mathbf{A} \wedge *d\mathbf{A} + \mathbf{A} \wedge *\mathbf{J}), \tag{4.69}$$

and variation,

$$\delta \mathcal{S} = \int_V (-d\delta\mathbf{A} \wedge *d\mathbf{A} + \delta\mathbf{A} \wedge *\mathbf{J}) = B(V, \delta\phi) + \int_V \delta\mathbf{A} \wedge (-d * d\mathbf{A} + *\mathbf{J}). \tag{4.70}$$

---

[3] Think of $\delta\phi^a$ as $X(\phi^a)$ for X an element of the Lie algebra of G.

The term,

$$B(V, \delta\phi) = \int_{\partial V} \delta\mathbf{A} \wedge *d\mathbf{A} \qquad (4.71)$$

is often referred to as the "topological" part, since it depends on the topology involved in the transition between a volume and its boundary. Under the usual assumption that B must be zero, this variational principle gives (4.59) where the physically observable field, $\mathbf{F}$, is given by (4.58).

At this point we must pause, however, to note that **the interaction term has removed gauge invariance.** That is, in (4.69), the free part, $\mathcal{S}_{EM} = \int_V (-\frac{1}{2} d\mathbf{A} \wedge *d\mathbf{A})$, containing only $\mathbf{A}$ terms, is invariant under the gauge transformation, (4.61), while the interaction part, $\mathcal{S}_I = \int (\mathbf{A} \wedge *\mathbf{J})$, is not as it stands. Of course, we have only a theory of the electromagnetic field, $\mathbf{A}$, and must also construct one for the **source** matter/fields that contribute the current, $\mathbf{J}$. Thus, we need another Lagrangian and action, $\mathcal{S}_J = \int \mathcal{L}_J$, so that the total action is

$$\mathcal{S} = \mathcal{S}_{EM} + \mathcal{S}_J + \mathcal{S}_I = \int (-\frac{1}{2} d\mathbf{A} \wedge *d\mathbf{A} + \mathbf{A} \wedge \mathbf{J} + \mathcal{L}_J). \qquad (4.72)$$

It is in the analysis of equations of this form, (4.72), that the physics, and later mathematics, of **gauge theory** has been formed.

To explore this further, we begin by looking at the transition motivated by quantum theory and build **field** theories (models) for what were classically described as **particles.** In the first steps, these fields were "probability densities," e.g., the Schrödinger wave function, $\psi(x^\mu)$ was such that $|\psi(x^\mu)|^2 dx dy dz$ is proportional to the probability of finding the classical particle in the spatial volume, $dx dy dz$ at time $t = x^0$. This originally provided a workable tool for investigating certain basic problems such as the behavior of an electron moving through slits, or being in a stationary states in a Hydrogen atom, etc. However, it soon became clear that this was inadequate to solve more complicated problems and to include special relativity. Most importantly however, this original approach to quantum fields had to be modified because it involved a hybrid approach in which some fields, such as the electromagnetic ones, described forces, and others, such as ψ described probability. In fact, Planck's first steps toward quantum theory involved "quantizing" the electromagnetic field, making it the field associated with "particles," later called **photons.** So, in some sense, Schrödinger's ψ is the quantum field for electrons, and $\mathbf{F}$ is the field for photons, etc. **Quantum field theory,** currently the most well established formalism for quantum phenomena, developed from this point. Fortunately, we need go no further into this complex subject, but merely sketch the mathematics of the subject which are important for gauge theory.

From special relativistic mechanics we identify the classical mechanical quantities **energy, momentum**, with certain time/space components of a four vector p^μ. For example, for a point particle,

$$p^\mu = m\frac{dx^\mu}{d\tau},$$

with $p^\mu = (E, \vec{p})$, relating the four-vector relativistic notation to pre-relativistic energy and three-momentum. From the definition of $d\tau$, we get

$$p^\mu p_\mu + m^2 = 0, \qquad (4.73)$$

or,

$$E^2 = |\vec{p}|^2 + m^2, \qquad (4.74)$$

Now we make the huge leap to **quantum theory**, replacing the notion of a point particle with that of a (generally complex) field, ϕ.

Remark 4.4.

We do not have time or space to give more than a passing idea of the physical interpretation of quantum fields. Suffice it to say that originally they provided **probability** information about outcomes to experiments asking classical questions such as "where-when, what are the values of momentum, energy," etc., but in the later development these fields, such as **A** for electromagnetism, and our sample ϕ, become **operators** on a Hilbert space whose rays represent states describing number of particles, such as photons for **A**, or some idealized scalar particle (Higgs?) for ϕ, having specific properties. In particular, these field-operators "create/annihilate" particles.

In making the transition from classical to quantum physics various rules were formulated, replacing classical variables (numbers) with quantum **operators**, and then imposing classical conditions, such as (4.73). In particular, the four-momentum components become complex differentiation operators,

$$p_\mu \to -i\partial_\mu, \qquad (4.75)$$

where here $i = \sqrt{-1}$, and not an index. We are also using an abbreviation familiar in quantum physics,

$$\partial_\mu \equiv \frac{\partial}{\partial x^\mu}, \qquad (4.76)$$

So the classical (4.73) becomes a differential equation

$$\Box^2 \phi - m^2 \phi = 0. \tag{4.77}$$

This is the **Klein-Gordon** equation which was first proposed as the relativistic generalization of the famous **Schroedinger** equation. In quantum mechanics, we associate particles to fields, or, more precisely, classes of particles to classes of fields. Thus, for (4.77) the field is a spacetime scalar (albeit complex) and with further analysis turns out to have zero internal angular momentum, or **spin** in the physical sense. We thus describe the particle as a scalar, spin zero, particle. Such particles are also called **Bosons**, and satisfy "Bose-Einstein statistics." A complete explanation of this fact would lead us far away from our purposes here. Suffice it to say that such particles have never been directly observed, but speculation on their existence continues because they might resolve certain theoretical difficulties.

It turns out that most matter in the everyday world, and probably in the local cosmological neighborhood, consists of particles of spin 1/2, called **Fermions**, because of their statistical behavior. We refer to the familiar classes of electrons, protons, neutrons, although there is increasing evidence that the latter two are not elementary but rather bound states of **quarks**. So, perhaps we should say that almost all familiar matter consists of electrons and quarks, both spin 1/2 fermions. It was **Dirac** who successfully developed the appropriate relativistic equation to describe such particles/fields.

We have always understood causality or time evolution in terms of equations of first order in time derivative, whereas (4.77) is obviously of second order. So Dirac searched for a "square-root" of this equation, postulating

$$(i\gamma^\mu \partial_\mu + m)\psi = 0, \tag{4.78}$$

such that the application of $(-i\gamma^\mu \partial_\mu + m)$ to this would result in the quantum energy-momentum relation of (4.77). Of course, this is not possible if the γ^μ are scalars, so he suggested that they be matrices/operators, satisfying

$$\gamma^\mu \gamma^\nu + \gamma^\nu \gamma^\mu = -2\eta^{\mu\nu}. \tag{4.79}$$

Introducing the very useful notation

$$\not{\partial} \equiv \gamma^\mu \partial_\mu,$$

(4.78) becomes

$$(i\not{\partial} + m)\psi = 0. \tag{4.80}$$

Mathematicians will recognize (4.79) as the defining relation for the **Clifford algebra** leading to the definition of the Spin group, covering the isometry group of the metric $\eta^{\mu\nu}$. In mathematics, the indefinite $\eta^{\mu\nu}$ might be replaced by a positive definite one. The book by Elie Cartan[Cartan (1966)] presents a definitive classical summary, including both definite and indefinite metrics. A more brief summary of Spin from the mathematical viewpoint is given in the first chapter of the book by Morgan[Morgan (1996)].

For the case of physical interest, the metric is indefinite, so the multiplicative group of Clifford algebra units should be labeled Spin(1,3). In any event, it provides a covering 2-1 representation of the Lorentz group, the isometry group of $\eta_{\mu\nu}$, as follows. If v^μ are the components of a real spacetime vector, define $v = v^\mu \gamma_\mu$. Then if $u_1, ..., u_k$ are Clifford algebra units,

$$v \to v' = u_1 \cdots u_k v u_k \cdots u_1, \qquad (4.81)$$

is a Lorentz transformation. Because of the indefinite character of the metric, Spin(1,3) cannot be a unitary representation[4], so $\psi^\dagger \psi$ will not be a spacetime scalar. However, $\bar\psi \psi$ is, where $\bar\psi \equiv \psi^\dagger \gamma_0$. Furthermore $\bar\psi \gamma^\mu \psi$ transform as the components of a spacetime vector. Finally, we note that the lowest dimensional representation (if $m \neq 0$) of ψ is a complex four-dimensional one, so the Fermion field can be thought of as a cross section of a spinor bundle, with fiber $\mathbb{C}^4$, and group (apparently, see below) Spin(1,3).

Of course we should point out that we have been dealing only with the **free** particle case. The problem of understanding **interactions** for such fields will then lead us to our goal of gauge theory. The key to this will be the observation above that in this quantum representation, with complex field, the group of the spinor bundle is actually Spin$(1,3) \otimes U(1)$. This extended group is the quantum **symmetry** group of this spin 1/2 model.

Remark 4.5.

The extension of the group from the spacetime symmetry of the Lorentz group, or its cover, Spin(1,3) to Spin$(1,3) \otimes U(1)$ is a noteworthy construction. The symmetry group of the theory is now a direct product group consisting of an **internal** part, $U(1)$, and **external**, or spacetime part, Spin(1,3). Later we will mention the natural extension of the internal $U(1)$ to larger groups. The suggestion that the rather artificial "direct product," internal cross external, nature of this symmetry group of the theory points to an extension of the group to

---
[4]The Lorentz group is not compact.

one in which the two parts are "mixed" in some ways. A famous "no-go" theorem however, showed that this cannot be done in a non trivial way for arbitrary Lie groups and led to the introduction of **supersymmetries**. We will refer to this in Chapter 5.

Return to the program of representing the theory by an action. Note that (4.80) could be obtained by extremising the free Fermion action

$$\mathcal{S}_\psi = \int \bar{\psi}(i\partial\!\!\!/ + m)\psi * \mathbf{1}. \qquad (4.82)$$

Recalling that we have been assuming special relativity, with flat spacetime and constant metric components, $\eta_{\mu\nu}$, we see that this action is clearly invariant under Spin(1,3) transformations. Furthermore it is also invariant under **constant** $U(1)$ transformations on ψ. The constancy of these transformations is summarized by saying that we are considering **global** $U(1)$ symmetry. As mentioned earlier, the Noether theorem implies that this symmetry is associated with a conserved current, (4.67). For $U(1)$, the group action on the fiber, $\mathbb{C}^4$, is multiplication by $e^{i\alpha}$, that is,

$$\psi \to e^{i\alpha}\psi, \quad \delta\psi = i\delta\alpha\psi. \qquad (4.83)$$

The conserved form, (4.67), becomes

$$\mathbf{J} = iC\bar{\psi}\gamma_\mu\psi dx^\mu. \qquad (4.84)$$

Following Planck's discovery of photons, we are led to a "quantization" of **F**, taking over the field equations or action formulation. If we omit the source terms, we have the wave equation for **A**, (4.60), of the same form as (4.77) for ϕ, if we set $m = 0$. This fact results in the statement that the particle associated to the electromagnetic field, the **photon**, is massless. But now we must describe the source, **J**, in quantum terms such as ψ or others, as well as the **interaction** of these two fields. Classically, this was described in the electromagnetic Lagrangian, (4.68). The problem of understanding how to do this in general is at the foundations of the physical subject of gauge theory and the intimately related topics of **symmetries** and **conservation**.

Recall the manner in which equations of physical theories can be expressed in terms of an action principle defined from a Lagrangian for each field, the electromagnetic, **F**, and ψ, for example. When we consider each field alone, we speak of them as "free," and have

$$\mathcal{L}_{EM} * \mathbf{1} = -\frac{1}{2}d\mathbf{A} \wedge *d\mathbf{A} = -\frac{1}{2}\mathbf{F} \wedge *\mathbf{F}, \qquad (4.85)$$

and

$$\mathcal{L}_\psi * \mathbf{1} = \bar{\psi}(i\partial\!\!\!/ + m)\psi * \mathbf{1}, \qquad (4.86)$$

for (4.82).

Non-trivial physics only arises in interactions, which in this case is represented by the $\mathbf{A} \wedge \mathbf{J}$ term in (4.69). However, as noted earlier, this interaction term breaks the electromagnetic gauge invariance. Similarly, we note that (4.86) is not invariant under **local** $U(1)$ transformations, for which the parameter α in (4.83) is variable. Gauge theory then balances, or cancels these two non-invariances in a striking, elegant and very inventive manner.

From bundle theory, we are naturally led to look at this phenomenon using bundles. Thus the physical field ψ is a cross section of a complex line bundle, with group $U(1)$, over spacetime. The coordinate differentiation, ∂_μ, applied to fields in the Lagrangian will **not** be invariant under general actions by $U(1)$ along the fiber, so, we must replace it by bundle **covariant** differentiation by introducing a $U(1)$ **connection**. Since the Lie algebra of this group is simply the additive group of $i\mathbb{R}$, we can describe the $U(1)$ connection in terms of a purely imaginary one-form, Γ. We can then replace the $\slashed{\partial}$ in (4.86) with $\slashed{D}$, defined by

$$\slashed{D} = \gamma^\mu(\partial_\mu - \Gamma_\mu), \quad \text{where} \quad \Gamma = \Gamma_\mu dx^\mu. \tag{4.87}$$

From the local transformation expressions for the connection described in (4.28), with the group action multiplication by $e^{i\alpha}$, the combined actions

$$\psi \to \psi' = e^{i\alpha}\psi, \quad \Gamma \to \Gamma' = e^{-i\alpha}\Gamma e^{i\alpha} + id\alpha, \tag{4.88}$$

leave the expression

$$\bar{\psi}(i\slashed{D} + m)\psi$$

invariant.

So, $U(1)$ gauge theory leads us to replace (4.86) by

$$\mathcal{L}_{\psi,\Gamma} * \mathbf{1} = \bar{\psi}(i\slashed{D} + m)\psi * \mathbf{1} = \bar{\psi}(i\slashed{\partial} - i\gamma^\mu \Gamma_\mu + m)\psi * \mathbf{1}, \tag{4.89}$$

or, in terms of Lagrangians

$$\mathcal{L}_{\psi,\Gamma} * \mathbf{1} = \mathcal{L}_\psi * \mathbf{1} - i\mathbf{\Gamma} \wedge *\mathbf{J}_\psi, \tag{4.90}$$

where $\mathbf{\Gamma} = \Gamma_\mu dx^\mu$, and $\mathbf{J}_\psi = \bar{\psi}\gamma_\gamma \psi dx^\gamma$ is (up to a multiplicative constant) the conserved ψ probability current density one form described in (4.84). Now, the surprising and interesting fact is that the last term in (4.90) is exactly of the same form as the ψ-EM interaction term in (4.72), if we relate the $U(1)$ bundle connection form to the electromagnetic potential, $\mathbf{A}$, up to constants. Specifically,

$$\mathbf{\Gamma} = iq\mathbf{A}, \tag{4.91}$$

where q is the electromagnetic **charge** of the particle corresponding to the ψ field.[5] Furthermore, we then have that the pure EM Lagrangian, (4.85) is simply proportional to the **second Chern class of the curvature of the gauge connection field**. That is, the curvature form is simply $d\mathbf{A}$ in this (abelian) ($U(1)$) case, and $\Omega \wedge *\Omega$ is clearly $U(1)$ invariant. We can then summarize what we have seen so far by describing the ψ field as a cross section of a $U(1)$ bundle, with connection form Γ, and curvature $\Omega = d\Gamma$. The full EM interaction of this field of charge q is then fully described by the gauge-geometric action,

$$S_g = \int (\frac{1}{2q^2}\Omega \wedge *\Omega + \bar{\psi}(i\slashed{D} + m)\psi * \mathbf{1}). \qquad (4.92)$$

Remark 4.6.

Note that we have full local $U(1)$ invariance, but still only **global** Lorentz or Spin(1,3) invariance. To accommodate an expansion to **local** Lorentz invariance, we would have to incorporate general relativity, again leading us far afield from our current topics.

4.5 Physical Generalizations, Yang-Mills, etc.

What we have just described for the electromagnetic-fermion interaction has formed the model for much of the progress of contemporary elementary particle physics, at least in its pre-string, pre-supersymmetry stages. We can summarize this procedure as follows:

- **"Elementary" particles:** In the Standard models the elementary particles, including the familiar electrons, neutrinos, quarks, etc., are all **fermions** of half odd spin, but with some internal structure, described by a Lie symmetry group, say G. Mathematically this means that the field is a cross section of a G bundle, generally, $G = SU(n)$. We then form an action for each such field in an **external**, Spin(1,3), cross **internal**, G, invariant manner.
- **Connection/gauge fields:** For non-trivial bundle structure, we will need a **connection** form, $\mathfrak{g}$-valued. This connection form then describes another physical field, the **gauge** or force field, and, in keeping with "wave-particle duality," is associated with its own class of gauge

---

[5]q^2 is a measure of the strength of the interaction of the ψ field with the electromagnetic one, and is sometimes referred to as the electromagnetic **coupling constant**. In natural, atomic units, it is referred to as the **fine structure constant** and has the dimensionless value of about 1/137.

particles. Because this connection field is a one-form, or vector, it has spin one and is a **boson**. Roughly, quantum field theory then describes the gauge particles as the exchange particles for the force defined by the symmetry group G.

- **Actions and field equations:** The total action is formed from the free particle action, but with gauge covariant derivatives. This provides the **interaction** between the particle and the gauge, or force, field. The gauge field equation is derived from an action formed from the second Chern class of Lie algebra trace of the curvature form of the gauge field.
- **Standard model(simplified):** The three basic forces, **electromagnetic, weak, strong** are associated with three classical groups $U(1), SU(2), SU(3)$, respectively. The corresponding particles are called **photons, W-bosons, gluons** respectively.

4.6 Yang-Mills Gauge Theory: Some Mathematics

Many of the mathematical applications of the physics of gauge theories come from analysis of the differential equations for the gauge field. However, as is usual, mathematics is more comfortable with a **positive definite** metric, replacing $\eta_{\mu\nu}$ with euclidean $\delta_{\mu\nu}$, and Spin(1,3) with Spin(4). The resulting positive definite metric contains no trace of physical time, so the pseudo-particles associated to the fields are called **instantons**. Of special importance is the first non-abelian gauge group, $SU(2)$, leading to the **Yang-Mills** equations. The resulting theory closely parallels the electromagnetic, $U(1)$, theory, but now the connection form is $\mathfrak{su}(2)$ valued, and the curvature form is the non-linear curvature form

$$\mathbf{F} = d\mathbf{A} + \mathbf{A} \wedge \mathbf{A} \in \Omega^2(M; \mathfrak{su}(2)). \tag{4.93}$$

The field equations, generalizing the electromagnetic case are obtained by extremising an action

$$\mathcal{S}_{YM} = -\int_M Tr(\mathbf{F} \wedge *\mathbf{F}). \tag{4.94}$$

The resulting generalization of the Maxwell equations leads to non-linear partial differential equations, of **elliptic** type, since the metric is definite. The analysis of these equations for closed M, such as S^4, leads to the famous **Atiyah-Singer** theorems. For a survey of this topic, see the lecture notes of Atiyah[Atiyah (1979)]. However, in the following chapter we also will look more closely into these mathematical applications arising from the physics of gauge theory.

Chapter 5

Gauge Theory and Moduli Space

5.1 Introduction

In this chapter we concentrate on the mathematics of **bundle theory**, Chapter 4, corresponding to the physics of **gauge theory**, expanding on the introductory discussions of the previous chapter. After reviewing the definition of vector bundles and their isomorphism classes, we proceed to the **classification** problem for vector and principal bundles over (paracompact, smooth) manifolds, using the tools of **K-theory, universal bundles** and **classifying spaces.** This leads to the section on **characteristic classes**, cohomology elements of the base space defined by a bundle over that space. Then we apply these characteristic classes to the existence question for Spin and $Spin_C$ structures on given manifolds. We close the chapter with discussions of various gauge theories and their resulting **moduli spaces**, i.e. the parameter spaces of gauge-equivalent connections with respect to particular field equations. The special cases of Donaldson and Seiberg-Witten moduli spaces so important to the study of smoothness questions bring the chapter to a close.

The reader may well ask why we need to discuss this vast machinery of bundle classifications. The reason is the central role such formalisms play in moduli spaces, one of the main components in the study of exotic smoothness. To build the moduli spaces of various connections over bundles, we start with a gauge group, G, a base manifold, M, and a G-principal bundle, $E(M, G)$. The moduli space for this $E(M, G)$ is the space whose points are classes of $\mathfrak{g}$-valued connections on E, modulo gauge transformations. Somewhat surprisingly, such spaces turn out to be finite dimensional smooth manifolds (with isolated singular points) for cases of interest to us. Since M and G by themselves do not uniquely determine the bundle E, any bundle equivalent to E (which necessarily has the same M, G) will have the same moduli space, since the equivalence map of bundles lifts to the moduli spaces. So a G-moduli space over M depends on the equivalence class of G-bundles over M. Such classes of bundles are studied with the tools discussed in this chapter. Thus a given G-moduli space over M can be identified by whatever tool we use to identify the corresponding equivalence class of G-bundles over M.

For example, Chern classes and numbers are bundle classification tools, and Donaldson's work describes the dimension of the moduli space of $SU(2)$ bundles over a given M as a function of k, the second Chern number. If we are looking at $SU(2)$ bundles over a 4-manifold, M, which has second Chern number, $k = 1$, the moduli space of $SU(2)$ connections over any representative bundle from this class will be of dimension 5 if M is simply connected. For $M = S^4$ Atiyah, Hitchin and Singer [Atiyah et al. (1978)] solved this problem for a wide class of groups as we discuss later in this chapter.

Assumptions in this chapter: Unless otherwise noted, we will be dealing with smooth, paracompact manifolds and Lie groups.

5.2 Classification of Vector and Principal Fiber Bundles

In 4.2 we presented a brief overview of bundle theory, an important mathematical underpinning for the expression of physical theories. In this chapter we will delve into certain tools for investigating bundles, in particular their isomorphism classes and other characterizations.

We begin with vector bundles and their isomorphism classes. Recall that we can regard a vector bundle as a family of vector spaces parameterized by an underlying smooth manifold, the base space. A section of the bundle can then be thought of as a generalization of a function over a manifold. For the purposes of this chapter we re-state the *coordinate* definition of a real m-dimensional vector bundle over M starting with the local transition function representation:

(1) a smooth manifold M with a given open covering $M = \bigcup_{\alpha \in A} U_\alpha$,
(2) (smooth) transition functions $g_{\alpha\beta} : U_\alpha \cap U_\beta \to Gl(m, \mathbb{R})$ for each $\alpha, \beta \in A$ satisfying the so-called "cocycle" condition:

$$g_{\alpha\beta} \cdot g_{\beta\gamma} = g_{\alpha\gamma} \quad \text{on } U_\alpha \cap U_\beta \cap U_\gamma \neq \emptyset \quad (5.1)$$

where $\cdot$ denotes the group operation in $Gl(m, \mathbb{R})$.

Now consider the set $\tilde{E} = \{(\alpha, p, v) \in A \times U_\alpha \times \mathbb{R}^m\}$ together with the equivalence relation:

$$(\alpha, p, v) \sim (\beta, q, w) \longleftrightarrow \begin{cases} p = q \in U_\alpha \cap U_\beta \\ v = g_{\alpha\beta}(p)w. \end{cases}$$

From this define the **total space** E of the bundle, $E = \tilde{E}/\sim$ as the set of equivalence classes $E = \{[\alpha, p, v]\}$. The naturally defined map, the **projection**, $\pi : E \to M$ is defined by $\pi([\alpha, p, v]) = p$. This definition of E is based on the set of local trivializations of E, i.e. the bijective maps $\pi^{-1}(U_\alpha) \to U_\alpha \times \mathbb{R}^m$. The space $\pi^{-1}(p)$ is called the **fiber** over p. Locally, that is in each U_α containing p, there is an isomorphism of the fiber to $\mathbb{R}^m$. Thus, each fiber is a real m-dimensional vector space, but in general with no *natural* or *canonical* isomorphism to $\mathbb{R}^m$. However, if there is one preferred isomorphism, so that all fibers $\pi^{-1}(p)$ can be canonically identified with $\mathbb{R}^m$, then the bundle is said to be **trivial**, or a **product bundle**, that is, $E = M \times \mathbb{R}^m$. A central problem of bundle theory, and its physical applications, is determining effectively whether or not a given bundle is trivial. Recall that what this procedure defines is a **coordinate bundle** and clearly contains the explicit choice of covering and transition functions.

Now consider a smooth map $f : E_1 \to E_2$ between the total spaces of two vector bundles (E_1, π_1) and (E_2, π_2) over M. f is a **bundle map** if it commutes with the projections and is linear on fibers, that is $\pi_2 \cdot f = \pi_1$, and the restriction of f to $\pi_1^{-1}(p)$, denoted by f_p, is a linear map of fibers. If this linear map is actually an isomorphism, then the bundle map, f is said to be a (vector) **bundle isomorphism** between E_1 and E_2. In terms of the transition functions, $g_{\alpha\beta}^{(1)}$ and $g_{\alpha\beta}^{(2)}$ of E_1 and E_2, respectively, the bundle isomorphism condition can be expressed by requiring

$$g_{\alpha\beta}^{(1)} = (f_p)^{-1} \cdot g_{\alpha\beta}^{(2)} \cdot f_p. \tag{5.2}$$

Now define the equivalence set under (5.2), that is

$$\hat{H}^1(\{U_\alpha\}, Gl(m, \mathbb{R})) = \{E : g_{\alpha\beta} : U_\alpha \cap U_\beta \to Gl(m, \mathbb{R}) \mid g_{\alpha\beta} \cdot g_{\beta\gamma} = g_{\alpha\gamma}\}$$
$$\text{mod (5.2)}.$$

$\hat{H}^1$ then describes all equivalence classes of coordinate vector bundles with respect to a fixed covering $\{U_\alpha\}$ of M, i.e. the construction depends on the covering of the manifold. By means of a limiting process of refinements of the coordinate patch covers, we can obtain a covering-independent definition by the process called "the inverse limit." Thus define the set resulting from this limit,

$$Vect_{\mathbb{R}}(M,m) = \lim_{\leftarrow \{U_\alpha\}} \hat{H}^1(\{U_\alpha\}, Gl(m,\mathbb{R})).$$

This then is nothing but the set of equivalence classes of m-dimensional coordinate vector bundles over M. One element of this set is a vector example of the general fiber bundle as defined by Steenrod[Steenrod (1951)], page 9.

Remark 5.1.

As an example, consider the space $Vect_{\mathbb{R}}(S^2, 2)$. It turns out that this set has an infinite number of elements, defined by the Euler class discussed later. At this point we merely point out that the product bundle, $S^2 \times \mathbb{R}^2$, and the tangent bundle, TS^2 must be distinct elements of $Vect_{\mathbb{R}}(S^2, 2)$ since the first has a non-zero global cross section, while the second does not.

Pullback bundle and homotopy theory of bundles

Homotopy again proves itself to be very useful in studying the dependence of $Vect_{\mathbb{R}}(M,m)$ on the base M. Consider a change of the base space and a map $F : N \to M$. The **pullback** of the vector bundle E over M is a vector bundle F^*E over N constructed by the following steps,

- the induced covering $\{F^{-1}(U_\alpha) : \alpha \in A\}$ on N is defined naturally from the covering $\{U_\alpha : \alpha \in A\}$ on M,
- similarly, the transition functions $\{g_{\alpha\beta} \circ F : \alpha, \beta \in A\}$ come from the transitions functions $g_{\alpha\beta}$ on E and,
- the total space is defined $F^*E = \{(p,v) \in N \times E : F(p) = \pi(v)\}$ with the projection $\pi : E \to M$ inducing a new projection $\pi^*(p,v) = p$.

In other words, we use the same transition functions from the original bundle, but "pulled back" by F, to define those of the new bundle. The following theorem illustrates an important feature of pullbacks:

Theorem 5.1. *If $F_0, F_1 : N \to M$ are smoothly homotopic maps and E is a vector bundle over M, then F_0^*E and F_1^*E are isomorphic vector bundles over N.*

The problem of determining whether or not two bundles over a given N are isomorphic, the **bundle classification problem,** is an important one in differential topology and physical field theories. Theorem 5.1 shows that this classification problem can be reduced to finding whether or not the two bundles are pullbacks of homotopic maps.

Let us briefly review why the classification tools are important. Mathematically one coordinate bundle may be constructed by one technique, and a second by another. A basic question then is whether or not the two structures are truly different, or are they in fact isomorphic. For example, if the bundles are both pullbacks by homotopic maps from the same bundle, then we know they are isomorphic from the preceding theorem. An analogous physical problem might go something like this. Two theories involve the construction of bundles, $E_i, i = 1, 2$, and each asserts that a certain cross section of the corresponding bundle, E_i, say f_i, describes a given physical situation. Are the predictions of the two theories physically "equivalent?" To answer this, we first need the mathematical machinery to see whether or not E_1 is isomorphic to E_2. If so, the isomorphism, ϕ, then carries f_1 into $\phi \cdot f_1$, which can then be directly compared to f_2, to answer the physical question. Of course, in this example, we assume the same base space, a priori.

A sketch of the proof of Theorem 5.1 using a bundle connection follows. A more general proof, without use of a connection can be found in §11 of Steenrod's book[Steenrod (1951)].

Remark 5.2.
Let $J_0, J_1 : N \to N \times [0, 1]$ be the smooth maps defined by: $J_0(p) = (p, 0)$, $J_1(p) = (p, 1)$. If F_0 is smoothly homotopic to F_1, there exists a smooth map $H : N \times [0, 1] \to M$ such that $H \circ J_i = F_i$ for $i = 0, 1$. Thus it suffices to show that if E is a vector bundle over $N \times [0, 1]$, then $J_0^* E$ is isomorphic to $J_1^* E$. Give E a connection and let $\tau_p : E_{(p,0)} \to E_{(p,1)}$ denote parallel transport along the curve $t \mapsto (p, t)$ in the total space which goes along the fiber over $p \in M$. We can then define a vector bundle isomorphism $\tau : J_0^* E \to J_1^* E$ by

$$\tau(p, v) = (p, \tau_p(v)), \qquad \text{for} \quad v \in E_{(p,0)} = J_0^* E_p,$$

which completes the proof.

An important special application is to a bundle over a **contractible** base space. Since such a space is homotopic to a point, and any bundle over a point is trivial, we arrive at the very important fact, particularly important in our later discussion of universal bundles:

Fact: *Any bundle over a contractible space is trivial.*

For the most part we will be concerned with vector bundles. However, many results, such as the homotopy theorem above, carry over to the associated principal bundles. Recall that a bundle is **principal** if its fiber is equal to its group, with left multiplication as group action. For any bundle with group G the **associated principal** bundle is constructed as the bundle over the same base space, with the same group and transition functions, but the fiber is G itself. The construction of a bundle from transition functions is spelled out in coordinate detail in Steenrod[Steenrod (1951)], §3. Conversely, given a principal fiber bundle and a space F on which G is an effective transformation group, we can construct an associated bundle with fiber F by similar techniques.

Although we have been discussing general vector bundles, an obviously important special case is that for which the vector space fiber is actually "soldered" to the base space as is the case for the tangent vectors to a smooth manifold. An obvious decomposition of each matrix of $Gl(m, R)$ into column vectors allows us to identify each such matrix as a **frame**, or basis, of m independent tangent vectors. Thus, for the tangent vector bundle to an m dimensional manifold, the bundle of linearly independent sets of vectors at each point can be identified with a principal $Gl(m, R)$ bundle and is sometimes referred to as the **principal bundle of frames**. Refer to our discussion of tangent bundles in chapter 4.

Remark 5.3.

We note that Milnor has generalized the notion of a tangent bundle with smooth manifold base space, to the more general TOP category through his construction of **microbundles**.

Looking again at the basic classification question of whether or not a bundle is trivial, we have the important "cross section" theorem,

Theorem 5.2. *A principal bundle with group G is trivial, that is, isomorphic to a product, if and only if it has a global cross section.*

From this it follows as a corollary that any bundle is trivial if its associated principal bundle has a global cross section. On the other hand, there are many examples of non principal fiber bundles which have global cross sections, but whose associated principal bundle has no global cross section. For example, every vector bundle admits a global cross section of zero vectors, but not all associated bundles are trivial. A particular example is the tangent vector bundle to S^2.

We note that the questions associated with constructions and extensions of cross sections were historically central to the development of **characteristic classes** which we will treat later in some detail. For now, we restrict ourselves to some preliminary results as explored in Steenrod[Steenrod (1951)], §12. Of particular interest is his Theorem 12.2 which establishes that it is always possible to construct a cross section of a bundle with fiber F which is "solid," a topological definition which includes the more familiar cases of open or closed cells, sufficient for our purposes. From elementary matrix analysis it is easy to see that the subset of $Gl(m, R)$ consisting of real, positive definite, symmetric matrices is convex and thus an open cell. From this Steenrod's theorem 12.2 establishes the important result

Fact: *It is always possible to construct a Riemannian metric on a real vector bundle.*

Another important class of questions is related to the **reduction of the bundle group** to a subgroup. If H is a subgroup of the bundle group, G, then we can ask whether or not a coordinate representation of the bundle exists in which all transition functions lie in H. Before looking at the linear group case of interest for our vector bundles, let us review especially important examples of bundles provided by groups and their subgroups, namely,

Coset bundles: *If H is a subgroup of G with a local cross section, then G is a bundle over G/H with group H.*

The condition that H have a local cross section is satisfied if G is a Lie group, which is always the case in our considerations. For more details on this topic see §7.4 in Steenrod[Steenrod (1951)].

Returning to vector bundles, let E be one with structure group $Gl(m, \mathbb{R})$ over M. An important sequence of theorems outlined in Steenrod clarifies the situation when $H = O(m, R)$. A basic result in real matrix theory ensures that we can write $Gl(m, R)$ as the Cartesian product, $O(m, R) \times A(m, R)$, where $A(m, R)$ is the set of real symmetric positive definite matrices. One way to see this is to consider a non-singular matrix as defining a basis. We can then choose a metric, an element of $A(m, R)$, and the Gram-Schmidt procedure then results in an orthonormal basis, an element of $O(m, R)$. At any rate, this decomposition proves that the coset space, $Gl(m, R)/O(m, R)$ reduces to $A(m, R)$. But, in discussing Riemannian metrics above, we showed that this latter space is a cell, and thus solid, even contractible. It is also the base space of $Gl(m, R)$ as an $O(m, R)$ bundle. Thus, this coset bundle is trivial, and has a global cross section, and, from Steenrod's Corollary 9.5, we arrive at the general linear group reduction theorem, Steenrod, page 57,

Theorem 5.3. *Any bundle with group $Gl(m, R)$ is reducible to one with*

group $O(m, R)$.

Operations on vector bundles

The fibers of a vector bundle are vector spaces of equal dimension so we can easily extend a fiber-wise operation to the whole bundle over a given base space. Consider the following operations on vector bundles E_1 and E_2 over the same base space:

- direct sum bundle $E_1 \oplus E_2$,
- tensor product bundle $E_1 \otimes E_2$,
- exterior product bundle $E_1 \wedge E_2$,
- operator bundle $Hom(E_1, E_2)$ (the homomorphisms between E_1 and E_2, or, the linear maps).

These operations, and others, are extensively used in differential geometry. For example, a map $g : TM \otimes TM \to \mathbb{R}$ defines a **metric tensor**, sections in the bundle $\bigwedge^k TM$ are k-differential form fields etc. Here, we use these operations to introduce more structure on the set $Vect_\mathbb{R}(M, m)$.

In general, of course, the operations of direct sum and tensor product change the dimension of the bundle. So, we form the union of spaces $Vect_\mathbb{R}(M, m)$ to get $Vect_\mathbb{R}(M)$ the set of all vector bundles on M. The space $Vect_\mathbb{R}(M)$ forms a semi-group with respect to the tensor product $\otimes$.

We will now discuss two techniques for studying the structure of classes of isomorphic vector bundles.

- First, we impose additional structure on $Vect_\mathbb{R}(M)$ to define a purely algebraic object, **K-theory**, which is a cohomology like structure (theory) for vector bundles. The third edition of the book by Husemoller[Husemoller (1994)] contains two chapters on the subject, and can serve as an excellent reference. We also note that recently the tools of **K-theory** have been of increasing interest in string theory to classify "D-brane charges"[Witten (1998)].
- The second approach describes $Vect_\mathbb{R}(M)$ using homotopy theory of the base space and constructing a **universal bundle** into which all vector bundles can be embedded.

Now we take up the K-theory approach.

K-theory of vector bundles

By way of introduction, we review how the direct sum procedure can be used to simplify the structure of a non-trivial vector bundle. For example, consider the non-trivial tangent bundle of the sphere TS^2. By using the standard embedding $S^2 \to \mathbb{R}^3$ we can define the bundle NS^2 of all normal vectors to the sphere. Obviously this bundle is trivial. The direct sum of the two bundles

$$TS^2 \oplus NS^2 = TS^2 \oplus \xi_1(S^2) = \xi_3(S^2) = \xi_2(S^2) \oplus \xi_1(S^2),$$

where $\xi_k(M)$ denotes the trivial vector bundle of rank k over M, is trivial as we expected by the embedding. In other words, even though TS^2 is not trivial, its sum, $TS^2 \oplus NS^2$ is.

Now we review a generalization of this approach as another tool for the vector bundle classification problem. First, a vector bundle over a manifold is said to be **stable** if the dimension of the fiber is greater than the dimension of the base space, M. Next, let E, F be two vector bundles over M of rank k and m, respectively. E **is stably equivalent to** F denoted by $E \sim_s F$ if there are trivial bundles $\xi_l(M)$ and $\xi_h(M)$ with

$$E \oplus \xi_l(M) \equiv F \oplus \xi_h(M)$$

and $k + l = m + h$. Here $\equiv$ means bundle isomorphism. In the example above we have shown that $TS^2 \sim_s \xi_2(S^2)$ but the tangent bundle TS^2 itself is non-trivial.

Now consider 3 vector bundles E, F and G over M with $E \oplus G \equiv F \oplus G$. Then [Husemoller (1966)] there is a vector bundle G' so that $G \oplus G' \equiv \xi_n(M)$ for some n. From $E \oplus G \equiv F \oplus G$ it follows that $E \oplus G \oplus G' \equiv F \oplus G \oplus G'$ or $E \oplus \xi_n(M) \equiv F \oplus \xi_n(M)$, so $E \sim_s F$. Thus,

Theorem 5.4. *If E, F and G are vector bundles over M satisfying $E \oplus G \equiv F \oplus G$ for some vector bundle G, then E and F are stably equivalent, i.e., $E \sim_s F$.*

The direct sum process for vector bundles, $\oplus$, makes the set of all complex vector bundles $Vect_\mathbb{C}(M)$ into an algebraic structure known as a **semigroup**, a set with associative combining operation, but not necessarily having an identity or inverses. Using an idea of Grothendieck we can then complete such an Abelian semi group to a full group, as follows. Suppose, that S is any semi-group. The corresponding Abelian group associated to S is the product $S \times S$ modulo the following equivalence relation $\sim$

$$(s_1, t_1) \sim (s_2, t_2) \leftrightarrow \text{ for some } u \in S, s_1 + t_2 + u = s_2 + t_1 + u .$$

Thus, although S may not have inverses, the equivalence relation could be formally restated as saying $s_1 \text{``}-\text{''} t_1 \sim s_2 \text{``}-\text{''} t_2$. In fact, the equivalence classes $[(s, t)]$ are often denoted by the formal differences $[s] - [t]$. A standard example starts with the natural numbers $\mathbb{N}$ (the additive positive integers) as an Abelian semi-group. The completion $\mathbb{N} \times \mathbb{N}/\sim$ is nothing else than the full set of integer numbers $\mathbb{Z}$, with $[(x,y)] \leftrightarrow x - y$, so $[(y,x)] \leftrightarrow y - x$.

For our application, start with complex vector bundles, and the semi-group is provided by $(Vect_\mathbb{C}(M), \oplus)$. Its completion is
$$K(M) = Vect_\mathbb{C}(M) \times Vect_\mathbb{C}(M)/\sim \qquad (5.3)$$
called the (complex) **K-theory of** M. Thus two pairs (E, F) and (G, H) of complex vector bundles over M are equivalent $(E, F) \sim (G, H)$ if and only if there is a complex vector bundle K over M with $E \oplus H \oplus K \equiv G \oplus F \oplus K$. Then, every element of $K(M)$ can be written as formal difference $[E] - [F]$ or $[E \ominus F]$ of two complex vector bundles. Notice that any element $[E] - [F]$ of $K(M)$ can be written in the form $[H] - [\xi_n(M)]$ for some bundle H and some integer n. The proof is very simple: Consider the bundle F' with $F \oplus F' \equiv \xi_n(M)$ for some n. Then,
$[E] - [F] = ([E] + [F']) - ([F] + [F']) = [E \oplus F'] - [F \oplus F'] = [H] - [\xi_n(M)]$
with $H \equiv E \oplus F'$. Define the virtual dimension of this element to be the difference $rk(E) - rk(F)$ of the ranks of the two complex vector bundles. Reduced K-theory $\tilde{K}(M)$ consists of all equivalence classes of complex vector bundles of virtual dimension 0, that is, the stable equivalence pairs of complex vector bundles of the same dimension. In fact, the elements of $\tilde{K}(M)$ are exactly the stable equivalence classes of complex vector bundles over M. Let E_m and F_n complex vector bundles over M of rank m and n, respectively. These bundles induce the elements $[E_m] - [\xi_m(M)]$ and $[F_n] - [\xi_n(M)]$ in $\tilde{K}(M)$. First, assume that these classes are in fact equal, i.e. $[E_m] - [\xi_m(M)] = [F_n] - [\xi_n(M)]$. So $E_m \oplus \xi_n(M) \oplus K \equiv F_n \oplus \xi_m(M) \oplus K$ for some bundle K. By the theorem 5.4 we get $E_m \oplus \xi_n(M) \sim_s F_n \oplus \xi_m(M)$ or equivalently $E_m \sim_s F_n$. For the converse, start with $E_m \sim_s F_n$ or $E_m \oplus \xi_j(M) \equiv F_n \oplus \xi_k(M)$ and write down the corresponding elements in $\tilde{K}(M)$, i.e. $[E_m \oplus \xi_j(M)] - [\xi_{m+j}(M)] = [F_n \oplus \xi_k(M)] - [\xi_{n+k}(M)]$ where the right hand side is equal to $[E_m] - [\xi_m(M)]$ and the left hand side is $[F_n] - [\xi_n(M)]$. Thus we obtain $[E_m] - [\xi_m(M)] = [F_n] - [\xi_n(M)]$, so

Fact: $\tilde{K}(M)$ *consists of all stable equivalence classes of complex vector bundles over* M.

We should remind the reader that the classification of vector bundles provided by K-theory, both $K(M)$ and $\tilde{K}(M)$, is only up to **stable** equivalence, which is, of course, weaker than full equivalence under bundle isomorphisms.

As a concrete example we will calculate the K-theory of a point $\{*\}$. The set $Vect_{\mathbb{C}}(\{*\})$ consists of all complex vector bundles over one point. So, for every possible rank we get only one complex vector bundle over that point, the trivial bundle. Thus, $Vect_{\mathbb{C}}(\{*\}) = \mathbb{N}$ and we can use the Grothendieck construction above, both for the integers and bundles, to show $K(\{*\}) = \mathbb{Z}$. For $\tilde{K}(\{*\})$ we can use the correspondence to the stable equivalence classes of complex vector bundles to show that all complex vector bundles over a point belong to the same element in $\tilde{K}(\{*\})$, i.e $\tilde{K}(\{*\}) = 0$. This calculation motivates the isomorphism $K(M) = \tilde{K}(M) \oplus \mathbb{Z}$ as explicitly shown in the next remark.

Remark 5.4.
The isomorphism $K(M) = \tilde{K}(M) \oplus \mathbb{Z}$ can be shown by abstract algebra. From above, all vector bundles over a point are classified by their rank, thus $K(\{*\}) = \mathbb{Z}$. Then we have a map $r : M \to \{*\}$ mapping all points of M to one point. Of course there is also a reversed map $i : \{*\} \to M$ where one point is mapped to one fixed point in M(inclusion). The induced maps of K-theory are:

$$i^* : K(M) \to K(\{*\}) = \mathbb{Z} \qquad r^* : K(\{*\}) \to K(M)$$

The first map can be seen as the rank map, i.e. $\ker i^* = \tilde{K}(M)$. Then we obtain the exact sequence:

$$\tilde{K}(M) \xrightarrow{I} K(M) \xrightarrow{i^*} K(\{*\}) = \mathbb{Z} \to 0$$

where I is the inclusion of $\tilde{K}(M)$ into $K(M)$ and the last map i^* has a (right) inverses r^*. Thus the sequence splits, i.e.

$$K(M) = \ker i^* \oplus K(\{*\}) = \tilde{K}(M) \oplus \mathbb{Z}$$

the required relation.

Remark 5.5.
Now consider a few examples of K-theory for some simple complex line bundles. Since every complex line bundle over S^1 is trivial, we have $\tilde{K}(S^1) = 0$. For higher dimensional spheres, recall that every $S^n, n > 1$ is coverable by two balls, intersecting in set homotopic to S^{n-1}. Thus, the bundle can be defined in terms of transition functions mapping $S^{n-1} \to U(k)$. If $k > n$ the bundle is stable and we can build the stable unitary group U which is something like $U(\infty)$ by the process of the "projective limit"

$$U = \lim_{\substack{U(n) \to U(n+1) \\ n \to \infty}} U(n).$$

Finally we arrive at

$$\tilde{K}(S^n) = \pi_{n-1}(U),$$

which identifies the K-theory and (stable) homotopy groups of the unitary group pointwise. With much more effort one can prove this isomorphism also for the algebraic structures. With the help of the Bott periodicity theorem one can determine these groups.

Remark 5.6.
Now we can calculate the groups $\tilde{K}(S^n)$. For that purpose we need the definition of the smash product

$$X \wedge Y = \frac{X \times Y}{X \vee Y}$$

of spaces X, Y, where $X \vee Y$ is the one point union of X and Y, i.e. we identify one point x_0 in X with one point y_0 in Y. Consider the case $S^1 \wedge S^1$ which is the torus $T^2 = S^1 \times S^1$ with all one point unions of circles $S^1 \vee S^1$ "contracted." To visualize this process, we consider the torus as a square with opposite sides identified. Each identified side in the torus is a circle S^1. Thus, the contraction of $S^1 \vee S^1$ is equivalent to the contraction of the boundary of the square, which is equivalent to S^2. This motivates the identification $S^1 \wedge S^1 = S^2$ and also the general result $S^k \wedge S^n = S^{k+n}$. The definition of the smash product can also be described by considering the exact sequence

$$X \vee Y \to X \times Y \to X \wedge Y$$

of spaces leading to a corresponding exact sequence

$$0 \to \tilde{K}(X \vee Y) \to \tilde{K}(X \times Y) \to \tilde{K}(X \wedge Y) \to 0$$

in K-theory. The Bott periodicity states that $K(X) \otimes K(S^2)$ is isomorphic to $K(X \times S^2)$. From the relation $K(X) = \tilde{K}(X) \oplus \mathbb{Z}$ we obtain the exact sequence

$$K(X) \otimes K(S^2) = (\tilde{K}(X) \otimes \tilde{K}(S^2)) \oplus \tilde{K}(X) \oplus \tilde{K}(S^2) \oplus \mathbb{Z}$$
$$\downarrow \text{isomorphism}$$
$$K(X \times S^2) \quad = \quad \tilde{K}(X \wedge S^2) \oplus \underbrace{\tilde{K}(X \vee S^2)}_{\tilde{K}(X) \oplus \tilde{K}(S^2)} \oplus \mathbb{Z}$$

where we have used the obvious relation $\tilde{K}(X \vee Y) = \tilde{K}(X) \oplus \tilde{K}(Y)$ (a vector bundle over a point union of two spaces always splits into two subbundles). Thus we obtain the isomorphism:

$$\tilde{K}(X) \otimes \tilde{K}(S^2) \to \tilde{K}(X \wedge S^2)$$

as another formulation of Bott periodicity (the KO-case is similar). Finally we obtain for $X = S^{2n}$ with $S^{2n} \wedge S^2 = S^{2n+2}$

$$\tilde{K}(S^{2n}) = \underbrace{\tilde{K}(S^2) \otimes \ldots \otimes \tilde{K}(S^2)}_{n \text{ times}}$$

and for $X = S^{2n+1}$

$$\tilde{K}(S^{2n}) = \tilde{K}(S^1) \otimes \underbrace{\tilde{K}(S^2) \otimes \ldots \otimes \tilde{K}(S^2)}_{n \text{ times}} .$$

Without proof we state that $\tilde{K}(S^1) = 0$ (all complex vector bundles over S^1 are trivial) and $\tilde{K}(S^2) = \mathbb{Z}$ (generated by the Hopf bundle $\pi : S^3 \to S^2$). Thus, we have

$$\tilde{K}(S^n) = \begin{cases} 0, n = \text{odd}, \\ \mathbb{Z}, n = \text{even}, \end{cases}$$

getting the "classical" Bott periodicity theorem:

$$\pi_n(U) = \pi_{n+2}(U)$$

or $\pi_{2n}(U) = 0$ and $\pi_{2n+1}(U) = \mathbb{Z}$. In some sense this periodicity in the homotopy groups of Lie groups is the source of all periodicities in topology (4-periodic L-groups in surgery etc.).

Note the tensor product map

$$K(X) \otimes_\mathbb{Z} K(X) \to K(X) ,$$

extends $K(X)$ to a ring. With this ring structure K-theory satisfies all of the Eilenberg-Steenrod axioms for a cohomology theory, *except for the dimension axiom*. In fact the K-theory of a point does not vanish. This is why K-theory is described as a generalized cohomology theory. Although it does not satisfy the dimension axiom, the rest of the cohomological structure makes K-theory very useful in (stable) classification problems. The ring structure means that K-theory of a manifold can be calculated by breaking up the manifold into simpler pieces. This is in contrast to the (full, not just stable) classification tool defined by the homotopy equivalence set of maps, $[M, BG]$, with classifying space, BG, which we define later.

Although we defined K-theory in terms of complex bundles, the real case proceeds similarly, resulting in the ring $KO(M)$. We now state without proof (see Chapter 11 of Husemoller[Husemoller (1994)]) one of the highlights of algebraic topology,

Theorem 5.5. *(Periodicity theorem of Bott)*
For a compact space, M, the tensor product map induces the isomorphism
$$K(M) \otimes K(S^2) \cong K(M \times S^2)$$
in complex K-theory and
$$KO(M) \otimes KO(S^8) \cong KO(M \times S^8)$$
in real KO-theory.

The consequences of this theorem for so many of the topics central to our study, such as group theory, surgery on manifolds, algebraic geometry and especially differential topology (Atiyah-Singer index theorem) are very important. Although we have constructed K-theory for the complex linear groups, it can be extended to more general groups. In the next subsection we remark that every (generalized) cohomology theory, and thus K-theory, can be represented by the map (homotopy functor): $M \mapsto [M, BG]$ for a suitable space BG. In particular, we will now look at the construction of this (classifying) space for (complex group) K-theory $K(M) = [M, BU]$.

The Universal bundle over the classifying space

The homotopy theory of bundles leads us to suspect that there might be a space containing all the information about vector bundles of given rank, at least those over manifolds. In fact, every homotopy class of a map from the base manifold into such a "universal" (or "classifying") space defines a specific vector bundle, up to isomorphism, over that base manifold. Good sources for this subject include the bundle theory book of Steenrod[Steenrod (1951)], especially §19, the book of Husemoller[Husemoller (1966)], especially Chapter 4, and the characteristic classes book of Milnor and Stasheff,[Milnor and Stasheff (1974)], especially their §5. In broad terms we can identify two approaches to the construction of these classes.

Informally, we begin with the following definition:

A universal bundle: *For dimension n and group G, a universal bundle, EG, (our notation) is a bundle into which any other bundle over a manifold of dimension n and group G can be embedded.*

Steenrod then points out that the base space, BG, of such a bundle is a **classifying space** in the following sense

Classifying space: *The homotopy classes of maps of any n manifold M into BG, denoted by $[M, BG]$, precisely describes (classifies) the equivalence classes of principal G bundles over M.*

He then proves a theorem that characterizes universal bundles as those for which the bundle space, EG, is n−connected that is, $\pi_i(EG) = 0$, for all $1 \leq i \leq n$. In particular, this is satisfied if EG is **contractible**. This can then be used as an effective test to determine whether or not a bundle is universal with respect to principal G bundles over all manifolds of dimension $\leq n$.

On the other hand, Husemoller starts from the classifying space viewpoint. He defines a universal bundle as one, EG, for which the pullbacks of homotopy classes of maps of the M into BG construct all bundles over M. Of course, an immediate consequence of this is that all bundles over M can be embedded in EG. He relegates the proof that EG is contractible to an excercise.

Milnor and Stasheff restrict themselves to vector bundles and emphasize a more constructive approach. Using a local patch presentation of a real vector bundle of rank m over a manifold of dimension n allows us to regard the bundle locally as the product of $\mathbb{R}^n \times \mathbb{R}^m$. Thus, informally, the total space of the bundle can be thought of as a union of subsets of Euclidean space of sufficiently high dimension. For complex vector bundles, the second factor is $\mathbb{C}^m$. In the following we will focus the discussion on this complex case. Recall we are restricting the base space to be a paracompact manifold.

In more detail, the base space of the universal bundle is constructed from the linear subspaces of $\mathbb{C}^n$. This set of k-planes in $\mathbb{C}^n$, appropriately topologized and smoothed, is called the **Grassmannian** manifold, $Gr_{k,n}(\mathbb{C})$. Each point in this manifold is a k-plane, which we identify with the fiber over that point. This results in a (canonical) complex vector bundle over $Gr_{k,n}(\mathbb{C})$ with fiber $\mathbb{C}^k$ and group $Gl(k, \mathbb{C})$. The total space of the associated principal bundle is called the **Stiefel manifold**.

Remark 5.7.
Given a k-plane lying in $\mathbb{C}^n$, choose some (not unique) complementary $n - k$ plane. Then the full action of $Gl(n, \mathbb{C})$ can be decomposed into two subgroups leaving each subspace invariant. Since the full $Gl(n, \mathbb{C})$ takes any k plane into any other, the family of all such planes, the Grassmannian manifold, can be identified with

$$Gr_{k,n}(\mathbb{C}) = \frac{Gl(n, \mathbb{C})}{Gl(k, \mathbb{C}) \times Gl(n - k, \mathbb{C})}.$$

It is easy to see that the imposition of some Hermitian metric on $\mathbb{C}^n$ can be used to reduce the groups to their unitary subgroups, so

$$Gr_{k,n}(\mathbb{C}) = \frac{U(n)}{U(k) \times U(n - k)}.$$

Similarly, the full bundle space, the Stiefel manifold, is

$$V_{k,n}(\mathbb{C}) = \frac{Gl(n, \mathbb{C})}{Gl(n - k, \mathbb{C})} = \frac{U(n)}{U(n - k)}.$$

From our earlier result on coset spaces and bundles, it follows that the Stiefel manifold, $V_{k,n}(\mathbb{C})$, is indeed a principal bundle over the Grassmannian, $Gr_{k,n}(\mathbb{C})$, with group $U(k)$. Further analysis shows that for sufficiently large n, $V_{k,n}(\mathbb{C})$ satisfies the connectivity requirements in order for any vector bundle over some fixed compact manifold with group $U(k)$ to be mapped into it. For this reason, Milnor and Stasheff [Milnor and Stasheff (1974)], page 61, refer to this Stiefel manifold as a "universal bundle," with Grassmannian, $Gr_{k,n}(\mathbb{C})$, the classifying space. However, the dimension n in the Stiefel manifold depends on the particular compact M, so that this usage of the term "universal" is not so widespread currently. In fact, this construction can be generalized to the general paracompact, not necessarily compact, case by letting n go to infinity. This can be done rigorously as described in some detail later in this same book. Formally, we have a chain of embeddings

$$\hookrightarrow Gr_{k,n}(\mathbb{C}) \hookrightarrow Gr_{k,n+1}(\mathbb{C}) \hookrightarrow \ldots \hookrightarrow Gr_{k,\infty}(\mathbb{C}) = Gr_k(\mathbb{C}).$$

The limiting bundle is obtained similarly, $V_{k,\infty}(\mathbb{C}) = V_k(\mathbb{C})$, resulting in the Stiefel manifold as the total space of the **universal bundle** over the **classifying space** $Gr_k(\mathbb{C})$, with projection $\pi : V_k(\mathbb{C}) \to Gr_k(\mathbb{C})$.

We can now use these tools to classify vector bundles in terms of maps from the base manifold to the Grassmannian, the base of the universal bundle. Given a map $f : M \to Gr_k(\mathbb{C})$ we can use the pullback construction to get a vector bundle $E = f^*V_k(\mathbb{C})$ of rank k. From the homotopy bundle theorem, 5.1, we know that any other bundle lying in the homotopy class $[f]$ of f is isomorphic to E. Thus

Fact : *The set of equivalence classes, $Vect_{\mathbb{C}}(M, k)$, of complex vector bundles of rank k over M is given by*

$$Vect_{\mathbb{C}}(M, k) = [M, Gr_k(\mathbb{C})]_{homotopy}.$$

For the real case we get a corresponding result.

But how does the classifying space change as we look at an arbitrary group G instead of the $U(k)$ or $O(k)$ subgroups of the general linear one? In a pair of papers, [Milnor (1956a)] and [Milnor (1956b)], Milnor provides the answer by constructing a space BG depending only on the group G, generalizing $BU(k)$ and $BO(k)$. Here we sketch his ideas.

First recall the topological operation, the join $X \star Y$ between two arbitrary topological spaces X, Y as the quotient space of the product $X \times Y \times [0, 1]$ with respect to the following equivalence relation. For each point $x \in X$ identify all points of the set $\{x\} \times Y \times \{0\} \sim \{x\}$ and for each point $y \in Y$ identify all points of the set $X \times \{y\} \times \{1\} \sim \{y\}$. For example the join $S^n \star S^m$ is homotopic (and even homeomorphic) to S^{n+m+1} for all n, m. Now consider the k-join

$$G^{\star k} = \underbrace{G \star G \star \cdots \star G}_{k}$$

of a topological group G. Then the union

$$EG = \bigcup_{k=1}^{\infty} G^{\star k}$$

equipped with the direct limit topology is a contractible space denoted by EG. On this space there is a natural action of G and the quotient EG/G is the classifying space BG. Thus every countable vector bundle with group G action on each fiber, and so every principal G-bundle, can be classified by an appropriate map into BG.

Remark 5.8.
The homotopy properties of EG, in particular its contractibility, are key to its role as a universal bundle. From the exact sequence of homotopy groups associated to the fibration $EG \to BG$ we get from the contractibility of EG:

$$\pi_{k+1}(BG) = \pi_k(G) .$$

Consider the set $\Omega(BG)$ of maps $S^1 \to BG$, the loop space of BG. The relation above shows that there is a (weak) homotopy-equivalence $\Omega(BG) \sim G$. So let G be a discrete group then $BG = K(G, 1)$ is an Eilenberg-MacLane space $K(G, 1)$, i.e. (universal) covering described by the action of the discrete group G can be classifying by the fundamental group.

Finally, we remark that an important example of a universal space occurs in quantum mechanics. The space $BU(1) = BS^1$ is $\mathbb{C}P^\infty$ which is the space of all non-vanishing states in the Hilbert space. In physical terms, a state is a "ray" in Hilbert space, that is, an equivalence class of unit vectors up to multiplication by a $U(1)$ phase factor. We will discuss this example in more detail below.

Summary review of K-theory and classifying spaces. K-theory provides a cohomology-like structure whose ring properties make it a very useful tool for classifying stable equivalence classes of vector bundles. The complete, not just stable, classification is provided by classifying spaces and bundles, BG, EG, for principal G bundles over M by identifying the equivalence classes of such bundles with the homotopy classes $[M, BG]$ of maps $M \to BG$. For the groups $U(k), O(k)$ and $Sp(k)$, we obtain spaces $BU(k), BO(k)$ and $BSp(k)$ classifying the bundles with structure group $U(k), O(k)$ or $Sp(k)$, respectively. While the information in these vector bundle classifying spaces is complete they are often harder to compute than K-theory. In fact, stable classifying spaces are obtained by taking a limit $k \to \infty$ giving U, O and Sp, for complex vector spaces. The corresponding classifying spaces BU, BO and BSp classify the stable equivalence classes of bundles with structure group the unitary, orthogonal and symplectic group (of arbitrary dimension). As expected, $K(M) = [M, BU]$ for complex vector bundles, etc. The incompleteness of stable equivalence is shown by the example of the tangent vector bundle to S^2 which is stably equivalent to the trivial bundle, but not actually isomorphic to it.

5.3 Characteristic Classes

Characteristic classes are algebraic constructions, elements of the cohomology groups of a base space, M, defined using a given bundle, E, over that space. Thus each class depends on the base space and the bundle over it. Originally, these structures were investigated iteratively as **obstructions** to continuing a bundle cross section from the base space simplicial skeleton of dimension n to that of dimension $n + 1$. Steenrod [Steenrod (1951)] discusses this early approach extensively. Later, they developed into important tools for both the **bundle classification** and **cobordism** problems. Milnor and Stasheff devoted a classic book [Milnor and Stasheff (1974)] entirely to characteristic classes. Various approaches are characterized by the type of vector bundle, the choice of cohomology coefficient group, and the role of K-theory, universal bundles and classifying spaces.

In particular, we will be dealing with:

- **Chern classes**, defined for integer cohomology coefficients, $\mathbb{K} = \mathbb{Z}$, and complex bundle groups.
- **Chern classes** with real cohomology coefficients, $\mathbb{K} = \mathbb{R}$, using deRham cohomology representations,
- **Pontrjagin Classes**, defined for integer cohomology coefficients, $\mathbb{K} = \mathbb{Z}$, and real bundle groups. These are defined in terms of Chern classes by complexifying the real bundle,
- **Stiefel-Whitney classes**, defined for integer mod 2 cohomology coefficients, $\mathbb{K} = \mathbb{Z}_2$, and arbitrary vector bundles.

In all of the approaches, a *universal characteristic class* associates to every G-principal fiber bundle $\pi : E \to M$ an element $c(E) \in H^*(M, \mathbb{K})$ in the cohomology ring of M with coefficient group $\mathbb{K}$ being one of $\mathbb{R}, \mathbb{Q}, \mathbb{Z}$ or $\mathbb{Z}_2$. Obviously the choice of group for cohomology coefficients is important. The classes are required to satisfy certain naturality conditions, including invariance under maps. Thus from $f : M \to N$ we have $f^*c(E(N)) = c(f^*E(M))$. Note the reversal of direction for cohomology maps. This naturality means that equality of characteristic classes can be used as a necessary condition for bundle equivalence. In fact, characteristic classes of G-bundles are uniquely determined by the classes $c(EG) \in H^*(BG)$ of the universal bundle $\pi_G : EG \to BG$.

Complex Case

We start with complex vector bundles, or equivalently $U(n)$-principal fiber bundles for which the universal bundle is $\pi_{U(n)} : V_n(\mathbb{C}) \to Gr_n(\mathbb{C})$, where V_n is the Stiefel manifold and Gr_n is the Grassmannian classifying space discussed above. As we explained there: $EU(n) = V_n(\mathbb{C})$ and $BU(n) = Gr_n(\mathbb{C})$, the space of all k-planes in $\mathbb{C}^\infty$. In this section, we will restrict the coefficient group to be the integers. The first step is to find the cohomology $H^*(BU(n), \mathbb{Z})$. Without proof we state ([Milnor and Stasheff (1974)] Theorem 14.5):

Theorem 5.6. *The cohomology ring $H^*(BU(n))$ is a polynomial ring over $\mathbb{Z}$ freely generated by n generators $\{c_1, \ldots, c_n\}$ with $c_k \in H^{2k}(BU(n), \mathbb{Z})$.*

So, to investigate $U(n)$ bundles over a manifold M, start with a classifying map $f : M \to BU(n)$. From this, the pull back produces the corresponding bundle $E(M, U(n)) = f^*(EU(n))$. From the universal bundle classes $c_k(EU(n)) \in H^{2k}(BU(n), \mathbb{Z})$ lying in the even dimensional cohomology groups we obtain

$$c_k(E) = f^* c_k(EU(n)) \in H^{2k}(M, \mathbb{Z})$$

called the **k-th Chern characteristic classes** of the $U(n)$ bundle E over the base space M. We will study these classes in more detail below, both as bundle classifiers and as obstructions to cross sections and other structures. Also, we will review the original presentation of these classes by Chern in terms of de Rham cohomology, with real rather than integer coefficients.

As an introductory example, consider the (generalized) Hopf bundle $\pi_k : S^{2k+1} \to \mathbb{C}P^k$. Each of these is a $U(1)$-principal fiber bundle with associated complex line bundle. Note that in the construction of the universal $U(1)$ bundle each $Gr_1(k)$ is just $\mathbb{C}P^k$. The limit process $k \to \infty$ then results in the universal bundle $\pi_\infty : S^\infty \to \mathbb{C}P^\infty$, with S^∞ contractible.

There is an interesting interpretation of this bundle in terms of quantum mechanics. First, recall that historically Schroedinger described the state of a physical particle in terms of a complex valued function over space (time is a parameter in this presentation), $\psi(x)$. All such realistic representations must also be non-zero and square integrable, or in physics terminology, "normalizable." In the development of the physical interpretation of quantum theory, it became evident that all physically observable properties of this representation were actually encoded in the absolute value of $\psi(x)$. That is, the physics is invariant under

$$\psi(x) \to e^{i\alpha} \psi(x), \qquad (5.4)$$

where α is real. These transformations of representation leaving the physics unchanged are called **gauge** transformations. This set of transformations obviously constitutes a group, in this case, the $U(1)$ gauge group. For now we leave unsettled the matter of whether or not α is a constant. As we noted earlier, and will review later, looking at the generalization to non-constant α provides deep insights into the physics of electromagnetism, and opens the door to much of modern high energy theory.

Shortly after physicists began these initial steps formulating quantum theory, von Neumann [von Neumann (1955)] undertook to make the formalism more rigorous. In particular, he pointed out that the space of physical, that is, normalizable, $\psi(x)$ could naturally be identified as, topological S^∞, in which a Schroedinger wave function is a **vector** in a Hilbert space. Dirac introduced the notation, $|\psi\rangle$, to represent the Schroedinger state function as a (ket) vector. In this formalism, each point of space, $x \in M$, is assigned a state vector (its eigenstate), denoted by $|x\rangle$, and the physical state, $|\psi\rangle$, corresponding to the Schroedinger wave function, $\psi(x) \in \mathbb{C}$, by the formal sum,

$$|\psi\rangle = \sum_x \psi(x)|x\rangle. \tag{5.5}$$

Because of physical invariance under $U(1)$ gauge, (5.4), the physical state ought properly be regarded as a one-dimensional "projection operator," an element of $\mathbb{C}P^\infty$ rather than S^∞. For the moment, let us restrict to the case α a constant in (5.4). Thus, the proper mathematical representation of all physical states is the complex projective space, $\mathbb{C}P^\infty$. As a bundle, we have

$$\begin{array}{ccc} U(1) & \longrightarrow & S^\infty \\ & & \pi \downarrow \\ & & \mathbb{C}P^\infty \end{array} \tag{5.6}$$

In fact, this displays S^∞ as the $U(1)$ universal bundle, $EU(1)$, over $\mathbb{C}P^\infty$ as the classifying space, $BU(1)$. The cohomology of $\mathbb{C}P^\infty$ is a polynomial algebra over $\mathbb{Z}$ generated by one generator lying in $H^2(\mathbb{C}P^\infty, \mathbb{Z})$. This element has a dual homology class represented by the space $\mathbb{C}P^1 \subset \mathbb{C}P^\infty$. Thus the classical Hopf bundle $\pi_H : S^3 \to S^2 = \mathbb{C}P^1$ is the generator of all other $U(1)$-principal fiber bundles and we obtain the first Chern class $c_1 \in H^2(S^2, \mathbb{Z})$. Now consider a map, $D_\psi : M \to S^\infty$, defined by

$$D_\psi(x) = \psi(x)|x\rangle \quad \text{no sum over } x.$$

Now combine this with the bundle projection in (5.6) to get another map,

$$f_\psi = \pi \cdot D_\psi : M \to \mathbb{C}P^\infty.$$

As a representative of a class in $[M, BU(1)]$ the wave function classifies a $U(1)$ principal bundle over M. The corresponding characteristic class is $c_1(E) = f_\psi^* c_1$.

Thus we can say that the homotopy class of a wave function, expressed by f_ψ, classifies a $U(1)$ bundle over space. However, in general, there is no obvious physical interpretation for this homotopy class of wave functions.

We now look at the application of the K theory, $K(M)$, for a manifold M, to the study of characteristic classes of vector bundles over M. The sum of all Chern classes

$$c(E) = 1 + c_1(E) + c_2(E) + \cdots + c_n(E)$$

of an n-dimensional, complex vector bundle E over M defines the **total Chern class**. For a complex line bundle L_1 we obtain $c(L_1) = 1 + c_1(L_1)$. In $K(M)$, we can use the so-called splitting principle[Hirzebruch (1973)] as an aid in studying an n-dimensional, complex line bundle $E(M)$. This principle asserts that for every such bundle there exists another manifold, M_1 and a map, $f : M_1 \to M$ such that the pullback bundle, $E_1(M_1) = f^*E(M)$ splits into a direct sum of n complex line bundles $L_1 \oplus L_2 \oplus \cdots \oplus L_n$ and $f^* : H^*(M) \to H^*(M_1)$ is a monomorphism. For more details see pages 251ff in [Husemoller (1994)].

With the help of the relation (see [Nakahara (1989)], [Milnor and Stasheff (1974)]) $c(L_1 \oplus L_2) = c(L_1)c(L_2)$ we obtain

$$c(E) = c(L_1 \oplus L_2 \oplus \cdots \oplus L_n)$$
$$= \prod_{i=1}^{n} c(L_i) = \prod_{i=1}^{n}(1 + x_i), \quad \text{where } x_i = c_1(L_i) \,.$$

Thus the $c_i(E)$ are elementary symmetric functions in the x_i. This splitting principle then facilitates algebraic manipulations with the generators. Define the **Chern character** $ch(E)$ of E by

$$ch(E) = \sum_{i=1}^{n} e^{x_i} \,.$$

The development of the sum leads to

$$ch(E) = n + c_1(E) + \frac{1}{2}(c_1^2(E) - 2c_2(E)) + \cdots$$

The properties of the Chern character are summarized in the following theorem:

Theorem 5.7. *The Chern character induces the following isomorphism*

$$ch : K(M) \otimes \mathbb{Q} \to \bigoplus_{i \geq 0} H^{2i}(M, \mathbb{Q})$$

between the K-theory and the even dimensional rational cohomology of M. Furthermore, given two complex vector bundles E, F then the relations:

$$ch(E \oplus F) = ch(E) + ch(F)$$
$$ch(E \otimes F) = ch(E)ch(F)$$

extends the isomorphism ch to a ring isomorphism.

Of course because this is expressed in terms of rational rather than integer coefficients the torsion part of K theory does not appear in this isomorphism. That is why characteristic classes do not give a complete answer to the classification problem, i.e. two bundles with identical characteristic classes need *not* be isomorphic.

Real Case

Now consider other groups including real ones. The case of an $SO(n)$ principal fiber bundle leads to classifying spaces denoted by $BSO(n)$. These spaces have a much more complicated structure than $BU(n)$ or $BSU(n)$ although with appropriate modifications real results parallel the complex ones sketched above. Corresponding to 5.6, we combine Theorem 15.9 and 7.1 from [Milnor and Stasheff (1974)] to get:

Theorem 5.8. *The (rational) cohomology ring $H^*(BO(n), \mathbb{Q})$ divides into two parts. The (rational) cohomology ring $H^*(BO(2k+1), \mathbb{Q})$ is a polynomial ring over $\mathbb{Q}$ freely generated by k generators $\{p_1, \ldots, p_k\}$ with $p_m \in H^{4m}(BO(2k+1), \mathbb{Q})$. Otherwise, the (rational) cohomology ring $H^*(BO(2k), \mathbb{Q})$ is a polynomial ring over $\mathbb{Q}$ freely generated by $k - 1$ generators $\{p_1, \ldots, p_{k-1}\}$ with $p_m \in H^{4m}(BO(2k), \mathbb{Q})$ and an additional class $e_{2k} \in H^{2k}(BO(2k), \mathbb{Q})$. The $\mathbb{Z}_2$ cohomology ring $H^*(BO(n), \mathbb{Z}_2)$ is a polynomial algebra over $\mathbb{Z}_2$ freely generated by n generators $\{w_1, \ldots, w_n\}$ with $w_m \in H^m(BO(n), \mathbb{Z}_2)$.*

We remark that instead of using the full rationals, $\mathbb{Q}$, as coefficient group it is enough to consider a ring over $\mathbb{Z}$ containing $1/2$. The classes described in this theorem are those mentioned at the beginning of this section.

From a map $f : M \to BSO(n)$ we obtain the corresponding classes on M by pullback. That is, let E be a real vector bundle over M defined by the homotopy class of the map f, then $p_k(E) = f^*p_k \in H^{4k}(M, \mathbb{Z})$ etc. These classes, $p_k(E)$ are called the **Pontrjagin** classes of the bundle E.

There are several notable relations between the characteristic classes. Let E be an n-dimensional, real vector bundle and $E_C = E \otimes \mathbb{C}$ its complexification. The expression $c(E_C)$ and $p(E)$ denotes the sum over all Chern and Pontrjagin classes, respectively.

- Let F be another vector bundle, then $p(E \oplus F) = p(E)p(F)$.
- $p_k(E) = (-1)^k c_{2k}(E_C)$
- Let P be a complex n-dimensional vector bundle. Using a real representation of the complex structure we obtain a real $2n$-dimensional

vector bundle P_r with

$$c((P_r)_C) = 1 - p_1(P_r) + p_2(P_r) - \cdots$$
$$= [1 + c_1(P) + c_2(P) + \cdots][1 - c_1(P) + c_2(P) - \cdots]$$

In particular, $p_1(P_r) = (c_1^2 - 2c_2)(P)$.
- Let E be an even-dimensional real vector bundle with $n = 2k$ then the highest Pontrjagin class is $p_k(E) = (e(E))^2$.
- For complex bundles we have $(e(P_r))^2 = p_k(P_r) = (c_k(P_r))^2$.

The corresponding map from KO-theory (K-theory for real vector bundles with orthogonal group) to cohomology is much more complicated and depends on the torsion in cohomology. The interested reader is referred to the standard literature [Husemoller (1994); Karoubi (1978)] where the necessary relationship to Clifford algebras is worked out.

The Weil homomorphism

To this point our discussion of characteristic classes has been rather abstract and formal. We now shift gears and review an approach pioneered by **Chern**, which makes use of a bundle connection, and thus is at least formally related to the physics of **gauge theory.** In this original form, the Chern classes are defined as elements of the deRham cohomology, with real coefficient group, in distinction to the integer classes we used in and following Theorem 5.6. First recall some facts from group theory. We know that every G-principal bundle over the manifold M is given by a map $f : M \to BG$ and the isomorphism classes are given by the homotopy classes $[f]$. Furthermore as shown in Steenrod[Steenrod (1951)] or Husemoller [Husemoller (1994)], homotopy groups $\pi_k(G)$ and $\pi_{k+1}(BG)$ are isomorphic, since the bundle space, EG, is contractible. The Hurewicz theorem provides a relation between the cohomology of $H^*(BG)$ and $H^*(G)$. Since the cohomology of a Lie group is determined by the cohomology of the corresponding Lie algebra we are led to seek a map "$H^*(\mathfrak{g}) \to H^*(M)$" corresponding to $f^* : H^*(BG) \to H^*(M)$.

Specifically, let G be a Lie group with Lie algebra $\mathfrak{g}$. Let $I^k(G)$ be the set of symmetric multilinear mappings

$$f : \underbrace{\mathfrak{g} \times \cdots \times \mathfrak{g}}_{k} \to \mathbb{R}$$

such that $f(at_1a^{-1}, \ldots, at_ka^{-1}) = f(t_1, \ldots, t_k)$ for $t_1, \ldots, t_k, a \in G$. The original Chern construction is based on a mapping from $I(G) = \sum_{k=0}^{\infty} I^k(G)$ to the $2k$-forms defined on the total space P of a principal fiber bundle over

M with group G. Choose a connection form ω in P with curvature form Ω. For each $f \in I^k(G)$, let $f(\Omega)$ be the following $2k$-form on P:

$$f(\Omega)(X_1,\ldots,X_{2k}) = \frac{1}{(2k)!} \sum_\sigma \epsilon_\sigma f(\Omega(X_{\sigma(1)},X_{\sigma(2)}),\ldots,\Omega(X_{\sigma(2k-1)},X_{\sigma(2k)}))$$

for tangent vectors $X_1,\ldots,X_{2k} \in T_u P$ in $u \in P$ where the summation is taken over all permutations σ of $(1,2,\ldots,2k)$ and ϵ_σ denotes the sign of the permutation σ. The following theorem defines the Weil homomorphism $I(G) \in H^*(M,\mathbb{R})$ and its properties.

Theorem 5.9. *Let P be a principal fiber bundle over M with group G and projection $\pi : P \to M$. Choosing a connection in P, let Ω be its curvature form on P.*

(1) For each $f \in I^k(G)$, the $2k$-form $f(\Omega)$ on P projects to a (unique) closed $2k$-form, say $\tilde{f}(\Omega)$, on M, i.e., $f(\Omega) = \pi^*(\tilde{f}(\Omega))$;

(2) If we denote by $w(f)$ the element of the deRham cohomology group $H^{2k}(M,\mathbb{R})$ defined by the closed $2k$-form $\tilde{f}(\Omega)$, then $w(f)$ is independent of the choice of a connection and $w : I(G) \to H^*(M,\mathbb{R})$ is an algebra homomorphism.

Detailed proofs are available in Husemoller[Husemoller (1994)], Milnor and Stasheff[Milnor and Stasheff (1974)] and Nakahara[Nakahara (1989)]. The critical issue in this theorem is the independence of the deRham cohomology element, $w(f)$, from the specific choice of connection. Informally, this is a consequence of an argument along these lines. First, we can reduce any change of connection to an "infinitesimal" one, $\delta\omega$, a $\mathfrak{g}$ valued one form. Then, to first order in $\delta\omega$, $f(\Omega)$ changes by an exact differential plus terms corresponding to the adjoint action of $\mathfrak{g}$ on Ω. The differential term vanishes since the map is into the deRham cohomology group and the adjoint action is trivial because of the invariance of I. This is indeed a beautiful result, relating the geometry and physics of gauge theory to topology. We should again point out that the Chern cohomology classes as just defined are all elements of deRham cohomology, which is necessarily based on real coefficients, rather than the integers as we used in the earlier introduction of Chern classes. This can lead to some confusion in the meaning of "Chern" class unless care is taken.

More concretely, we now investigate the algebra $I(G)$ (or commutative ring) for the classical (semi-simple) Lie groups $U(n)$, $SO(n)$ and $Sp(n)$. It turns out that the full set of invariant polynomials can actually be generated from determinant evaluations reminiscent of the eigenvalue problem. See [Milnor and Stasheff (1974)], page 299 ff. The polynomial functions $f_1, \ldots, f_n$ on the Lie algebra $\mathfrak{g}$ of G for these groups are, respectively:

(1) For $G = U(n)$,

$$\det(\lambda\, Id_n + iX) = \lambda^n + \sum_{k=1}^{n}(-1)^n f_n(X)\lambda^{n-k} \text{ for } X \in \mathfrak{u}(n)$$

These f_n are algebraically independent and generate the algebra of (invariant) polynomial functions of $\mathfrak{u}(n)$.

(2) For $G = O(m)$ (where $m = 2n$ or $m = 2n+1$),

$$\det(\lambda\, Id_n - X) = \lambda^n + \sum_{k=1}^{n} f_n(X)\lambda^{n-k} \text{ for } X \in \mathfrak{o}(m)$$

These f_n are algebraically independent and generate the algebra of (invariant) polynomial functions of $\mathfrak{o}(m)$.

(3) For $G = Sp(n)$,

$$\det(\lambda\, Id_{2n} + iX) = \lambda^{2n} + \sum_{k=1}^{n}(-1)^n f_n(X)\lambda^{2(n-k)} \text{ for } X \in \mathfrak{sp}(n)$$

These f_n are algebraically independent and generate the algebra of (invariant) polynomial functions of $\mathfrak{sp}(n)$.

(4) For $G = SO(m)$ and $m = 2n+1$, we have the functions $f_1, \ldots, f_n$ analogous to the $O(m)$ case. But for $m = 2n$ there exists a polynomial function g such that $f_n = g^2$ and the functions $f_1, \ldots, f_{n-1}, g$ are algebraically independent and generate the algebra of (invariant) polynomial functions of $\mathfrak{so}(2n)$.

The algebra $I(G)$ for the exceptional Lie groups F_4, G_2, E_6, E_7, E_8 is also generated by a finite number of polynomial functions i.e. 2 functions for G_2, 4 for F_4, 6 for E_6, 7 for E_7 and 8 for E_8. We will not go into the details here.

Chern-Weil theory

Now let P be a $Gl(n, \mathbb{C})$-principal fiber bundle over M with curvature form Ω defined on P. By theorem 5.9 there is a unique closed $2k$ form γ_k such that $\pi^*(\gamma_k) = f_k(\Omega)$. Thus

$$\det\left(Id_n + \frac{i}{2\pi}\Omega\right) = \pi^*(1 + \gamma_1 + \cdots + \gamma_n)$$

leading to the kth Chern class as $c_k \in H^{2k}(M, \mathbb{R})$ generated by γ_k. Let F be the curvature form on M projected to P, i.e. $\pi^* F = \Omega$ then the total Chern class $c(P)$ is

$$c(P) = c(F) = \det\left(Id_n + \frac{i}{2\pi}F\right) = 1 + c_1(F) + c_2(F) + \cdots. \qquad (5.7)$$

As a simple example, consider a 2-dimensional complex vector bundle P over a 4-dimensional manifold M with structure group $G = SU(2)$. The curvature form F on M can be chosen locally to be:

$$F = F^a(\sigma_a/2i) = \frac{1}{2}F^a_{\mu\nu}(\sigma_a/2i) dx^\mu \wedge dx^\nu$$

where $\sigma_a, a = 1, 2, 3$ are the Pauli matrices, a basis of $\mathfrak{su}(2)$. From this

$$\begin{aligned}
c(F) &= \det\left(Id_2 + \frac{i}{2\pi} F^a(\sigma_a/2i)\right) \\
&= \det\begin{pmatrix} 1 + (i/2\pi)(F^3/2i) & (i/2\pi)(F^1 - iF^2)/2i \\ (i/2\pi)(F^1 + iF^2)/2i & 1 - (i/2\pi)(F^3/2i) \end{pmatrix} \\
&= 1 + \frac{1}{4}(i/2\pi)^2(F^3 \wedge F^3 + F^1 \wedge F^1 + F^2 \wedge F^2)
\end{aligned}$$

or $c_1(F) = 0$ and $c_2 = (i/2\pi)^2 \, tr(F \wedge F) = \det(iF/2\pi)^1$. For a general k-dimensional complex vector bundle the Chern classes are

$$\begin{aligned}
c_1(F) &= \frac{i}{2\pi} tr(F) \\
c_2(F) &= \frac{1}{8\pi^2}[tr(F \wedge F) - tr(F) \wedge tr(F)] \\
&\cdots \\
c_k(F) &= (i/2\pi)^k \det(F).
\end{aligned}$$

Because of the tracelessness of the Pauli matrices, this agrees with our previous example.

Now shift to the real case. Consider a real k-dimensional vector bundle P over M with structure group $O(k)$ (by introducing a fiber metric leading to a Riemannian metric on M). The curvature form F on M defines the **Pontrjagin classes** (recall theorem 5.8) by

$$p(F) = \det\left(Id + \frac{F}{2\pi}\right) = 1 + p_1(F) + p_2(F) + \cdots$$

---

[1] The trace and determinant are always understood to be taken with respect to a suitable representation of the structure group.

which leads to the concrete formulas

$$p_1(F) = -\frac{1}{2}\left(\frac{1}{2\pi}\right)^2 tr(F^2)$$

$$p_2(F) = \frac{1}{8}\left(\frac{1}{2\pi}\right)^4 [(tr(F^2))^2 - 2tr(F^4)]$$

$$p_1(F) = \frac{1}{48}\left(\frac{1}{2\pi}\right)^6 [-(tr(F^2))^3 + 6(tr\,F^2)(tr\,F^4) - 8(tr\,F^6)]$$

......

$$p_{[k/2]}(F) = \left(\frac{1}{2\pi}\right)^k \det F.$$

We will now look more closely at the **Euler** class, referred to in theorem 5.8. Let M be a $2m$-dimensional (if the dimension is odd, the Euler class vanishes) orientable Riemannian manifold with tangent bundle TM. Denote the curvature form by R. By introducing an orthonormal frame reduce the structure group of TM down to $SO(2m)$. Let p_m be the highest Pontrjagin class of M. The curvature R is a 2-form and is thus commutative under the wedge product $\wedge$. The form R is also $\mathfrak{so}(2m)$-valued, that is, its value is a skew-symmetric $2m \times 2m$ matrix. Note that the determinant of a $2m \times 2m$ skew-symmetric matrix A can be written as a square of a polynomial called the Pfaffian $Pf(A)$,

$$\det A = (Pf(A))^2$$

The Pfaffian is given by

$$Pf(A) = \frac{(-1)^m}{m!} \sum_P sgn(P) A_{P(1)P(2)} A_{P(3)P(4)} \cdots A_{P(2m-1)P(2m)} \quad (5.8)$$

where the sum is over all permutations P weighted by the sign of the permutation, $sgn(P)$. For more details, see pages 309ff of [Milnor and Stasheff (1974)]. In our notation above the Euler class $e(R)$ is then given by

$$e(R) = Pf\left(\frac{R}{2\pi}\right) \quad (5.9)$$

By way of illustration consider the example of the (unit-)sphere $M = S^2$ with tangent bundle TS^2. Recall that theorem 5.9 assures us that the classes will be independent of choice of connection, so we choose the simplest, spherical geometry. The Lie algebra of the $SO(2)$ group is of course the one dimensional set of antisymmetric 2X2 matrices. The Lie algebra-valued curvature form then has only one component, given by

$$R_{12} = \sin\theta d\theta \wedge d\phi$$

in the coordinate system (θ, ϕ). Formally we obtain for the Pfaffian with $m = 1$

$$Pf(A) = -A_{12} = A_{21}$$

and thus for the Euler class

$$e(R) = Pf\left(\frac{R}{2\pi}\right) = \frac{1}{2\pi}R_{12}\,.$$

In the coordinate system of the sphere we obtain the Euler class,

$$e(S^2) = \frac{1}{2\pi}\sin\theta d\theta \wedge d\phi\,.$$

It is interesting to note that

$$\int_{S^2} e(S^2) = \frac{1}{2\pi}\int_0^{2\pi} d\phi \int_0^{\pi} d\theta \sin\theta = 2$$

which agrees with the topologically defined Euler characteristic, or number, of the sphere, $\chi(S^2) = 2$. This agreement is the content of the **Gauss-Bonnet theorem**:

$$\int_M e(M) = \chi(M), \qquad (5.10)$$

for a compact orientable manifold M without boundary. If M is odd-dimensional both e and χ vanish.

As a second example we will write down the Euler class for an orientable four-manifold M. The structure group of TM is $SO(4)$ and we obtain for the Pfaffian of a 4×4 skew-symmetric matrix A:

$$Pf(A) = A_{12}A_{34} - A_{13}A_{24} + A_{14}A_{23} = \frac{1}{4!}\epsilon^{ijkl}A_{ij}A_{kl}$$

and finally for the Euler class

$$e(M) = \frac{1}{2(2\pi)^4}\epsilon^{ijkl}R_{ij} \wedge R_{kl} \qquad (5.11)$$

with the totally anti-symmetric tensor ϵ^{ijkl}.

Remark 5.9.

This Euler characteristic, or number, for a compact space, M, $\xi(M)$ has a long and fruitful history connecting topology, differential topology, and geometry. It was originally defined as the signed sum of the dimensions of homology groups of M. See for example, Vick [Vick (1994)], p 63. If M is smooth, it is related to the existence of fixed points for flows generated by vector fields [Milnor (1965b)]. In turn, this implies that it is an "obstruction" to the continuation of a global non-zero vector field. Finally, the Gauss-Bonnet theorem ties it to geometry, in a manner presaging Chern classes.

Stiefel-Whitney classes

Stiefel-Whitney classes are characteristic classes, generally denoted by $w_i(E(M))$, lying in the cohomology group $H^i(M, \mathbb{Z}_2)$ but dependent on the choice of vector bundle, E over M. Because the coefficient group is $\mathbb{Z}_2$, these cohomology groups cannot be described by the differential forms of deRham cohomology over $\mathbb{K} = \mathbb{R}$. However, the w_i have their own special applications as we will shortly see. In the book by Milnor and Stasheff[Milnor and Stasheff (1974)], dedicated to characteristic classes, Stiefel-Whitney classes are the first ones introduced and then using an axiomatic presentation. Their construction in a later chapter makes use of the techniques of Steenrod squares and the Thom isomorphism, [Milnor and Stasheff (1974)], page 90, as we will briefly review below. Here we will comment on a few salient points. First, the fact that the coefficient group is $\mathbb{Z}_2$ rather than $\mathbb{Z}$ leads to the rather mysterious looking equations

$$1 + 1 = 0, \quad 1 = -1. \quad \text{over } \mathbb{Z}_2 \tag{5.12}$$

Next, a multiplicative operation, the cup product in cohomology (see, for example, [Vick (1994)]) can be used to convert $H^*(M, \mathbb{K})$ into a ring. This product generalizes the wedge product for forms in deRham cohomology and shares the antisymmetry properties of latter. However, in the Stiefel-Whitney case of $\mathbb{K} = \mathbb{Z}_2$, all antisymmetry is replaced by symmetry since $-1 = 1$, noted in (5.12).

An important property of this cup product for Stiefel-Whitney classes is the Whitney "Theorem," which is actually presented as an "Axiom" by Milnor and Stasheff,

$$w_k(E \oplus F) = \sum_{i=0}^{k} w_i(E) \cup w_{k-i}(F), \tag{5.13}$$

where E, F are two vector bundles over the same base space. When applied to the total class, defined by

$$w(E) = 1 + w_1(E) + w_2(E) + w_3(E) + \cdots$$

we get
$$w(E \oplus F) = w(E) \cup w(F)$$
For the special case of $E = TM$, the tangent vector bundle, it turns out that the total class is generated by one element. More precisely, we paraphrase Corollary 11.15 to Wu's theorem as stated by Milnor and Stasheff[Milnor and Stasheff (1974)],

Corollary: *Let M be a manifold of dimension km for which $H^*(M, \mathbb{Z}_2)$ is generated by one generator $a \in H^k(M, \mathbb{Z}_2)$ $1 \leq k$. Then the total Stiefel-Whitney class is given by*
$$w(TM) = (1+a)^{m+1} .$$

Note that the condition on $H^*(M, \mathbb{Z}_2)$ will be satisfied for $m = 2$ and $k = 1, 2, 4, 8$.

We again refer the reader to Milnor and Stasheff[Milnor and Stasheff (1974)], §8, for the explicit construction of Stiefel-Whitney classes using the Steenrod squaring operation. We close this section by noting some applications of Stiefel-Whitney classes for the tangent bundle TM of a manifold M.

- w_1 is an obstruction to orientability of M. That is, M can be oriented if and only if $w_1(TM) = 0$.
- w_2 is an obstruction to the definition of a (global) spin structure on an oriented manifold M. That is, such a spin structure exists on oriented M if and only if $w_2(TM) = 0$.

We will discuss these points further in the next section.

Because of the importance of a spin structure in physical theories and their implications for mathematics, the obstruction to a spin structure presented by w_2 limits the range of useful base manifolds. However, it turns out that a generalization of Spin to Spin_C, is not subject to this limitation, at least for four manifolds.

5.4 Introduction of Spin and Spin$_C$ Structures

Einstein's principle of (special) relativity requires the invariance of the spacetime distance measured by the Minkowski (flat) metric to insure that all reference frames see the same speed of light. This bedrock principle is based on the observations of Michelson and Morley in the late 19th century showing the independence of light speed on direction as the earth moves through its orbit. Mathematically, such reference frame transformations are described by the group $SO(3,1)$ also known as the (proper, homogeneous) Lorentz group.[2] Quantum mechanics or quantum field theory incorporates Einstein's relativity by requiring that the presentation of its theories also be invariant under the Lorentz group. Since quantum states are represented by complex vectors, this means that quantum theory requires complex representations of this group. Elements of the Lie algebra, described physically as "infinitesimal" transformations, turn out to be related to physical observables, and invariance of a theory under elements of a transformation group can be shown to correspond to the conservation (time constancy) of some corresponding physical quantity. This deep and important connection between invariance and conservation laws is summarized in **Noether's Theorem** as we discussed in the previous chapter. For more insights into this subject, refer to standard quantum field theory texts, such as [Ramond (1990)],[Hatfield (1992)].

First, recall a fundamental fact of Lie theory: A Lie group determines its algebra, but its algebra only determines the local structure of the group. One obvious difference occurs when the group is not connected. But even if connected it may not be simply connected. In fact, this is the case for $SO(3,1)$ for which $\pi_1(SO(3,1)) = \mathbb{Z}_2$. If the physics resides in the Lie algebra, then we must search for a uniquely related group. This is where the physics of spin appears.

The "unentangling" of a non simply connected space is accomplished by the mathematical structure known as a **covering** space. For the Lorentz group, $SO(3,1)$, this is provided by $SL(2,\mathbb{C})$, the spin group of Minkowski spacetime.

The indefinite nature of the Lorentz metric deprives it of an essential feature for many mathematical applications. In particular, the Lorentz metric distance cannot be used to generate the locally Euclidean topology of spacetime manifolds. For this, and other reasons, physicists often investigate formal theories resulting from the replacement of the metric of signature $(-,+,+,+)$ by the Euclidean one, $(+,+,+,+)$.

---

[2]The full group representing all possible reference frame transformations satisfying the Einstein's principle of special relativity includes dilations, translations and inversions. However, its structure does not play a role in our considerations here.

Remark 5.10.
Since the Hodge operator equation (generalization of ∇^2) for the Minkowski metric produces a **hyperbolic** equation, with wave motion of finite speed, while the Euclidean metric produces an elliptic equation ("infinite" speed), quantum particle models for Euclidean signature metrics are referred to as **instantons**. While the physics of such structures is certainly dubious, they have proven to be useful in exploring theoretical constructs in both quantum physics and related mathematics, such as, especially, the Seiberg-Witten formalism.

For now we deal with Euclidean space with group $SO(4)$ for which the spin group can be represented as $\text{Spin}(4) = SU(2) \times SU(2)$[3]. Note, however, that it is standard terminology to use the term "spin" to refer to the covering groups of the special orthogonal groups for metrics of both definite and indefinite signature.

First, we note that for a manifold to support a spin structure, it must at least be orientable. Let TM be the tangent bundle of a manifold M with structure group $O(m)$. To TM we associate a frame bundle LM with transition functions $g_{ij} : U_i \cap U_j \to O(m)$. Referring to Nakahara[Nakahara (1989)], pages 402ff, we can define

$$f(U_i, U_j) = \det g_{ij} = \pm 1, \text{ so, } f \in C^1_{\check{C}ech}(M),$$

as a Čech chain. In fact, it is a cocycle since for $U_i \cap U_j \cap U_k \neq 0$,

$$g_{ij} g_{jk} g_{ki} = Id_{O(m)}.$$

As a cocycle it has a projection into an element of $H^1_{\check{C}ech}$ which turns out to be the first Stiefel-Whitney class, w_1. On the other hand, this class is trivial, i.e., equal to one, if and only if f is a cobound. In the Čech sense, this would mean

$$f(U_i, U_j) = \det g_{ij} = \lambda(U_i) \lambda(U_j),$$

for some λ as a function of coordinate patch, which can be taken as its orientation. Thus, the orientability of M, i.e., the existence of a covering for which every $f(U_i, U_j) = +1$, is precisely the condition that $w_1 = 1$ using multiplicative representation, or 0 for the additive one.

Remark 5.11.
At this point we should clarify the relationship between two possible representations of the coefficient group $\mathbb{Z}_2$, namely as the additive group with elements $\{0, 1\}$, or equivalently as the multiplicative group of elements $\{1, -1\}$. The map $x \to \exp(i\pi x)$ can serve as the isomorphism. In the orientability discussion above we use the multiplicative representation.

---

[3]Recall that $SU(2)$ is isomorphic to the multiplicative group of unit quaternions, so $SU(2)$ is diffeomorphic to S^3.

From here on assume an orientation of M.[4] Furthermore, we will be interested only in the case of a compact, even closed, M. Introducing a Riemannian metric on TM, we can construct a principal bundle, represented as LM, the frame bundle with transition functions $g_{ij} \in SO(m)$. Now we look into the question of building a Spin bundle over M. Denote the covering map by $\phi : \text{Spin}(m) \to SO(m)$. We look to lift the $SO(m)$ frame bundle, LM, to a spin bundle, $\text{Spin}(M, m)$. Such a lift is called **Spin structure**. This requires the existence of transition functions $\tilde{g}_{ij}$ in $\text{Spin}(m)$ such that

$$\phi(\tilde{g}_{ij}) = g_{ij}, \tag{5.14}$$

with bundle consistency conditions,

$$\tilde{g}_{ij}\tilde{g}_{jk}\tilde{g}_{ki} = Id_{\text{Spin}(m)}, \quad \tilde{g}_{ii} = Id_{\text{Spin}(m)}. \tag{5.15}$$

Assuming a "good covering" [Bott and Tu (1995)] all intersections $U_i \cap U_j$, are contractible, so the local lifting in (5.14) is ensured. However, the "cocycle" condition in (5.15) may not be satisfied. An argument extending that above for orientability is described both in Nakahara[Nakahara (1989)] and Moore[Moore (2001)] and leads to the following theorem.

Theorem 5.10. *Let M be a manifold. The tangent bundle of M admits a spin bundle structure if and only if $w_1(M)$ and $w_2(M)$ are trivial (i.e. M is orientable and a unique transport of spinors is possible). The spin structures are in one-one correspondence with the elements of $H^1(M, \mathbb{Z}_2)$.*

From Clifford algebra arguments, the covering group $\text{Spin}(m)$ can be represented as a matrix group acting on $\mathbb{C}^k$ where $k = 2^{[m/2]}$ and $[x]$ is the integral part of x. Associated to the principal bundle, $\text{Spin}(m, M)$ then is a vector bundle, $\text{Spinor}(m, M)$, with fiber $\mathbb{C}^k$. Sections of this bundle are called **spinors**. If we choose a non-trivial spin structure, corresponding to a non-trivial element of $H^1(M, \mathbb{Z}_2)$, then the sections of $\text{Spinor}(M, m)$ associated to this structure are called exotic spinors.

For realistic physics we need a pseudo-Riemannian manifold with structure group $SO(n, k)$. Specifically, this group is the set of real $(n+k) \times (n+k)$ matrices leaving invariant a "metric," represented by a diagonal matrix with n positive and k negative diagonal entries. For a thorough study of these groups using Clifford algebras, see [Morgan (1996)] and [Moore (2001)]. The double cover of this group is the spin group $\text{Spin}(n, k)$ of the manifold defining the corresponding spin bundle. Manifolds which can carry

---
[4]It should be mentioned that even in the non-oriented case there is a precursor to Spin, the so-called Pin structure defined by a map $\phi : Pin \to O(m)$ with kernel $\ker \phi = \mathbb{Z}_2$.

a pseudo-Riemannian structure and a spin structure are described by the following theorem.

Theorem 5.11. *Let M be a m-dimensional manifold. M can be given a pseudo-Riemannian structure of type (n,k) if and only if the tangent bundle admits a decomposition $TM = \zeta^n \oplus \eta^k$ into a time-like and space-like subbundle. M can be given a spin structure if and only if*
$$w_2(M) = w_1(\zeta) \cup w_1(\eta)$$
in $H^2(M, \mathbb{Z}_2)$. The distinct spin structures are in one-one correspondence with the elements in $H^1(M, \mathbb{Z}_2)$.

The preceding discussion points out that not every manifold will support a Spin structure. However, a generalization, Spin_C can be provided for any oriented 4-manifold M. Furthermore, this is precisely the structure we need to formulate Seiberg-Witten theory.

First, the group $\text{Spin}_C(n)$ can be defined by the exact sequence
$$0 \to \mathbb{Z}_2 \to \text{Spin}_C(n) \to SO(n) \times U(1) \to 0 \ .$$
In dimension four, of interest to us, the Spin group can be written $\text{Spin}(4) = SU(2) \times SU(2)$ with explicit representation as the set of 4×4 matrices of type
$$\text{Spin}(4) = \left\{ \begin{pmatrix} A_+ & 0 \\ 0 & A_- \end{pmatrix} \right\}$$
with $A_\pm \in SU(2)$. This Lie group of real dimension six is contained in a Lie group of dimension seven,
$$\text{Spin}_C(4) = \{B\} = \left\{ \begin{pmatrix} \lambda A_+ & 0 \\ 0 & \lambda A_- \end{pmatrix} : A_\pm \in SU(2), \lambda \in U(1) \right\} \ .$$
This results in a homomorphism $\alpha : \text{Spin}_C(4) \to U(1)$ with $\alpha(B) = \det(\lambda A_+) = \det(\lambda A_-) = \lambda^2$.

Remark 5.12.
Another representation of Spin_C is based on the space of spacetime quaternions, V, complex 2×2 matrices of the form
$$Q = \begin{pmatrix} t+iz & -x+iy \\ x+iy & t-iz \end{pmatrix} \ .$$
Define $\rho : \text{Spin}_C(4) \to Gl(V)$ by
$$\rho \begin{pmatrix} \lambda A_+ & 0 \\ 0 & \lambda A_- \end{pmatrix} (Q) = (\lambda A_-) Q (\lambda A_+)^{-1}.$$

The Euclidean spacetime metric is given by $\det Q$ which is preserved by ρ. However, the $U(1)$ component, λ, factors out of this, giving the map of Spin_C onto $SO(4) \times U(1)$ described above.

A Spin$_C$ structure on a 4-manifold M is the extension of the structure group $SO(4)$ of TM to Spin$_C(4)$. Let $g_{ij} : U_i \cap U_j \to SO(4)$ be the transition function of TM and $\tilde{g}_{ij} : U_i \cap U_j \to$ Spin$_C(4)$ the extension of g_{ij} known to exist in the contractible $U_i \cap U_j$. The obstruction presented by a nontrivial w_2 to the global extension of this map for Spin is not present for Spin$_C$ precisely because of its additional $U(1)$ component, so that

Theorem 5.12. *Every compact, oriented four-manifold possesses (at least one) Spin$_C$ structure.*

For details of the proof of this see [Moore (2001)] and [Hirzebruch and Hopf (1958)]. The proof rests on the fact that if M^4 is a compact, oriented 4-manifold, then $w_2 \in H^2(M^4, \mathbb{Z}_2)$ is characterized by the equation $x^2 = w_2 \cup x$ for all $x \in H^2(M^4, \mathbb{Z}_2)$. In fact, w_2 is the $\mathbb{Z}_2$-reduction of an integer class in $x \in H^2(M, \mathbb{Z})$, i.e. $x^2 = w_2 \cup x$ mod 2.

In general, an oriented vector bundle, ξ^n, over a manifold M possesses a spin structure if and only if $w_2(\xi)$ is the mod 2 reduction of an integral class.

5.5 More on Yang-Mills Theories

In the previous chapter we discussed an important insight into the significance of local phase shifts of the wave function and a resulting $U(1)$ gauge theory, which in its simplest form is formally identical to electromagnetic theory derived from macroscopic experience. Here we will briefly review this argument from the time evolutionary viewpoint. First, recall that the quantum state is properly represented by a ray in Hilbert space, that is, a point in $\mathbb{C}P^\infty$. However, practical equations are generally expressed in terms of "wave functions." For brevity in presentation and ease of comparison with the physics literature, we divide space from time, replacing $x \to (t, \mathbf{r})$ so the wave function is typically something like $\psi(t, \mathbf{r})$ and its time evolution expressed in terms of a Hamiltonian operator, H, by the Schroedinger equation,

$$i\hbar \frac{\partial \psi}{\partial t} = H\psi, \qquad (5.16)$$

where $\hbar$ is Planck's constant. Also, we will assume a constant spacetime metric to replace forms by vectors. Typically H is an operator which contains spatial derivatives, say ∇. Now, recalling that the physical state is actually the equivalence class of all $\{e^{i\alpha}\psi\}$, we are naturally led to investigate the question of invariance of (5.16) under *local* phase transformations, $\psi \to e^{i\alpha(\mathbf{r},t)}\psi$. If we can do this we will have constructed a $U(1)$ **gauge theory**. In fact, the replacement

$$\Box = (\frac{\partial}{\partial t}, -\nabla) \to \Box - \mathcal{A}, \qquad (5.17)$$

where

$$\mathcal{A} = \frac{ie}{\hbar}(\phi, \mathbf{A}), \qquad (5.18)$$

and where the transformation $\psi \to e^{i\alpha(\mathbf{r},t)}\psi$ is accompanied by

$$\mathcal{A} \to \mathcal{A} - \frac{i}{\hbar}\Box\alpha, \qquad (5.19)$$

achieves two important results:

- a formulation describing the effect of an external (independent of ψ) field, represented by potentials $(\phi, \mathbf{A})$, on the particle, ψ, and,
- a $U(1)$ gauge theory, if $\mathcal{A}$ is regarded as the $U(1)$ connection, with standard gauge transformations (rewritten from exterior derivative form) (5.19).

Of course, before we can claim to have a complete theory, we must write the equations for the connection. If we choose the most "natural" and "simple" we arrive a formalism which we can, at least *formally*, identify with electromagnetism, as discussed in the previous chapter.

Our next step is to generalize the gauge group from the abelian $U(1)$ acting on a state representation contained in the single ψ, to the more general non-abelian case, for example, $SU(n)$ acting on matrices (ψ^α), $\alpha = 1...n$.

Non-abelian Yang-Mills theory

Prompted by the $U(1)$ physical model, consider now a principal fiber bundle P over the smooth manifold M with structure group G. Let A and F be the (local, base space) connection 1-form and corresponding curvature 2-form. The covariant (exterior) differentiation is defined by $D\omega = d\omega + A \wedge \omega + (-1)^k \omega \wedge A$ for a k-form ω with values in the Lie algebra $\mathfrak{g}$ of G. Recall that $F = DA = dA + A \wedge A$ and $DF = dF + A \wedge F - F \wedge A = 0$ (Bianchi identity). Also note that the wedge product of Lie algebra valued forms implicitly involves the Lie bracket operation.

Our task now is to obtain field equations for the connection. Using the models in the previous chapter, we look to **variational principles** as the means to this end. That is, we construct a functional from an integral, called the **action**, and obtain field equations by extremizing this action, subject to boundary conditions, which are not needed, of course, if the manifold is closed. If the manifold is open, these boundary conditions are generally of the form "vanishing sufficiently rapidly at infinity." For an action, we look at the G-invariant expressions such as $tr(F)$, $tr(F \wedge *F)$ etc. The simplest non-trivial one of these is $tr(F \wedge *F)$. We now select the integral

$$S = -\frac{1}{4} \int_M tr(F \wedge *F) \quad (5.20)$$

as the **action** of the (free) **Yang-Mills theory** for group G. Extremizing S with respect to variations of A with suitable boundary conditions, gives the (free) **Yang-Mills** equations

$$DF = 0 \qquad D*F = 0 \quad (5.21)$$

where the first equation is nothing but the well-known Bianchi identity. We now note some facts about these equations. For the definition of the action we need a metric on the manifold encoded into the Hodge operator $*$ and a representation of the group G needed for the definition of the trace. We can either assume a pre-given metric, or look to Einstein's general relativity to give us one. The group representation is *a priori* undetermined, but we are guided by how interesting the resulting theory is, either in physics or mathematics, or ideally by both.

The action in (5.20) and equations (5.21) are incomplete because they do not include a coupling to matter. To set this question in context, let us review the physics of matter $\leftrightarrow$ connections, assuming a spacetime model of four dimensions.

- "**Matter**" is represented by a matrix of wave functions, say

$$\Psi = (\psi^\alpha), \ \alpha = 1...n.$$

- From physics we have reason to suspect the existence of a symmetry, a group G acting linearly on Ψ.
- Again from physical considerations, we construct a "free" matter action for Ψ,

$$S_{matter} = \int \mathcal{L}_{matter}(\Psi, \nabla \Psi, ...).$$

- From spacetime invariance of S_{matter} we obtain a one form, J, whose dual is closed. This is a particular application of Noether's theorem with respect to spacetime translation invariance. We interpret J as the matter current, and $d * J = 0$ describes conservation of matter (represented by Ψ in this case).
- Gauge, G, invariance of S_{matter} can be accomplished by replacing ∇ by a gauge covariant derivative, involving a connection, A, with values in the Lie algebra of G.
- We choose some appropriate "free" action for A, such as in (5.20).
- The replacement of ∇ by gauge covariant derivative in $\mathcal{L}_{matter}$ typically results in a replacement of this "free" Lagrangian,

$$\mathcal{L}_{matter} \to \mathcal{L}_{matter} + \mathcal{L}_{interaction}(J, A).$$

- Finally, the total action is

$$S = S_{matter} + S_{connection} + S_{interaction}.$$

Physical Gauge Theories: *The physical assumption of G gauge invariance has led to an explicit form for the interaction Lagrangian and action, which in turn describes how the matter produces the connection field, and then how the connection field effects the matter field. This is the standard quantum field theoretic model for understanding the classical notion of force interaction between particles.*

Returning to our explicit Yang-Mills model, excluding the action for the matter itself, we arrive at field equations

$$S = -\frac{1}{4} \int_M tr(F \wedge *F + A \wedge *J) \tag{5.22}$$

$$DF = 0 \qquad D * F = *J. \tag{5.23}$$

Recall our extensive discussion of this procedure for the $U(1)$ representation of electromagnetism starting with equation (4.87) in the previous chapter. For a given matter action (like a Dirac field action) we can construct such a current by changing the partial derivative to a covariant derivative.

We now note a few additional points related to physical gauge theories. The strength of the connection (force) field is properly measured by the curvature form, F. For quantum gauge theories derived from classical theories, this form measures the macroscopically observed classical "force per unit charge." Another, specifically quantum quantity is the Wilson loop which is the exponential of the integral of connection A around a closed curve. This is also called holonomy. For example, if a stream of particles represented by this quantum field is split into two paths, corresponding to closed loop, and then recombined the Wilson loop measures the phase difference and can be directly measured by the interference pattern. this is known as the Aharonov-Bohm effect, originally described (and experimentally detected) for the $U(1)$ electromagnetic field.

We should briefly discuss possibly confusing usage of the term "gauge group." Originally this term applied to G itself, corresponding to the physical notion of a **global** group of transformations acting on a fixed vector representation space. However, in more recent applications G is the structure group of a bundle over a base manifold, M, providing the mathematical underpinnings for the physical notion of **local** gauge transformations, where the group element varies with the point in M. For a topologically trivial manifold (contractible) for which the bundle is simply a product, the group action is on sections, provided by functions from the manifold to the structure group G acting on each section point by point. This is often referred to as the **local gauge group.** For non-trivial bundles we have to consider the smooth automorphisms $Aut(P)$ of the principal bundle P together with the projection $\pi : P \to M$ which projects $Aut(P)$ to the diffeomorphism group $Diff(M)$ by way of a group homomorphism $h : Aut(P) \to Diff(M)$. The kernel $\ker h$ of this homomorphism forms a group again, $\mathcal{G}$, which is the proper representation of the physical notion of the local gauge group.

Remark 5.13.

In more detail, consider the principal bundle P and form its associated adjoint bundle $Ad\, P$. The bundle $Ad\, P$ is the associated bundle over M where each fiber is an isomorphic copy of the group G but the action of the group is conjugation ($g \mapsto a \cdot g \cdot a^{-1}$) rather than right translation; we write this as $Ad\, P = P \times_{AdG} G$. Then one can now show that $\mathcal{G}$ is given by all smooth sections $\Gamma(M, Ad\, P)$ of $Ad\, P$.

Finally, we note that a non-abelian structure group G, leads to $A \wedge A$ (A is the connection form) terms in F (curvature form) in the Lagrangian density. In turn this leads to non-linear field equations. Physically, this is described as **self-interaction,** which is, of course, absent in the electromagnetic interaction, which is linear.

5.6 The Concept of a Moduli Space

In a general sense, a moduli space is a space labelling a particular set of structures on manifolds, topological spaces, etc. The etymology of the term "moduli" lies in the fact that generally we are not interested in particular structures, but in equivalence classes of such structures under some group action. Thus, we are interested in a structure "modulo" this group action. Again we find similarity to the relativity principle in physics.

Formally, if $\mathcal{A}$ is the set of structures, and $\mathcal{G}$ is the group, then

$$\text{moduli space} = \mathcal{A}/\mathcal{G}.$$

As an example, consider the set of all complex structures on a compact Riemannian surface. Here the appropriate group is the set of biholomorphisms. The resulting moduli space is called Teichmüller space.

In general these spaces are singular (having a varying dimension) and can be described as stratifying spaces. Some special cases, such as the moduli space of Donaldson and Seiberg-Witten theory which we discuss in detail below, are manifolds with singularities. Another example is the universal space BG which can be regarded as the moduli space of all possible G principal fiber bundles under the group of bundle isomorphisms. In physical applications, a solution space of a gauge theory (possibly coupled to a matter field), is the moduli space. This space is a parameter space (sometimes called *moduli*) for the solutions of the field equations, modulo gauge action.

We now provide more detail in an important example, a complex line bundle L over a compact manifold M. This is the associated bundle to a $U(1)$ principal fiber bundle over M. A connection (unitary because of $U(1)$) on L can be written as a purely imaginary differential form $A \in \Omega^1(M, i\mathbb{R})$ on M. Thus the space of unitary connections $\mathcal{A}$ is the space of one-forms on M. In this case of $U(1)$-bundles, a *gauge transformation* of L is described locally by smooth maps over neighborhoods, $g : V \subset M \to U(1) = S^1$ inducing a vector bundle isomorphism $g : L \to L$, g acting by scalar multiplication. Let $\mathcal{G}$ denote the gauge group (understood as this group of local gauge transformations) and $\mathcal{G}_0$ be the subgroup of based gauge transformations. That is, for some base point $p_0 \in M$ and $g \in \mathcal{G}$ with $g(p_0) = 1$. Every gauge transformation can be decomposed into a based and a constant (global) gauge transformation, $\mathcal{G} = \mathcal{G}_0 \times S^1$. Locally, the action of such an element of $\mathcal{G}$ on a connection is described by

$$A \to A - id\alpha, \text{ where } g(p) = e^{i\alpha(p)}. \tag{5.24}$$

This leads to the space of equivalence classes of connections on L as $\mathcal{B} = \mathcal{A}/\mathcal{G} = \mathcal{A}/\mathcal{G}_0$. Clearly, the curvature form, $F = dA$, is gauge invariant.

Interpreting these fields as electromagnetic, our previous action principles lead to the gauge invariant, source free Maxwell equations

$$dF = d(dA) = 0 \qquad d * F = 0$$

for the unitary curvature $F = dA$. Thus F is a closed and co-closed 2-form with values in the Lie-algebra of $U(1)$ isomorphic to $i\mathbb{R}$, and $-iF$ represents a cohomology class in $H^2(M, \mathbb{R})$. Vice versa, a cohomology class Ω is the curvature $i\Omega$ of a unitary connection of a line bundle when this cohomology class lies in the image of the coefficient homomorphism: $H^2(M, \mathbb{Z}) \to H^2(M, \mathbb{R})$.

Remark 5.14.
For the case of a two sphere, it is easy to see this in detail. Cover S^2 with two hemispheres, $B_\pm$, intersecting in the equatorial sphere, S^1. From (5.24) above, the difference between the connection form on these two regions is $A_+ - A_- = -id\alpha$. The evaluation of the cohomology class generated by $-iF$ on the top homology class, S^2 itself, gives

$$-F[S^2] = -i\int_{S^2} F = -i(\int_{B_+} dA_+ + \int_{B_-} dA_-) = -i\int_{S^1} (A_+ - A_-) = \alpha(2\pi) - \alpha(0) \in \mathbb{Z},$$

because of the single valued requirement on $g(p)$ in (5.24).

From the physical point of view, this result can be interpreted as the quantization of the magnetic monopole charge, if this is ever non-zero in our universe. The full argument requires the expansion of the connection from S^2 to (punctured) space and time, $\mathbb{R} \times (\mathbb{R}^3 - \{0\}) = \mathbb{R}^2 \times S^2$. For more details on this see [Trautman (1984)] and [Moore (2001)].

Proceeding with the construction of the moduli space of connections modulo $U(1)$ gauge transformations, consider simply-connected and the non-simply-connected base spaces separately. For the first case, we have an isomorphism between the moduli space and the space of closed 2-forms $\mathcal{C}$ representing the first Chern class of the line bundle and given by the cohomology class of the curvature Ω. The other case of a manifold with non-trivial fundamental group is more complicated. Let $\gamma_1, \ldots, \gamma_n$ be the generators of the first homology group $H^1(M, \mathbb{R})$ represented by closed curves. A unitary connection A of the line bundle L defines a parallel transport of a section along the closed curve γ_i. Let L_p be the fiber of L above $p \in M$ which is the beginning and end point of the closed curve then $\tau_i : L_p \to L_p$ defines an isomorphism defined by

$$\tau_i(A) = \exp\left(i\oint_{\gamma_i} A\right),$$

the holonomy around γ_i. Note that because of the local nature of A, this expression makes sense only for closed paths, for which $\int d\alpha = 0$. Without proof we state that the moduli space is isomorphic to $\mathcal{C} \times S^1 \times S^1 \times \cdots \times S^1$. That is, the equivalence class $[A]$ of a unitary connection on L is mapped to $(F_A/(2\pi), \tau_1(A), \ldots, \tau_n(A))$ (see [Moore (2001)]).

In general, the space of connections $\mathcal{A}$ is an affine space and thus contractible. By dividing out the gauge group $\mathcal{G}$ we obtain a topologically non-trivial space $\mathcal{A}/\mathcal{G}$, the moduli space. This moduli space can be seen as a base space of a principal bundle $\mathcal{A} \to \mathcal{A}/\mathcal{G}$ with structure group $\mathcal{G}$. Because of the contractibility of the space $\mathcal{A}$, the moduli space is the classifying space $B\mathcal{G} = \mathcal{A}/\mathcal{G}$ of the gauge group $\mathcal{G}$, i.e., any principal bundle over M with structure group $\mathcal{G}$ can be classified by the homotopy classes $[M, B\mathcal{G}]$. Thus the homotopy properties of the moduli space are of central importance for any gauge theory. This classifying space also appears naturally in the physical context of functional integrals over gauge fields. Because of the gauge invariance of the action, we have to integrate over the classes of gauge-equivalent gauge fields (connections) instead over all possible gauge fields. Physically this is referred to as the Gribov ambiguity. We stop here with these general remarks. In the next section we will look at an important example of a moduli space which was of central importance in the early discovery and study of exotic $\mathbb{R}^4_\Theta$.

5.7 Donaldson Theory

Now we proceed to the gauge theory and resulting moduli space used by Donaldson [Donaldson (1983)] to get his famous characterization of smooth structures on simply-connected, oriented 4-manifolds. In a later chapter these results will be applied to obtain smoothness classifications for the base space itself. For details of the proof we refer to [Freed and Uhlenbeck (1990)] and to [Donaldson and Kronheimer (1990)] for the Donaldson polynomials.

Consider a Yang-Mills theory with respect to a compact Lie group G over a compact simply-connected 4-manifold M. Thus we are looking at connections on some G principal fiber bundle P over this manifold. The field strength of the Yang-Mills theory F is the curvature of P while the connection A is the gauge potential. After choosing some Riemannian metric to define the Hodge dual, the natural (i.e. simplest) action functional

is
$$S = \int_M tr(F \wedge \star F) = ||F||^2_M,$$

where tr is the trace with respect to the representation of G. Note the essential results will be independent of the particular metric choice. As usual, the variation of the functional with respect to the gauge potential A leads to the Yang-Mills equations,

$$DF = dF + A \wedge F - F \wedge A = 0, \qquad D \star F = 0. \qquad (5.25)$$

Solutions to this will lead to *local* minima of S, of which there may be more than one. Note that the first of these equations, $DF = 0$, is just the Bianchi identity, automatically satisfied by any curvature form. Thus, if F satisfies either of

$$F = \star F \qquad \text{or} \qquad F = - \star F, \qquad (5.26)$$

it is automatically a Yang-Mills solution. Because of the positive definite metric signature, these solutions are known as *instantons* in physics and as self-dual and anti-self-dual connections respectively in mathematics. From the splitting

$$F = \frac{1}{2}(F + \star F) + \frac{1}{2}(F - \star F) = F_+ + F_-.$$

into self-dual and anti-self-dual curvatures, the action becomes

$$S = ||F_+ + F_-||^2_M = ||F_+||^2_M + ||F_-||^2_M.$$

In section 5.3 we introduced various characteristic classes. From these the important **characteristic numbers** of Chern and Pontrjagin are evaluated as integrals of an invariant polynomial of the curvature. In this case of a 4-manifold the expression

$$k = \frac{1}{8\pi^2} \int_M tr(F \wedge F)$$

is exactly the second Chern or the first Pontrjagin number. The physics literature refers to such k as the **topological charge** defined by the bundle. The generalizations to higher dimensions and bundles are called **topological quantum numbers**, since they are integers, that is, discrete, a defining property of many observed quantities in quantum theory. Recall that these numbers depend not only on the base space, but also on the bundle over it and particularly the choice of the group G. Using the splitting above, we obtain

$$k = \frac{||F_+||_M^2 - ||F_-||_M^2}{8\pi^2},$$

and the important condition

$$8\pi^2|k| = |(||F_+||_M^2 - ||F_-||_M^2)| \leq ||F_+||_M^2 + ||F_-||_M^2 = S$$

where the middle "$\leq$" becomes the equality "$=$" only solutions which are either self-dual or anti-self-dual. This provides an important and productive bridge between topology and physics: the absolute minimum of the Yang-Mills action is a topological quantity, the Chern number.

Remark 5.15.
Consider self-duality in the context of Einstein theory. The vacuum equations of general relativity are the vanishing of the Ricci tensor, locally described by

$$0 = R_{\mu\nu} = e_\mu^a e_\nu^b R_{ab} = e_\mu^a e_\nu^b R_{acbd}\eta^{cd}$$

where e_μ^a are the vierbeins (solder forms), R_{acbd} are the Riemann tensor components and $\eta^{ab} = e_\mu^a e_\nu^b g^{\mu\nu}$ are those of the flat metric of the tangent space, simply the identity matrix in this Riemannian case. The torsion-free condition of the Levi-Civita connection can be expressed by

$$\epsilon^{abcd} R_{ebcd} = 0$$

where ϵ^{abcd} denotes the totally anti-symmetric symbol. The conditions for the curvature to be (anti-)self-dual are

$$R_{abcd} = \pm \frac{1}{2} \epsilon_{ab}^{\ \ mn} R_{mncd}.$$

Then together with the torsion-free condition we obtain:

$$0 = \epsilon^{abcd} R_{ebcd} = \pm \frac{1}{2} \epsilon^{abcd} \epsilon_{eb}^{\ \ mn} R_{mncd}$$
$$= \pm(R\eta_{ae} - 2R_{ae})$$

with the curvature scalar $R = R_{ab}\eta^{ab}$. Thus (anti-)self-dual curvatures are also solutions of Einstein vacuum equations. Of course, not all Einstein metrics are (anti-)self-dual.

Now, we focus in more detail on the solution space (moduli space) of the (anti-)self-dual equations. Let $\eta = P \times_G V$ be the vector bundle associated to a G principal bundle with respect to a representation $G \to Aut(V)$. The remarkable fact is that for a large class of interesting groups, such as $U(1)$, or $SU(2)$, and related Yang-Mills equations, this moduli space actually has the structure of a smooth finite dimensional manifold, at least locally. The details of this argument are spelled out in [Freed and Uhlenbeck (1990)]. Here we only review the high points.

First, the establishment of a finite dimensional manifold structure depends on developing a formalism to parameterize moduli space. Next, we note that certain families of connections may be more general than others. This is similar to the classification of metrics, those with symmetries, and those without. In particular, a connection is said to be **reducible** if $A = A_1 \oplus A_2$, where each A_i is the connection on a sub bundle, λ_i, and the bundle splits, $\lambda = \lambda_1 \oplus \lambda_2$. Clearly then reducibility depends on the reduction of the Lie algebra of G, $\mathfrak{g} = \mathfrak{g}_1 \oplus \mathfrak{g}_2$, so that A_i is $\mathfrak{g}_i$ valued. Otherwise, the connection is said to be **irreducible**. In some sense, irreducible connections are more general, and thus require more parameters, or dimensions. If we consider the parameterized space of all connections, the irreducible connections generate a smooth manifold part, with the reducible ones singularities where the dimension is lower. First, we note

Lemma 5.1. *Let $\mathcal{A}_k$ be the space of all irreducible, anti-self-dual connections with respect to the second Chern number equal to k. The group of gauge transformations $\mathcal{G} = C^\infty(Aut(\eta))$ act transitively on this space. Thus we can form a quotient $\mathcal{M}_k = \mathcal{A}_k/\mathcal{G}$ which is the moduli space of irreducible, anti-self-dual connections with second Chern number k.*

The next crucial step was first obtained by Atiyah, Hitchin and Singer [Atiyah et al. (1978)] who showed that this moduli space $\mathcal{M}_k$ admits a smooth manifold structure. The dimension is given by the following table with respect to the compact group G, the compact 4-manifold $M = S^4$, the second Chern number k and the irreducibility condition.

G	$\dim \mathcal{M}_k$	Irreducibility condition
$SU(N)$	$4Nk - N^2 - 1$	$k \geq N/2$
$Spin(N)$	$4(N-2)k - N(N-1)/2$	$k \geq N/4, N \geq 7$
$Sp(N)$	$4(N+1)k - N(2N+1)$	$k \geq N$
E_6	$48k - 78$	$k \geq 3$
E_7	$72k - 133$	$k \geq 3$
E_8	$120k - 248$	$k \geq 3$
F_4	$36k - 52$	$k \geq 3$
G_2	$16k - 14$	$k \geq 2$

The general formula of the dimension $\dim \mathcal{M}_k$ in terms of characteristic classes can be found in [Atiyah et al. (1978)]. In this article only so-called self-dual manifolds were considered but a result of Taubes [Taubes (1982)] leads to the extension of the formula to nearly all interesting smooth 4-manifolds. In particular, the results obtained for $\dim \mathcal{M}_k$ using $M = S^4$ agree for those for which M is an arbitrary 1-connected manifold as long as it has a positive definite intersection form, discussed more fully in the next chapter.

Now specialize to the standard Yang-Mills $G = SU(2)$. From obstruction theory we find that every principal $SU(2)$ bundle over a compact simply-connected 4-manifold is completely classified by the second Chern class. Let η be the complex rank-2-vector bundle associated to the principal $SU(2)$ bundle P. We now look for a condition for η to split in our special case. Assume it does, so $\eta = \lambda_1 \oplus \lambda_2$. This corresponds to a reduction of the structure group from $SU(2)$ to $U(1) \otimes U(1)$. First, note

$$c_1(\eta) = c_1(P) = 0 \qquad c_2(\eta) = c_2(P) \;.$$

Because of the relation $c_1(\lambda_1 \oplus \lambda_2) = c_1(\lambda_1) + c_1(\lambda_2)$ we obtain $c_1(\lambda_1) = -c_1(\lambda_2)$. Thus the two bundles are conjugated to each other and we write $\lambda_1 = \lambda$ and $\lambda_2 = \lambda^{-1}$. From $c_2(\eta) = c_1(\lambda) \cup c_1(\lambda^{-1}) = -c_1(\lambda) \cup c_1(\lambda)$, integration gives

$$k = -\int_M c_2(\eta) = \int_M c_1(\lambda) \cup c_1(\lambda),$$

where the sign is pure convention and depends on the orientation of M. *Here and in the following we assume that M is a compact, 1-connected smooth 4-manifold without boundary and has a positive definite intersection form.* The 1-connectedness condition means $\pi_1(M) = 0$ and so $H_1(M, \mathbb{Z}) = 0$. According to the universal coefficient theorem $H_2(M, \mathbb{Z})$ has no torsion and we can define the intersection form as a bilinear from over $\mathbb{Z}$. The manifold M is closed and so $H_2(M, \mathbb{Z})$ is dual to $H^2(M, \mathbb{Z})$ via the Poincaré duality map. Thus we can evaluate the intersection from also as a bilinear pairing between cohomology classes. Furthermore all complex line bundles over M generate the full cohomology group $H^2(M, \mathbb{Z})$ via the first Chern class.

Remark 5.16.

For more details, see [Mosher and Tangora (1968)] and [Moore (2001)]. Here we sketch the argument. Let λ be a complex line bundle over M. Such bundles are classified by the homotopy classes $[M, BU(1)]$ with the classifying space $BU(1)$. But $BU(1)$ is the infinite complex projective space $\mathbb{C}P^\infty$ with $\pi_2(\mathbb{C}P^\infty) = \mathbb{Z}$ and $\pi_n(\mathbb{C}P^\infty) = 0$ for all other n. Such a space is called an Eilenberg-MacLane space $K(\mathbb{Z}, 2)$. Such spaces are important because there is an isomorphism between $[M, K(\mathbb{Z}, 2)]$ and $H^2(M, \mathbb{Z})$. This completes the argument. Thus complex line bundles are completely classified by their first Chern class $c_1 \in H^2(M, \mathbb{Z})$.

Let Q_M be the intersection form of M and PD denote the Poincaré dual, then
$$k = Q_M(PD(c_1(\lambda)), PD(c_1(\lambda))) = Q_M(c_1(\lambda), c_1(\lambda)) \qquad (5.27)$$
where the first paring is defined in homology and the second in cohomology. This expression can be interpreted geometrically as the self-intersection of a surface representing $PD(c_1(\lambda))$. On the other hand, the classification property of $c_1 \in H^2(M, \mathbb{Z})$ for line bundles means that for each $\in H^2(M, \mathbb{Z})$ there is a line bundle, λ, such that $c_1(\lambda) = \alpha$ and $c_2(\lambda \oplus \lambda^{-1}) = -k$. So η is bundle equivalent to $\lambda \oplus \lambda^{-1}$, so we have

Lemma 5.2. *Let λ be a complex line bundle over M completely classified by its first Chern class $c_1(\lambda)$. The bundle η splits topologically into two line bundles $\eta = \lambda \oplus \lambda^{-1}$ if and only if the equation $k = Q_M(\alpha, \alpha)$ admits a solution for some $\alpha = \pm c_1(\lambda)$.*

To complete the relation between splittings of bundles η and reducible connections, we need the following lemma, [Freed and Uhlenbeck (1990)].

Lemma 5.3. *Suppose M has positive definite intersection form. Then for split $SU(2)$ bundles $\eta = \lambda \oplus \lambda^{-1}$ there is a unique self-dual Yang-Mills field*
$$F = \begin{pmatrix} f & 0 \\ 0 & -f \end{pmatrix}$$
which respects the splitting.

Thus we have the following situation: According to the table above, the moduli space of *irreducible* (anti-)self-dual $SU(2)$ Yang-Mills fields is $8k-3$-dimensional whereas the space of *reducible* fields is 1-dimensional (the only freedom is multiplication by a real number). As we have seen the singularities in the moduli space come precisely from these reducible connections, whose appearance is very important for studying smoothness conditions. In fact, (5.27) relates the existence of reducible connections to the intersection form. Following Donaldson we note the following facts:

- The moduli space of $SU(2)$ instantons has for $k = 1$ the dimension $\dim \mathcal{M}_1 = 5$ (see the table above).
- A (5-dimensional) cobordism between two oriented 4-manifolds is smooth if and only if the corresponding 4-manifolds each admit a differentiable structure.
- The signature $\sigma(M)$ of the intersection form is an invariant of the cobordism.

From these three points, one may ask whether the moduli space is itself the cobordism between two 4-manifolds leading to possible restrictions on the intersection form of smooth 4-manifolds. The following theorem of Donaldson [Donaldson (1983)] clarifies the structure of the moduli space.

Theorem 5.13. *Let η be a principal $SU(2)$-bundle with second Chern number $k = 1$ over a compact, 1-connected, oriented, smooth 4-manifold M with positive definite intersection form Q_M. Furthermore, let $\mathcal{M}_1$ be the moduli space of all (not just irreducible) anti-self-dual connections of η, then it follows:*

(1) Let m be half of the number of solutions of $Q_M(\alpha, \alpha) = 1$. Then there are points $p_1, \ldots, p_m \in \mathcal{M}_1$ for a (generic) dense set of metrics g of M such that $\mathcal{M}_1 \setminus \{p_1, \ldots, p_m\}$ is a smooth 5-manifold. The points p_i are in 1-1 correspondence to the topological splitting $\eta = \lambda \oplus \lambda^{-1}$.

(2) There is a neighborhood $\mathcal{O}_{p_i}$ of p_i so that $\mathcal{O}_{p_i}$ is homotopic to a cone in $\mathbb{C}P^2$.

(3) $\mathcal{M}_1$ is an orientable, possibly singular, manifold.

(4) $\mathcal{M}_1 \setminus \{p_1, \ldots, p_m\}$ is non-empty. In fact there is a collar $(0, \lambda_0] \times M \subset \mathcal{M}_1$ and $\overline{\mathcal{M}_1} = \mathcal{M}_1 \cup M \supset [0, \lambda_0] \times M$ which is a smooth manifold with boundary.

(5) $\overline{\mathcal{M}_1}$ is a compact manifold.

The Figure 5.1 visualizes the structure of this moduli space. A critically important point of this result is that the base manifold, M, itself is smoothly collared into the moduli space of connections, $\mathcal{M}_1$, so that the smoothness properties of M can be gleaned from a study of $\mathcal{M}_1$. For more details on this collar theorem, see §9 of [Freed and Uhlenbeck (1990)].

An easy, but important, corollary is

Corollary 5.1. *M is oriented cobordant (with respect to the cobordism $\mathcal{M}_1$) to the disjoint union $\underbrace{\pm \mathbb{C}P^2 \amalg \ldots \amalg \pm \mathbb{C}P^2}_{m}$.*

Now note the following algebraic lemma (see Chapter 6).

Lemma 5.4. *Let Q be a positive definite symmetric unimodular form of rank $r = r(Q)$, and let m be half the number of solutions α to $Q(\alpha, \alpha) = 1$. Then $m \leq r$ with equality if and only if Q is diagonalizable over the integers.*

A proof of this lemma can be given by an induction with respect to the rank r. Now we know from the above corollary, the definiteness of Q_M and

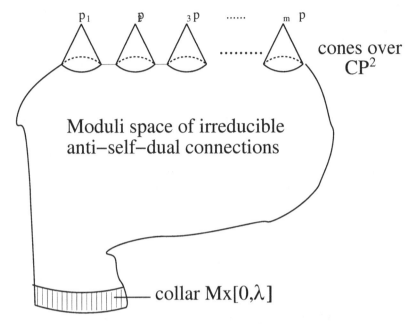

Fig. 5.1 Structure of the moduli space $\mathcal{M}_1$

the cobordism invariance of the signature $\sigma(M)$ that

$$r(Q_M) = \sigma(Q_M) \leq m \cdot \sigma(\mathbb{C}P^2) = m \,.$$

But we know from the preceding lemma that $m \leq r(Q_M)$. Hence $m = r(Q_M)$ and Q_M is diagonalizable over $\mathbb{Z}$. This provides a sketch of the proof of the all important **Donaldson's theorem**:

Theorem 5.14. *The intersection form of a compact, simply connected, oriented smooth 4-manifold M is diagonalizable over the integers $\mathbb{Z}$.*

An extension of these discussions to Chern numbers, $k > 1$, leads to the formalism of **Donaldson polynomials,** which we will briefly review now. Let P be an $SU(2)$-principal bundle over M with second Chern number k. The moduli space of irreducible, anti-self-dual, gauge-equivalent connections is denoted by $\mathcal{M}_k$ as usual. Assume now that this moduli space is even-dimensional, $\dim \mathcal{M}_k = 2d(k)$. Rather detailed considerations show that this dimension can be computed as $\dim \mathcal{M}_k = 8k - 3(1 + b_2^+(M))$ with $b_2^+(M)$ as the number of self-dual 2-cocycles (elements of $H^2(M,\mathbb{R})$). Thus, we can ensure that the dimension is even by assuming that $b_2^+(M)$

is odd. While this is a restriction on the manifold M it does include all interesting cases. Furthermore, it can be shown that the rational cohomology ring $H^*(\mathcal{M}_k, \mathbb{Q})$ of the moduli space is generated by $d(k)$ classes in $H^2(\mathcal{M}_k, \mathbb{Q})$ and one class in $H^4(\mathcal{M}_k, \mathbb{Q})$. Donaldson constructed a map $\mu : H_2(M, \mathbb{Q}) \to H^2(\mathcal{M}_k, \mathbb{Q})$ as follows.

Let $\mathcal{G}$ be the gauge group of P and $\mathcal{A}$ the space of irreducible connections. From the bundle P one can induce the tautological bundle $\underline{P} = \mathcal{A} \times P$. But the gauge group $\mathcal{G}$ does not freely act on $\mathcal{A}$, recall (4.26) in Chapter 4. So we must factor out the center of the $SU(2)$, given by the group $\mathbb{Z}_2$. This procedure results in a universal $SO(3)$ bundle $\mathcal{P}^{ad} = \underline{P}/\mathcal{G}$ over $\mathcal{M}_k \times M$ with the first Pontrjagin class $p_1(\mathcal{P}^{ad})$ generating $H^4(\mathcal{M}_k, \mathbb{Q})$. Next define the map $\mu : H_2(M, \mathbb{Q}) \longrightarrow H^2(\mathcal{M}_k, \mathbb{Q})$ by

$$\int_T \mu(\Sigma) = -\frac{1}{4} \int_{T \times [\Sigma]} p_1(\mathcal{P}^{ad}),$$

where Σ is a 2-dimensional submanifold of M generating an element of $H_2(M, \mathbb{Q})$ and $T \subset \mathcal{M}$ is a compact 2-dimensional submanifold of the moduli space. Another more useful realization of the μ-map is given by the following construction. We choose a Dirac operator $\partial\!\!\!/_\Sigma$ on Σ and the trivial $SU(2)$ bundle E over Σ. Next we twist the Dirac operator with a connection A of E to get $\partial\!\!\!/_{\Sigma,A}$. By $\mathcal{B}$ we denote the set of all (framed) connections and by $\mathcal{E}$ the bundle over $(\mathcal{B}/SO(3)) \times \Sigma$ induced by E. The family index $ind(\partial\!\!\!/_\Sigma, \mathcal{E})$ of the Dirac operator can be considered as a vector bundle over $\mathcal{B}/SO(3)$. Donaldson defined a line bundle by

$$\mathcal{L}_\Sigma = \bigcup_{A \in \mathcal{B}/SO(3)} (\Lambda^{max} \ker \partial\!\!\!/_{\Sigma,A})^* \otimes \Lambda^{max} \ker \partial\!\!\!/^*_{\Sigma,A},$$

i.e. as determinant bundle of the index bundle $ind(\partial\!\!\!/_\Sigma, \mathcal{E})$. This line bundle $\mathcal{L}_\Sigma$ has the following property

$$c_1(\mathcal{L}_\Sigma) = \mu([\Sigma]) \quad . \tag{5.28}$$

For more details see [Donaldson and Kronheimer (1990)].

Given this definition, consider $[\Sigma] \in H_2(M, \mathbb{Z})$, generated by a 2-dimensional submanifold Σ in M. With the help of the μ-map we get the corresponding cohomology element $\mu(\Sigma) \in H^2(\mathcal{M}_k, \mathbb{Q})$ and define the diffeomorphism invariant as the pairing

$$q_k(\Sigma) = \int_{[\mathcal{M}]} \mu(\Sigma) \cup \ldots \cup \mu(\Sigma), \text{ (d cup products)}. \tag{5.29}$$

Because of the non-compactness of the moduli space this definition may lead to convergence problems, but as Donaldson showed one can "repair" this defect by evaluating the cup-product $\mu(\Sigma) \cup \ldots \cup \mu(\Sigma)$ on a regular subset of $\mathcal{M}_k$. We then combine all the polynomials q_k into one analytic function or formal power series $q : H_2(M, \mathbb{R}) \to \mathbb{R}$ by

$$q(h) = \sum_k \frac{1}{k!} q_k(h) \qquad h \in H_2(M, \mathbb{R}) \qquad (5.30)$$

Using Donaldson's invariant of a smooth simply connected 4-manifold of the so-called simple type, Kronheimer and Mrowka [Kronheimer and Mrowka (1994)] proved the existence of p classes $K_1, \ldots, K_p \in H^2(M, \mathbb{Z})$ (also called basic classes) and non-zero rational numbers $a_1, \ldots, a_p$ such that

$$q = \exp\left(\frac{Q}{2}\right) \sum_{s=1}^{p} a_s \exp(K_s) \qquad (5.31)$$

where Q is the intersection form of M regarded as quadratic form. This representation of the Donaldson polynomials was the starting point for many investigations. Later [Fintushel and Stern (1995)], Fintushel and Stern established a relationship between the Donaldson invariants and the newly discovered Seiberg-Witten invariants. We will discuss this matter briefly in the next section and in chapter 9 more extensively.

Using these Donaldson polynomial techniques, we arrive at a crucial step in the discovery of the first exotic $\mathbb{R}^4$. In brief,

Theorem 5.15. *([Donaldson (1990)]) There is a 5-dimensional h-cobordism which is not a smooth product.*

This implies the existence of two h-cobordant 4-manifolds which are thus homeomorphic but not diffeomorphic. On the other hand, a result of Wall insures that two diffeomorphic 4-manifolds are also h-cobordant.

5.8 From Donaldson to Seiberg-Witten Theory

The so called "Seiberg-Witten" theory so important in recent studies of the differential topology of 4-manifolds has its roots Witten's formulation of Donaldson theory as a topological quantum field theory based on $N = 2$ supersymmetry. In this section we give an overview of these topics. Unfortunately, the amount of physics required to properly understand Witten's path to his equations is vast indeed so we will only be able to skim the surface,[Witten (1994)]. The book by Nash, [Nash (1991)], contains a more complete and pedagogical review of the relevant formalisms.

Quantum field theory and symmetry groups

The story begins in the 1930's when the founders of quantum mechanics began to develop a quantum theory respecting the Poincaré group, now called *quantum field theory* as a result of attempts to include the electromagnetic force field as a quantum field in its own right. The particles associated with this field, **photons,** are created/annihilated as exchange particles mediating the electromagnetic force between "stable" particles such as electrons, protons, etc. This formalism came to be know as **quantum electrodynamics, QED** and was generalized to what is now known as "quantum field theory." This theory extends the notion of "stable" particles to particles which are created/annihilated in interactions. The formalism uses new operators, a_k, a_k^+, called annihilation/creation operators for particles in states defined by quantum numbers, k, and the eigenvalues of the Hermitean operator $N_k = a_k^+ a_k$ correspond to the number of particles in state k. At this point the question of the spin of the particle turns out to be critical when special relativistic invariance is required[5]. For half-odd integer the particles are called **fermions** and for integer spin **bosons.** From the spin-statistics theorem, the a_k operators for bosons must satisfy commutation relations

$$[a_k, a_{k'}] = 0, \qquad [a_k^+, a_{k'}^+] = 0, \qquad [a_k^+, a_{k'}] = \delta_{kk'}$$

while for fermions, these are replaced by anti-commutators

$$\{a_k, a_{k'}\} = 0, \qquad \{a_k^+, a_{k'}^+\} = 0, \qquad \{a_k^+, a_{k'}\} = \delta_{kk'} .$$

As a result for fermions the operator N_k has only eigenvalues 0 or 1, i.e. one state k can only be occupied by at most one fermion ("Pauli exclusion principle"). For more details we refer the reader to the standard quantum field theory books, such as [Hatfield (1992); Kaku (1993)], or others.

High energy nuclear accelerator experiments in the 1950's and 1960's led to an explosive expansion of the list of known "particles." In fact, during this period Oppenheimer was reported to have suggested that the next Nobel prize should be awarded to someone who had **not** discovered a new particle. Attempts to systematically classify this new zoo of particles during this time led to a taxonomy based on observed ad hoc conservation laws. By assigning appropriate numbers to various particles participating in an interaction it was found that the totals before and after the interaction were the same, i.e.,

---

[5]This is the spin-statistics theorem of Pauli.

something was "conserved." As discussed in chapter 4, Noether's theorem derived from the Lagrangian formulation of field theories provides a crucial link between symmetries of the action and conservation laws. From this, theorists began to postulate symmetries, **internal symmetries**, that could be understood as acting on some **internal** particle space.

Remark 5.17.
Recall that the quantum mechanical state of a system can be represented by a one dimensional projection operator in a complex vector space (generally a Hilbert space). For our purposes, we can replace the projection operator by a single vector. A symmetry group action then defines a representation of itself as a linear action on such vectors. For example, the behavior of protons and neutrons led to their description as two states of the spin 1/2 representation of the internal **isotopic spin** group, $SU(2)$, with the proton as the "spin up" and the neutron as "spin down." The important point here is that this isotopic spin space is internal, and not at all related to physical space in which the angular momentum carrying spin, corresponding to the $SU(2)$ representation of $SO(3)$ resides. This early notion of isotopic spin was soon subsumed into larger groups, including the famous $SU(3)$ group for which the three base vectors were identified with the original **quarks**. Using these tools, the vast zoo of particles began to be tamed and organized according to irreducible representations of the symmetry groups. Of course, these early models have continued to evolve. We mention these points here only to give the reader an idea of the notion of **internal symmetry groups**, as opposed to **external, spacetime symmetries**, such as the Poincaré group for flat spacetime, or for the tangent vector space in general relativity.

Clearly, this artificial, but apparently rigidly observed, division between internal and external (spacetime) symmetries was not satisfactory to the natural desire for ultimate simplification. The theoretical search was for a unification of these two groups, internal and external, as parts of a larger group, in some non-trivial way, that is, not simply as a direct product. This hope was quickly destroyed by the work of Coleman and Mandula [Coleman and Mandula (1967)] in 1967, who showed that under certain reasonable assumptions, namely, Lorentz invariance, existence of finite presentations and some technical properties about the analyticity of the scattering amplitudes and the group representation via integral kernels, any connected symmetry *Lie* group G (of the S matrix) must be locally isomorphic to the direct product of an internal symmetry group and the Poincaré group. Note the emphasis on the Lie structure.

Another, apparently artificial, division of quantum theory pertained to the division of particles according to their spin representation, in particular the **fermion/boson** dichotomy, which was introduced above. Up to now, direct observation has detected "elementary" particles with spins of only 1/2 or 1, but the general classification would apply to higher values.

Remark 5.18.

As an aside, we note that fermions seem to refer to "permanent" particles, such as electrons, muons, neutrinos, perhaps quarks, etc., while the bosons correspond to the "exchange" particles of various forces, photons for electromagnetic, W-particles for the weak force, gluons for the strong force, and, very speculatively, gravitons for gravity. Furthermore, in quantum field theory the basic **creation/annihilation** operators for these two particle types satisfy **commutation** relations (bosons), and **anti-commutation** relations (fermions). This famous "spin-statistics" theorem is a result of a deep analysis of relativistic quantum theory. The anti-commutation properties of fermions will show up later in the supersymmetry algebra.

Shortly after the Coleman-Mandula result, Veneziano [Veneziano (1968)] introduced an important suggestion concerning the strong interaction. From mathematical sources, he recognized a striking correspondence between properties of scattering amplitudes in the strong interaction with the Beta function introduced by Euler. Later this model was extended by Nambo and Goto [Nambo (1970); Goto (1971)] to the first ideas of **string theory**, which, of course, has recently exploded as a field of theoretical interest. Except for the introduction of additional space dimensions, the Veneziano theory had another serious problem: there were no fermions in the theory. Of course, a natural objective might be to explore extending symmetries to include both fermions and bosons, resulting in what is known as a **supersymmetry, SUSY.** Wess, Zumino, et al., developed what is now called the Wess-Zumino model.

It is important to note that not only does supersymmetry mix fermions with bosons, but also internal with spacetime supergroups, in a sense unifying them in a way forbidden by the Coleman-Mandula theorem for ordinary internal and spacetime groups. Formally, the transition from a Lie algebra to a Lie superalgebra involves generalizing the Lie bracket operation (a commutator) in an ordinary Lie algebra to include some anti-commutators. We will sketch one such Lie superalgebra in (5.32) below.

Green, Schwarz and Scherk suggested that combining strings with supersymmetry, producing **superstrings**, could lead to a theory of quantum gravity. Later, 1984, Green and Schwarz published an advanced version of superstring theory which was anomaly-free and finite in the first orders of perturbation theory. As investigation proceeded it became clear that many superstring theories could be proposed, an "embarrassment of riches." This first stage of the "superstring revolution" was followed by a second one in 1995 when Witten proposed a unified theory of the five 10-dimensional superstring theories in 11 dimension with the low energy limit being a **supergravity** theory now generally called "M theory". One of the big problems in superstring theory is the 10 space-time (or 11 in case of M theory) dimensions instead of the 4 apparent to us. The obvious answer is to compactify 6 dimensions (or 7 dimensions in case of M-theory) to be left with the 4 observable dimensions. Here we use the word "compactify" in both the mathematical topological sense and in the physical sense of making "small." It turns out that a special kind of supersymmetric field theory, called the twisted $N = 2$ supersymmetric Yang-Mills theory will do this for us. The development of this theory motivates the construction of the so-called topological quantum field theories which we will define later. These are used to relate Donaldson [Witten (1988)] and Seiberg-Witten theory [Seiberg and Witten (1994a,b)], and of course, the SW equations of interest in our study of differential topology. First we will briefly introduce supersymmetry.

Supersymmetry

Recall that particles seem to be unnaturally segregated into the fermion/boson dichotomy. Fermions are classified by representations of internal symmetry groups. Our discussion in chapter 4 related how the $U(1)$ symmetry of charged particles could be extended to a local, bundle, symmetry with the addition of a connection corresponding to the quantum field for the photon, the electromagnetic potential. So, the group-bundle-connection model was used in attempts to explain other force fields in terms of internal symmetries, but unification of these symmetries with the basic spacetime Lorentz group was stymied by the Coleman-Mandula result. This obstacle could be overcome by extending the concept of the unification group. We know that symmetry groups in quantum field theories and spacetime physics are represented by Lie groups having a Lie algebra as infinitesimal generators. What was proposed was an extension to a "super" Lie algebra containing not only ordinary Lie brackets [,], but also a second product given by the anti-commutator $\{\,,\,\}$. Mathematically the result is a $\mathbb{Z}_2$-graded Lie algebra.

Remark 5.19.

In general such an algebra is vector space S which is the set-theoretic union of two subspaces $S = S_0 \cup S_1$ of dimensions N_0 and N_1, respectively. Every element of S has grade which is 0 if the element belongs to S_0 and 1 if it belongs to S_1. To complete S to an algebra we need a product [, } respecting the grading i.e. $[S_0, S_0\} \subset S_0$, $[S_1, S_1\} \subset S_0$ and $[S_0, S_1\} \subset S_1$

with some additional relations (symmetry/anti-symmetry and Jacobi relation with respect to grading).

Informally, we introduce operators, Q, which change

$$Q|\text{fermion of spin } s\rangle = |\text{boson of spin } s - \tfrac{1}{2}\rangle,$$

fermions to bosons and vive versa. As generators the Q fulfill anti-commutator relations among themselves resulting in the Poincaré group generators of a general form (indices and other details omitted),

$$\{Q, Q\} = \gamma^\mu P_\mu,$$

where P_μ is the generator of infinitesimal translation in the Poincaré group and γ^μ are Dirac's spin matrices.

In 1975, Haag, Łopuszański and Sohnius [Haag et al. (1975)] obtained all possible (infinitesimal) generators of supersymmetric theories. Let G be a generator of a supersymmetric transformation of S-Matrix. G acts on the Hilbert space of physical massless states which are the square-integrable functions $L^2(\mathcal{F})$ over the superspace $\mathcal{F}$. The operator G is assumed to fulfill two basic assumptions for massless states:(1) G commutes with the S-matrix, establishing the symmetry and (2) G acts additively on the states of several incoming particles. In addition, for a massive theory we require: (3) G connects only particles having the same mass. Consider the simplest non-trivial example: the supersymmetric extension of the Poincaré algebra with a single set of supergenerators. Use subscripts $a, b, \ldots = 1, 2$, for a self-conjugate (Majorana) spinor. The supergenerators must themselves transform as components of such a spinor, and so are denoted by Q_a. Let P^μ be the usual generators of displacement and $M^{\mu\nu}$ ($\mu = 0, 1, 2, 3$) be the homogeneous Lorentz transformations. The super Lie algebra is then defined by

$$\{Q_a, \bar{Q}_{\dot{b}}\} = -2(\gamma^\mu)_{ab} P_\mu \quad [Q_a, M^{\mu\nu}] = i(\sigma^{\mu\nu})_{ab} Q^b$$
$$[M^{\mu\nu}, P^\rho] = i(\eta^{\rho\mu} P^\nu - \eta^{\rho\nu} P^\mu) \quad [P_\mu, P_\nu] = 0 \quad [P_\mu, Q_\alpha] = 0$$
$$[M^{\mu\nu}, M^{\rho\sigma}] = i(\eta^{\mu\rho} M^{\nu\sigma} + \eta^{\nu\sigma} M^{\mu\rho} - \eta^{\mu\sigma} M^{\nu\rho} - \eta^{\nu\rho} M^{\mu\sigma})$$

in which $\sigma^{\mu\nu} = \tfrac{1}{4}[\gamma^\mu, \gamma^\nu]$, and $\eta^{\mu\nu}$ is the Minkowski metric.

Notice the desired mixing of internal and spacetime indices. Thus, the restrictions imposed by Coleman-Mandula on unifying internal and external symmetries have been eliminated by extending the Lie algebra to a super or graded one. Clearly, the next step is to include gravity (spacetime geometry) in supersymmetry, leading to supergravity. For more on this, see, for example, [West (1990)]. Later supergravity became part of superstring theory, but so far has not led to any significant contributions to differential topology.

The physical notion of a multiplet refers to a set of particles which are naturally grouped as a result of having similar quantum numbers, such as spin, internal group representation parameters, etc. Often in the first presentation of a theory members of a (super)multiplet have the same mass. However, physically they may not. The breaking of this symmetry from the original larger theoretical one to the smaller observed one is an important phenomenon in quantum field theories, including supersymmetric ones. What are the corresponding "supermultiplets"? In a supermultiplet fermions and bosons of the same mass are grouped together. The structure of the supermultiplets are strongly determined by the number N of supersymmetry generators Q. This number N is just the (generally complex) dimension of the internal symmetry group, with the generators, Q_a^I providing a representation of this internal group. In the case $N = 1$, described above, in each supermultiplet we must have one fermion of spin s and one boson of spin $s - 1/2$ or one boson of spin s and one fermion of spin $s - 1/2$. The next interesting case is $N = 2$ (for example, internal symmetry=$SU(2)$) where for instance a boson of spin 1, two fermions of spin $1/2$ and one boson of spin 0 are grouped in one supermultiplet. As before, all particles in one supermultiplet must have the same mass, which we do not expect to be realized in nature. Thus, supersymmetry must be broken.

Physical considerations lead to the following list of properties to be satisfied by all supersymmetric theories:

(1) The spectrum of a supersymmetric Hamiltonian contains no negative eigenvalues.
(2) Each supermultiplet must contain at least one boson and one fermion whose spins differ by $\frac{1}{2}$.
(3) All states in a multiplet of unbroken supersymmetry have the same mass.
(4) Supersymmetry is spontaneously broken if and only if the energy of the lowest lying state is not exactly zero.

The next step is the construction of meaningful field theories from the concept of supersymmetry. A natural approach for this is to introduce a superspace (supermanifold, see deWitt, [DeWitt (1984)]) on which the super group (the super Poincaré group for instance) acts. The superspace is an ordinary manifold with the addition of a vector space of anti-commuting numbers (also called Grassmann numbers) attached to each point of the manifold, in a bundle-like formalism. Fields over that superspace are functions which are to be used in some Lagrangian. A formal integration of the Lagrangian over these anti-commuting numbers produces a supersymmetric Lagrangian on the original real manifold. To supplement this obviously skimpy review, the reader should consult the, by now classical, books such as [West (1990); Wess and Bagger (1992)].

But what has any of this to do with the differential topology of manifolds?

Witten's way to the Topological Quantum field theory

In [Witten (1982)] Witten gave a physically inspired approach to Morse theory using supersymmetry. Consider the algebra $\Omega^*(M)$ of all differential forms on a manifold M, naturally divided into odd and even degree. In fact the forms of odd degree anti-commute, while those of even degree commute, i.e.

$$\eta \wedge \rho = -\rho \wedge \eta, \qquad \eta, \rho \in \Omega^{2p-1}(M)$$
$$\psi \wedge \xi = \xi \wedge \psi, \qquad \psi, \xi \in \Omega^{2p}(M)$$

for $0 < 2p < \dim M$. Formally, then, we identify forms of odd degree $2p-1$ with fermions of spin $(2p-1)/2$ and forms of even degree $2p$ with bosons of spin p. A supersymmetry provides a definition of the maps (generators) $Q_1 : \Omega^m(M) \to \Omega^{m+1}(M)$ and $Q_2 : \Omega^m(M) \to \Omega^{m-1}(M)$. Witten then suggested that the Hamiltonian of the theory be defined by $H = Q_1 Q_2 + Q_2 Q_1$. Returning to differential forms, there is a natural choice of Q_1, Q_2 by

$$Q_1 = d, \qquad Q_2 = \delta$$

where δ is the codifferential $\delta \sim *d*$, so that this Hamiltonian is just the Laplace operator $H = d\delta + \delta d$. To relate this to Morse theory, Witten introduced a smooth function $f : M \to \mathbb{R}$ with isolated critical points, a Morse function, and defined

$$Q_1 = e^{-tf} d e^{tf}, \qquad Q_2 = e^{-tf} \delta e^{tf}$$

with the real number t as parameter. Clearly, the Hamiltonian becomes much more complicated, but Witten was able to show that the spectrum of H encodes all relations between the number of critical points of f and the number of generators of the homology group, known from Morse theory.

That was the beginning of the use of supersymmetry in understanding the topology of manifolds. In 1988 Witten [Witten (1988)] inspired by the work of Atiyah [Atiyah (1988)] constructed a topological quantum field theory which is now conjectured to produce the Donaldson polynomials as a vacuum expectation value of some operator.

General structure of a topological quantum field theory

We will now review the highlights of the formalism of topological field theory, TQFT. Refer to the book by Nash, [Nash (1991)], for a more complete and pedagogical review of the relevant formalisms. As usual, our spacetime is modelled on an n-dimensional Riemannian manifold M endowed with a metric $g_{\mu\nu}$. In quantum field theory, the quantum fields themselves are subject to probability laws to answer the question: What is the probability that the spacetime evolution of a quantum field is given by a particular function, $\phi_i(x^\mu)$? In the Feynmann approach to quantum theory, sometime referred to as the **path-integral** formulation, the probability of a **path**, that is a spacetime evolution of such quantum fields is postulated to be proportional to something of the form $\exp(-iS(\phi_i))$ where S is the action integral, which is a functional of the "path" of ϕ_i, that is, the form of the functions, $\phi_i(x^\mu)$. For example, see (4.92) for the action representing the electromagnetic field interacting with a Dirac particle. In general, an **observable**, a quantity which could, in principle, be measured, is represented by an operator, say $\hat{O}$, quantum mechanically, and as a function of the fields, say $\mathcal{O}$, classically. From this probability for a path, use standard quantum techniques to obtain the **expectation** or **average** value of such an observable,

$$\langle\hat{O}\rangle = \int [D\phi_i]\mathcal{O}(\phi_i)\exp(-iS(\phi_i)/g) \qquad (5.32)$$

where g denotes the coupling constant. The left side is the standard notation for expectation value of the observable $\mathcal{O}$. To make sense out of the right side of this equation we really need to cover much more material than we can give here. The reader can refer to any standard quantum field theory book, such as Kaku's, [Kaku (1993)], especially chapter 8. Briefly, the measure for the integral, $[D\phi_i]$ is a measure on the space of all paths, that is functional expressions, for all of the fields. Making sense of this "space of paths," and finding its measure is the major difficulty in this approach. In addition to other mathematical problems, the notion of convergence must be addressed. A standard way of handling this issue involves "euclideanization," replacing the indefinite spacetime metric by the positive definite Euclidean one, and also replacing $i \to 1$. The assumption then is that some sort of analytic continuation will allow us to make physical sense of the result. Thus, (5.32) will be replaced by

$$\langle \hat{\mathcal{O}} \rangle = \int [D\phi_i] \mathcal{O}(\phi_i) \exp(-S(\phi_i)/g) \qquad (5.33)$$

The coupling constant, g, not to be confused with the metric determinant, was added to facilitate limits. Now, extend (5.33) to a product of observables,

$$\langle \hat{\mathcal{O}}_\infty \hat{\mathcal{O}}_\in \cdots \hat{\mathcal{O}}_\sqrt{} \rangle = \int [\mathcal{D}\phi_\rangle] \mathcal{O}_\infty(\phi_\rangle) \mathcal{O}_\in(\phi_\rangle) \cdots \mathcal{O}_\sqrt{}(\phi_\rangle) \exp\left(-\mathcal{S}(\phi_\rangle)/\}\right). \qquad (5.34)$$

A quantum field theory is defined as topological if the following relation is satisfied,

$$\frac{\delta}{\delta g^{\mu\nu}} \langle \hat{\mathcal{O}}_1 \hat{\mathcal{O}}_2 \cdots \hat{\mathcal{O}}_p \rangle = 0, \qquad (5.35)$$

i.e., if the expectation values of some set of selected operators is independent of the metric $g_{\mu\nu}$ on M, and thus dependent only on the topology of this space. If such is the case those operators are called "observables."

There are two ways to guarantee, at least formally, that condition (5.35) is satisfied.

(1) Schwarz type: $S(\phi_i)$, $\mathcal{O}_i$ are explicitly metric independent (for instance Chern-Simons gauge theory), or,
(2) Witten type: there exist a symmetry, which will be denoted by Q, satisfying the following properties:

$$Q\mathcal{O}_i = 0. \quad T_{\mu\nu} = QG_{\mu\nu}, \qquad (5.36)$$

Note that

$$T_{\mu\nu} = \frac{\delta S}{\delta g^{\mu\nu}}, \qquad (5.37)$$

is the standard definition for the energy-momentum tensor, but the requirement (2) above is that it can be derived from the action of some Q on some other tensor $G^{\mu\nu}$. We will see an example of this in the following.

The reader will note that the same symbol, Q, is used above as for supersymmetry generators and for the map Q in Witten's approach to Morse theory. This is no accident, as we will see.

We now skip over many technical details and formulate the result of interest to our study of differential topology, that is, **Witten's TQFT**. Suffice it to say that the tools and motivations of both supersymmetry and TQFT were used to arrive at the following formalism.

Let M be a compact oriented four-dimensional manifold endowed with a metric $g_{\mu\nu}$, and let us consider on it a principal fiber bundle P with group G (not to be confused with its use as a generator of a SUSY earlier) which will be assumed to be simple, compact and connected. Let E be the vector bundle associated to P via the adjoint representation, and let $\mathcal{A}$ be the space of G-connections on E. A connection in $\mathcal{A}$ will be denoted by A and its corresponding covariant derivative and self-dual part of its curvature by D_μ and F^+, respectively. Let us introduce the following set of fields:

$$\xi \in \Omega^{2,+}(M,\mathfrak{g}), \quad \psi \in \Omega^1(M,\mathfrak{g}), \quad \Phi, \lambda, \phi \in \Omega^0(M,\mathfrak{g}), \quad (5.38)$$

with local components $\xi_{\mu\nu}, \psi_\mu$ respectively. In (5.38) $\mathfrak{g}$ denotes the Lie algebra associated to G. Now we will choose $G = SU(2)$ with Lie algebra $\mathfrak{g} = \mathfrak{su}(2)$. The action of the theory has the form:

$$S_{WTQFT} = \int_M d^4x \sqrt{g}\, \text{Tr}\Big(\frac{1}{4}F^{+2} + \frac{1}{2}\phi D_\alpha D^\alpha \Phi - i\xi^{\mu\nu}D_\mu\psi_\nu + i\lambda D_\mu\psi^\mu \quad (5.39)$$

$$+\frac{1}{4}\phi\{\xi_{\mu\nu},\xi^{\mu\nu}\} + \frac{i}{4}\Phi\{\psi_\mu,\psi^\mu\} - \Phi D_\mu D^\mu \phi + \frac{i}{2}\phi\{\lambda,\lambda\} + \frac{1}{8}[\Phi,\phi]^2\Big).$$

This action was not just arbitrarily constructed, but in fact was motivated by the fact that it is a twisted $N = 2$ supersymmetric field theory. This fact and the motivation is discussed by Baulieu-Singer [Baulieu and Singer (1988)] (or Brooks-Montano-Sonnenschein [Brooks et al. (1988)]) and by Labastida-Pernici [Labastida and Pernici (1988)] (see [Birmingham et al. (1991)] for an overview of the whole theory). Furthermore, the TQFT nature of this theory follows from definition and action of the following operator Q:

$$QA_\mu = \psi_\mu, \qquad Q\xi = F + \frac{1}{2}\star F,$$
$$Q\psi = d_A\phi, \qquad Q\lambda = i[\Phi,\phi],$$
$$Q\phi = 0, \qquad Q\Phi = \lambda. \quad (5.40)$$

where d_A is the covariant derivative with respect to the connection A.

Remark 5.20.
In fact, we can construct the energy-momentum tensor $T_{\mu\nu}$ from $T_{\mu\nu} = QG_{\mu\nu}$,

$$T_{\mu\nu} = QG_{\mu\nu}$$
$$G_{\mu\nu} = \frac{1}{2}\text{Tr}(F_{\mu\sigma}\xi_\nu^\sigma + F_{\nu\sigma}\xi_\mu^\sigma - \frac{1}{2}g_{\mu\nu}F_{\sigma\tau}\xi^{\sigma\tau})$$
$$+ \frac{1}{2}\text{Tr}(\psi_\mu D_\nu \Phi + \psi_\nu D_\mu \Phi - g_{\mu\nu}\psi_\sigma D^\sigma \Phi) + \frac{1}{4}g_{\mu\nu}\text{Tr}(\lambda[\phi,\Phi]).$$

Thus, this is a TQFT according to (5.35) since the expectation values of the observables are thus topological invariants.

Now we take up the observables of the theory. For $SU(2)$, there is one expression ϕ^2 of the Lie algebra $\mathfrak{su}(2)$-valued scalar field ϕ commuting with all generators of the Lie algebra, the Casimir operator. Define the invariant operator

$$W_0 = \frac{1}{2}\text{Tr}(\phi^2) \tag{5.41}$$

leading to the observable

$$\mathcal{O}^{(0)} = \mathcal{O}(\gamma_0) = \int_{\gamma_0} W_0 = \frac{1}{2}\text{Tr}(\phi(\gamma_0)^2),$$

with $\gamma_0 \in H_0(M)$ which is a point in M. With the help of the descent equations

$$QW_i = dW_{i-1},$$

generate the operators

$$W_0 = \tfrac{1}{2}\text{Tr}(\phi^2) \qquad\qquad W_1 = \text{Tr}(\phi \wedge \psi),$$
$$W_2 = \text{Tr}(\tfrac{1}{2}\psi \wedge \psi + i\phi \wedge F), \qquad W_3 = i\text{Tr}(\psi \wedge F), \tag{5.42}$$

leading to the following observables:

$$\mathcal{O}^{(k)} = \mathcal{O}(\gamma_k) = \int_{\gamma_k} W_k, \tag{5.43}$$

where $\gamma_k \in H_k(M)$.

Remark 5.21.
As an example, we verify the equation $dW_0 = QW_1$, i.e.

$$\frac{1}{2}d\text{Tr}(\phi^2) = \text{Tr}(\phi \wedge d_A\phi)$$
$$Q\text{Tr}(\phi \wedge \psi) = \text{Tr}(\delta\phi \wedge \psi) + \text{Tr}(\phi \wedge \delta\psi) = \text{Tr}(\phi \wedge d_A\phi)$$

where we use the transformations (5.40) and the fact $d\text{Tr}(.) = \text{Tr}(d_A.)$.

Consequently, we have the important fact that every functional integral of the form

$$Z(\mathcal{O}) = \langle \hat{\mathcal{O}}^{(k)} \rangle =$$
$$= \int [DA\, D\xi\, D\Phi\, D\phi\, D\psi\, D\lambda]\, \mathcal{O}^{(k)} \exp(-S_{WTQFT}/g), \tag{5.44}$$

with respect to the action S_{WTQFT} (5.39) is a topological invariant, called the **vacuum expectation value of the observable**. Now, what is the topological invariant defined by such vacuum expectation values?

The relationship to Donaldson polynomials

In [Witten (1988)] Witten analyzed the functional integral (5.34) by using the property that the Lagrangian L of the action (5.39) is Q-exact, i.e.

$$L = QV$$
$$V = \frac{1}{4}\text{Tr}(F_{\mu\nu}\xi^{\mu\nu} + 2\psi_\mu D^\mu \Phi - \eta[\phi, \Phi])$$

which leads to the independence of the theory of the coupling constant g. Thus the limit $g \to \infty$[6] where perturbation methods apply is equivalent to the limit $g \to 0$ where non-perturbative methods are needed. The previous argument for $g \to \infty$ implies that the "semiclassical approximation" of the theory is exact. That means that in this limit the contributions to the functional integral are dominated by the field configurations which minimize S. These are precisely the solutions to the equations defined by $\delta S = 0$. Witten showed that these solutions are equivalent to the solutions of the anti-self dual equation $F^+ = 0$. Thus, the functional integral of the theory is equivalent to an integral over the moduli space $\mathcal{M}_k$ of irreducible solutions of the anti-self dual equation which is finite dimensional $\dim \mathcal{M}_k = d(k) = 8k - 3(\chi + \sigma)/2$, where χ is the Euler number and σ the signature of the 4-manifold M. This is an important result: every expectation value of an observable can be expressed by an integral over the moduli space $\mathcal{M}_k$. Now specialize to the operators W_0, W_2 (see 5.42) together with the homology cycles $p \in H_0(M), S \in H_2(M)$ leading to the observables

$$\mathcal{O}(p) = W_0(p)$$
$$\mathcal{O}(S) = \int_S W_2$$

Then the expectation value (see the definition (5.44)

$$Z(p, S) = \langle \exp(\hat{\mathcal{O}}(p) + \hat{\mathcal{O}}(S)) \rangle$$

is a topological invariant. By expanding the exponential we see that the expectation value $Z(S, p)$ is a polynomial with respect to the cycles S and p. Now we fix the cycle p and obtain the important

- **TQFT $\leftrightarrow$ Donaldson polynomials conjecture:** The map $Z : H_2(M) \to \mathbb{R}$ is the Donaldson polynomial.

The way to Seiberg-Witten theory
Thus, there is a reasonable conjecture that Donaldson theory can be formulated as a topological quantum field theory (TQFT). This TQFT is also

---
[6]Remember, the g is inverse in the definition of the functional integral (5.44).

referred to as **Donaldson-Witten theory.** In [Witten (1988)] and later in [Moore and Witten (1998)] Witten et.al. the structure of the functional integral (5.44) for the action (5.39) is analyzed in some detail. As we stated above, in the limit $g \to \infty$, the functional integral is dominated by the anti-self dual solutions $F^+ = 0$ and this led to the possible relation to the Donaldson polynomials. The other limit $g \to 0$ is not so fruitful because we need non-perturbative methods to get any results. In 1988, Seiberg discovered new techniques to obtain very explicit formulas for the perturbative expansion for $N = 2$. This approach related non-perturbative effects to instantons, the solutions of the anti-self dual equation. In 1994, Seiberg and Witten [Seiberg and Witten (1994a,b)] expanded on earlier work to obtain new results for the non-perturbative structure of $N = 2$ supersymmetric quantum field theories including the action (5.39). The main point of the Seiberg-Witten approach was the discovery of "S-duality" exchanging the couplings $g \leftrightarrow 1/g$, i.e. transforming a perturbative situation into a non-perturbative one and vice versa. So, what is the S-dual of Donaldson-Witten theory, i.e. the TQFT leading to the Donaldson polynomials as vacuum expectation values of special operators? The answer is provided by Seiberg-Witten theory, as we review next.

Recall that a possible $N = 2$ supermultiplet has one boson of spin 1, two fermions of spin 1/2 and one boson of spin 0, that is a scalar. The action of the corresponding field theory looks like a gauge theory with group G coupled to this scalar field, possibly a Higgs field. Higgs had introduced a complex scalar field (spin 0) ϕ as a mechanism for breaking symmetry and leading to mass in a previously massless theory. In this $N = 2$ supersymmetric theory, this scalar takes values in the group $SU(2)$, interacts with the gauge field as usual and in addition has a self-intersection with the potential $V(\phi) = -2|\phi|^2 + |\phi|^4$ where $|\phi|^2 = \phi^a \bar{\phi}_a$ and $a = 1, 2$ is the group index for the group $SU(2)$. The absolute minimum of the potential $V(\phi)$ is given by the non-zero values $|\phi| = 1$, i.e. we choose ϕ that the non-zero component is given by $\phi^1 = e^{i\psi}$. At these values, the full gauge symmetry is broken and only a reduced version given by the subgroup $U(1) \subset SU(2)$ remains. In other words, the $SU(2)$ gauge theory is broken to a $U(1)$ gauge theory by the Higgs potential $V(\phi)$. Now consider the energy of the whole theory and ask for the absolute minimum. Surprisingly, this minimum is also determined by $|\phi| = 1$. Thus the energy minimum (the classical vacuum) is parametrized by the phase of ϕ. It was Seiberg's [Seiberg (1988)] insight that a similar process occurs for the quantized version of a $N = 2$ supersymmetric field theory. Here the expectation value $u = \langle W_0 \rangle \in \mathbb{C}$ parameterizes the "moduli space of quantum vacua", i.e the set of quantum states with lowest energy. Call that parameter space the u-plane. Seiberg and Witten showed that

(1) Among $u = \infty$, there are two other singularities of the quantum theory at $u = \pm 1$.
(2) There is a symmetry $u \leftrightarrow -u$.
(3) The $SU(2)$ is broken at $u = \pm 1$ to $U(1)$.
(4) The coupling constant g is mapped to $-1/g$ at $u = \pm 1$.

In [Moore and Witten (1998)], Moore and Witten analyzed this $U(1)$ theory to show that the instanton equation $F^+ = 0$ must be modified at $u = \pm 1$. Also, they found that at $u = \pm 1$ there is a $U(1)$ gauge field A coupled to a spinor field Φ. Mathematically, this coupling is given by a $Spin_C$ structure on a 4-manifold M (see section 5.4). After a long argument, Moore and Witten derived the field equations of the $U(1)$ theory

$$F^+ = (\bar{\Phi}\Phi)^+ \qquad (5.45)$$
$$\not{\partial}_A \Phi = 0 \qquad (5.46)$$

which replace the anti-self dual equation $F^+ = 0$ of the $SU(2)$ theory.

• **These are the famous Seiberg-Witten equations.**

Remark 5.22.
In physics (explicit spinor) notation, these equations can be written as follows. Using Clifford matrices Γ_μ (with anticommutators $\{\Gamma_\mu, \Gamma_\nu\} = 2g_{\mu\nu}$), $\Gamma_{\mu\nu} = \frac{1}{2}[\Gamma_\mu, \Gamma_\nu]$ and the curvature in components $F_{\mu\nu} = \partial_\mu A_\nu - \partial_\nu A_\mu$. we obtain

$$F^+_{\mu\nu} = -\tfrac{i}{2}\bar{\Psi}\Gamma_{\mu\nu}\Psi$$
$$\Gamma^\mu D_\mu \Psi = 0.$$

In the second equation, $\Gamma^\mu D_\mu$ is the Dirac operator $\not{\partial}_A$ where D_μ denotes the covariant derivative corresponding to the Levi-Civita and to the $U(1)$ connection A.

The corresponding invariants and the moduli space of these equations will be discussed in chapter 9.

We close with some remarks about the open conjecture that Seiberg-Witten theory is related to Donaldson theory. The S-duality between Donaldson-Witten and Seiberg-Witten theory described above also predicts particular formulas that relate Donaldson invariants to Seiberg-Witten invariants. A rigorous proof of this result is still lacking, however, since Seiberg and Witten's work does not constitute a proof. There remain a number of non-rigorous arguments, ranging from assuming that no other factors arise from integration in the u-plane, to the whole notion of functional integration (which is still not founded on rigorous mathematics, even today). However, the conjectured relationship between the Donaldson and Seiberg-Witten invariants does actually occur in the many cases where the Donaldson invariants and the Seiberg–Witten invariants are both known. This "empirical" evidence may be convincing, but for mathematicians (and physicists) concerned with calculating these invariants, the lack of a rigorous proof is of course problematic.

In 1995, Victor Pidstrigach and Andrei Tyurin proposed a program to prove the relationship between the Donaldson invariants and the Seiberg–Witten invariants. Their approach is to consider a theory that contains both the scalar field Φ and the non-abelian gauge group $SO(3)$ (which is basically $SU(2)$, except it identifies $+I$ with $-I$). The theory they examine is analogous to the Seiberg–Witten equations (5.45), though slightly more complicated.

Carrying out their program involves a great deal of difficult mathematics, and this mathematics is being developed by Paul Feehan and Thomas Leness in a series of papers (see [Feehan and Leness (2003)] for instance). Meanwhile, with what they have accomplished so far, Feehan and Leness have proved Witten's conjectured relationship between Donaldson invariants and Seiberg–Witten invariants for a large class of manifolds, up to a certain number of terms. Given the impressive work so far, it is reasonable to hope that this program will eventually prove the equivalence of the Donaldson invariants and the Seiberg–Witten invariants.

Chapter 6

A Guide to the Classification of Manifolds

The topics discussed in this chapter can be compared to those that arise in the various "relativity principles" of physics. There we need to know whether or not two representations of spacetime and physical fields are essentially different, or whether they can be transformed from one to the other by coordinate transformations, within a class specified by the principle. For Galilean relativity, the transformations are restricted to time changes which are linear and space changes which are linear in cartesian spatial coordinates and time. Obvious extensions of this lead to special and general relativity principles respectively. For a thoughtful presentation of these ideas in a novel bundle setting, see [Trautman (1984)].

In this chapter we look at the mathematical version: the classification of manifolds. Note that we are using the term "manifold" in this introductory discussion to mean a space which is mapped locally into some standard archetype by coordinate patches. In its most general form this problem has not been completely solved. In fact, even at the topological level, it has been proved by A.A. Markov [Markov (1958)] that there *cannot* exist any algorithm (in the sense of the theory of recursive functions) that would allow one to determine if two Euclidian simplicial complexes X, Y of dimension greater than three are homeomorphic or not[1]. Thus, by extension, there can be no algorithm (or computer program) which would solve the topological classification for manifolds with dimension greater than three. Notwithstanding this pessimistic fact, much progress has been made for restricted types of manifolds. Because the notion of equivalence implies a class of transformation, we need to organize the study according to restric-

---

[1]The proof connects this problem with the so-called "word problem" in group theory. One has to decide whether the two fundamental groups $\pi_1(X)$ and $\pi_1(Y)$ are isomorphic or not. But it is known that no such algorithm exists.

tions on the transition functions between the charts defining the manifold. Specifically, we have:

- homeomorphisms (TOP case)
- piecewise-linear (PL case)
- or diffeomorphisms (DIFF case).

As mentioned above, the TOP case cannot be effectively solved. However, the additional structures in the last two are so restrictive that some classification results are known.

Actually, a more basic question than classification is that of existence. So we can ask under what conditions a TOP manifold can carry a PL or a DIFF structure. In fact, this extension problem from TOP to PL or DIFF has now been solved, except for our notorious dimension 4. In dimensions 1 and 2 the classification problem was solved at the end of the 19th century. The reader can find in [Rado (1925)] a proof that every topological manifold of dimension less than 3 admits a unique PL and DIFF structure. An extension of this result to dimension 3 for any TOP manifold can be found in [Moise (1952); Cerf (1968)].

In dimension 4 there is a complete classification for TOP manifolds for a certain restricted class of fundamental groups. See [Freedman and Quinn (1990)]. Furthermore, every PL-4-manifold admits a unique induced DIFF structure. However, the question of the transition from a topological (TOP) to smooth (DIFF) structure on an arbitrary 4-manifold is still open. Donaldson [Donaldson and Kronheimer (1990)] and Seiberg-Witten theory [Akbulut (1996); Morgan (1996)] provide some strong conditions on a topological 4-manifold for it to admit a smooth structure. In all other dimensions (≥ 5) fairly complete results are known concerning extensions from TOP to PL, PL to DIFF and TOP to DIFF, as described in Kirby and Siebenmann [Kirby and Siebenmann (1977)]. For dimension 3 the classification problem is closely connected to the famous Poincaré conjecture, which, as of this writing in 2004, is not yet completely resolved. However Thurston's "Geometrization Program"[Thurston (1997)] is being used as the basis for many current attacks on the problem. Thus again the physically interesting dimensions, three and four, stand out as uniquely important in differential topology, as well as for our studies in this book.

We begin with a look at Morse theory on smooth manifolds. This technique decomposes manifolds into simple pieces, handles, using the critical points of a real valued function on the manifold, called a Morse function. This procedure is appropriately called surgery. The initial and final level surfaces of an appropriate Morse function are then said to be cobordant. Next, we apply these and other techniques to study low dimensional manifolds, dimensions 1,2,3. For the first two cases, algebraic topology, that is, homology theory, is sufficient for classification. The theory of 3-manifolds is much more complicated and has recently been greatly enhanced by the work of Thurston and others. We temporarily skip in section 6.3 to higher-dimensional n-manifolds $n > 4$, using the h-cobordism theorem to explore wider questions of the relationship between the categories, TOP, PL, and $DIFF$. Two simply-connected higher-dimensional manifolds are diffeomorphic iff they are h-cobordant. Note that the h-cobordism theorem is valid only for compact manifolds,and so excludes $\mathbb{R}^n$. To get that, we need the engulfing theorem. See Stallings, [Stallings (1962)], discussed in our chapter 8. The next two sections are concerned with the case of most interest to us, four dimensional smooth manifolds. These methods are very specific to 4-dimensions: Casson handles and Kirby calculus. We discuss the failure of the higher-dimensional methods (h-cobordism etc.) in the 4-dimensional case. This failure is due to the failure of Whitney's trick used in finding an embedding of a disk for doing surgery in dimensions greater than four. the last two sections are concerned with the important tool, the intersection form, as applied to 4-manifold topology. In fact, it is possible to construct 4-manifolds with any given intersection form. But what is the TOP class of such constructions? This topic was explored successfully by Freedman. We close with a brief algebraic discussion of the (integer) equivalence class of quadratic forms.

Assumption: *Unless otherwise specified, all manifolds in the remainder of this chapter will be* **smooth, compact, and connected.**

6.1 Preliminaries: From Morse Theory to Surgery

6.1.1 *Morse theory and handle bodies*

Morse theory is, on the one hand, a theory about the relationship between the critical points of a function $f : M \to \mathbb{R}$ and the topology of the underlying space M, and, on the other hand, a theory about the decomposition of a space M into simple pieces called handle bodies (see Figure 6.1 for a description). Standard, very readable, references including many explicit local chart presentations are in two books based on lectures by Milnor,

[Milnor (1963)] and [Milnor (1965a)].

We begin with a review of classical dynamics of particle motion. In standard elementary cartesian non-relativistic form we have the equations of Newton for a particle of unit mass subjected to a conservative force with potential energy f,

$$\frac{d\vec{x}}{dt} = \vec{v},$$
$$\frac{d\vec{v}}{dt} = -\nabla f(\vec{x}(t)). \qquad (6.1)$$

From the physical point of view, the critical points of this dynamical system are the equilibrium, force-free, points. In mathematics, the solution of the system (6.1) generates *local* diffeomorphisms in $TM \ni (\vec{x}, \vec{v})$ between starting values, $(\vec{x}(0), \vec{v}(0))$, and ending values, $(\vec{x}(t), \vec{v}(t))$. Of course, the diffeomorphism breaks down precisely at the critical points, $\vec{v} = 0$ and $\nabla f(\vec{x}(t)) = 0$. A study of the number and nature of these critical points, and their relationship to the differential topology of M is precisely the subject matter defining Morse theory. [2]

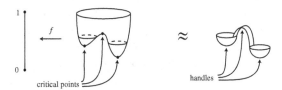

Fig. 6.1 From critical points of functions to handle decomposition (figure from [Gompf and Stipsicz (1999)] p. 104, Fig. 4.4)

Specifically, $p \in M$ is a *critical point* of a function $f : M \to \mathbb{R}$, if all first-order partial derivatives vanish at p. That is, the induced map $f_* : T_p M \to T_{f(p)} \mathbb{R}$ is the zero map. The real number $f(p)$ is the critical value of f. Let U be a coordinate patch neighborhood of the critical point p with coordinates $(x^1, \ldots, x^n)$. The matrix of second derivatives is known as the Hessian. The critical point is called degenerate if and only if the Hessian is singular at the point p, that is, the determinant of the matrix of second derivatives of f at p is zero. A smooth function $f : M \to \mathbb{R}$ which has only non-degenerate critical points is called *Morse function*. The important issue of the existence of Morse functions is settled in [Milnor (1963)]. From this,

---

[2] Of course, equations (6.1) have been presented in *global* Cartesian coordinates, so the consideration of manifolds with non-trivial topologies requires extensions of the formalism to take into account the local nature of the $\vec{x}$.

and elementary analysis, it is easy to see that any Morse function contains only isolated critical points.

The following important lemma clarifies the local structure in the neighborhood of the critical point of a Morse function.

Theorem 6.1. *(Lemma of Morse)*
Let p be a critical point of a Morse function, f, which, by definition, must be non-degenerate. Then there is a local coordinate system $(y^1, \ldots, y^n)$ in a neighborhood U of p with $y^i(p) = 0$ and in which

$$f = f(p) - (y^1)^2 - \ldots - (y^k)^2 + (y^{k+1})^2 + \ldots + (y^n)^2.$$

The integer k is called the *index of the critical point* p of f. The proof of this lemma is a straightforward exercise in real analysis. Introduce local coordinates, y^i, in a neighborhood of the critical point. The Taylor expansion of f can then be written

$$f = f(p) + \sum_{i \leq j} H_{ij} y^i y^j + O(y^3), \tag{6.2}$$

where H_{ij} is the symmetric Hessian matrix of second derivatives at p. First order terms do not appear because $\nabla_p f = 0$. A linear transformation of coordinates diagonalizes H and a rescaling reduces its eigenvalues to ± 1. Finally, the removal of third order terms requires a little more work. See Milnor's book, [Milnor (1963)].

Morse functions and topology

Consider the "landscape" over the manifold spanned by the graph of the function $f : M \to \mathbb{R}$. Using terminology appropriate for such a picture, the height of points on the torus manifold, as indicated in Figure 6.2, defines such a function, f. Also, note that in this case all critical points are non-degenerate, as required of a Morse function. Define $M^a = \{x \in M | f(x) \leq a\} = f^{-1}(-\infty, a]$. The behavior of these regions as a function of a leads to the conjectured relationship between critical points and topology. The heights $0, 1, 2, 3$ are precisely the critical values of the height function. Now, let us explore the region free of critical points, say $0 < a < b < 1$. Clearly the set $f^{-1}[a, b]$ is compact and does not contain any critical point of f. Now consider the orbits of solutions to

$$\frac{dx^i(t)}{dt} = \rho(x^i(t)) \frac{\partial f}{\partial x^i}(x^i(t)).$$

It is then possible to choose the normalization factor ρ so that $f(x^i(0)) = a$, $f(x^i(1)) = b$. The absence of critical points in this region then implies

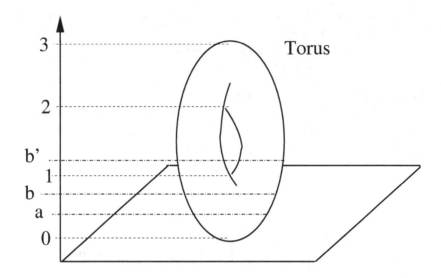

Fig. 6.2 Morse function of the torus $T^2 = S^1 \times S^1$ with critical points $0, 1, 2, 3$

that these curves generate a diffeomorphism between the two level sets, $f^{-1}(a)$ and $f^{-1}(b)$. Furthermore, the region between them is diffeomorphic to the product, $[0,1] \otimes f^{-1}(a)$. Finally, requiring that ρ vanish outside a compact neighborhood of $f^{-1}[a,b]$ leads to an argument showing that M^a is indeed a smooth deformation retract of M^b. Details are available in [Milnor (1963)],§3.

The other case $0 < a < 1 < b' < 2$, that is, the intervention of a critical point, is more interesting and is described in the theorem of Morse, Bott and Smale which we state as

Theorem 6.2. *Let $f : M \to \mathbb{R}$ be a Morse function with critical point p of index k. We set $f(p) = c$ and assume that $f^{-1}[c - \epsilon, c + \epsilon]$ is compact and does not contain any critical point except p for some $\epsilon > 0$. Then for sufficiently small ϵ the set $M^{c+\epsilon}$ is homotopy equivalent to $M^{c-\epsilon} \cup e_k$, that is, $M^{c-\epsilon}$ with a k-handle attached. To attach a handle, take a copy of $D^k \times D^{n-k}$ and embed $\partial D^k \times D^{n-k}$ in $\partial M^{c-\epsilon}$ with a map $\varphi \colon \partial D^k \times D^{n-k} \to \partial M^{c-\epsilon}$.*[3]

Furthermore, let $b_i = \dim H_i(M, \mathbb{F})$ for any field, $\mathbb{F}$, be the ith Betti number of the smooth, compact manifold M and C_k be the number of critical

---

[3] For details of this approach consider section 6.1.3.

values of f with respect to the index k. Then
$$b_i \leq C_i \quad \forall i$$
$$\sum_i (-1)^i b_i = \sum_k (-1)^k C_k.$$

It is important to emphasize that the equivalence discussed here is in the category of **homotopy**, not **topology**. But, within this realm of homotopy equivalence, this result shows that it is possible to decompose a given manifold into handle bodies (see section 6.1.3 for the details) by using a Morse function over the smooth manifold M.

Now, back to the example of a torus. Consider the levels a and b with $0 < a < 1 < b$ so that the critical value, 1, intervenes(see Fig. 6.3). It is now

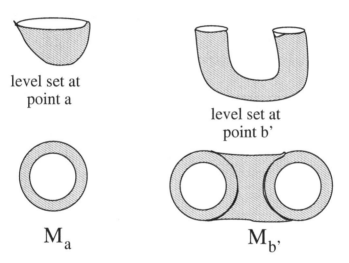

Fig. 6.3 Level sets of the points a and b

possible to visualize that the difference between M_a and M_b is the addition of a 1-handle $D^1 \times D^1$ (see Fig. 6.4)). Continuing this analysis to the other critical points, we see that the 2-torus is homotopically decomposable into a 0-handle, two 1-handles, and a 2-handle (see Fig. 6.5).

We close this section by mentioning some more recent extensions of Morse theory. One assumes that the critical points (now degenerate) are concentrated along a submanifold. Then we use the Poincaré polynomial of the submanifold instead of the Betti numbers to get the same relation as in the non-degenerate case (see [Nash (1991)]). Another generalization

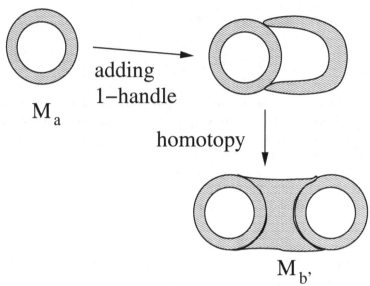

Fig. 6.4 Getting from level set M_a to M_b via handle adding

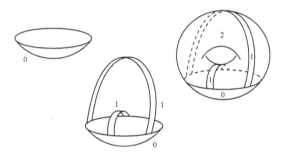

Fig. 6.5 Morse decomposition of the torus $T^2 = S^1 \times S^1$ (figure from [Gompf and Stipsicz (1999)] p. 105, Fig. 4.5)

involves the action of a Lie group on the manifold. This leads to equivariant cohomology with respect to this group (see [Nash (1991)]). Finally, there is a generalization of the manifold itself to a pseudo-manifold or more generally a stratified space. A good example of a pseudo-manifold is the cone, having a singularity on the peak. Examples of stratifying spaces occur often in moduli spaces of gauge theories where the dimension of the space jumps from area to area (i.e. area=strata). Morse theory for these kinds of space uses very complicated methods from algebraic geometry (sheaves, perversity of sheaves, intersection cohomology, etc.) to again obtain the

A Guide to the Classification of Manifolds   159

result (see [Goresky and MacPherson (1988)]) that the weighted sum of the Betti numbers of the intersection homology (see [Borel (1984)]) is given by the weighted sum of the critical points or subspaces of the Morse function. More details of these generalizations lie beyond the scope of this book.

6.1.2 Cobordism and Morse theory

Next we review the use of Morse function techniques to explore the relationship between different manifolds which, as a disjoint pair, constitute the boundary of a third manifold. Let X_-, X_+ be two n-dimensional, closed, oriented manifolds. If there exists an $(n+1)$-dimensional compact, oriented manifold W with boundary such that $\partial W = \overline{X}_- \amalg X_+$ ($\overline{X}_-$ is the manifold X_- with opposite orientation) then the two manifolds are said to be **(oriented) cobordant** to each other. Here $\amalg$ indicates the disjoint union. Denote this relationship by $X_- \sim_c X_+$. The triple (W, X_-, X_+) is called a *cobordism* between X_- and X_+. Figure 6.6 sketches an example of a cobordism. There is an obvious equivalence: Two cobor-

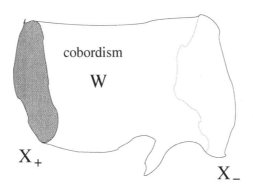

Fig. 6.6 Cobordism W between X_+ and X_-

disms W and W' are equivalent if and only if there is a diffeomorphism $g : (W, X_+, X_-) \to (W', X'_+, X'_-)$. Recall that in this part of the chapter we are assuming that manifolds are smooth, which we must to talk about Morse functions.

Remark 6.1.

It is easy to see that $\sim_c$ is an equivalence relation. The equivalence classes of n-dimensional cobordant manifolds form an Abelian group Ω_n with disjoint union $\amalg$ as addition. The equivalence class of the n-dimensional empty manifold $\emptyset$ plays the role of 0, and $\overline{X}$ (the manifold X with the opposite orientation) gives an inverse for X. The class represented by X will be denoted by $[X] \in \Omega_n$. Furthermore, we note that the Cartesian product $M_1 \times M_2$ defines a product in Ω_n which make it to a $\mathbb{Z}_2$-algebra. R. Thom [Thom (1954)] has investigated this very complicated algebra structure.

The manifolds cobordant to $\emptyset$ are called *null-cobordant*. So, a manifold X^n is null-cobordant if and only if there exists W^{n+1} such that $\partial W = X$.

We introduced cobordism using examples derived from Morse functions. Conversely, a cobordism (W, X_+, X_-) defines a Morse function $f : W \to [a, b]$ satisfying conditions:

(1) $f^{-1}(a) = X_+$, $f^{-1}(b) = X_-$,
(2) all critical points of f are in the interior of W.

Define the Morse number μ of (W, X_+, X_-) to be the minimum number of critical points of f with respect to all Morse functions f. According to [Milnor (1965a)] we obtain the following important and powerful result:

Theorem 6.3. *Every cobordism (W, X_+, X_-) admits a Morse function. Furthermore every cobordism can be decomposed into a union of cobordisms, each with Morse number 1.*

The simplest example of a cobordism with one critical point is given by the so-called "pants" surface, providing a cobordism between the disjoint union of two circles $S^1 \amalg S^1$ and single circle S^1 (see Fig. 6.7). The critical

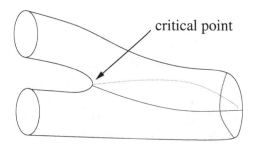

Fig. 6.7 The simplest cobordism with Morse number 1

point lies in the crotch of the pants. Thus, the cobordism of any two non-homeomorphic manifolds must have a non-trivial Morse number.

Remark 6.2.
A less trivial example confirming the assumption above can be given by later introduced surgery on a manifold X^n. Attach a $(k+1)$-handle to $I \times X^n$ along $\hat{\varphi} \colon S^k \times D^{n-k} \to \{1\} \times X \subset \partial(I \times X)$. As a result we get a cobordism between $X = \{0\} \times X$ and the surgered manifold $X_{\hat{\varphi}}$.

The tools from Morse theory and cobordism will be helpful for us in understanding the smooth classifications in higher dimensions in the later sections and chapters.

6.1.3 *Handle bodies and surgery*

In the preceding section we briefly reviewed Morse functions and cobordism providing an intuitive notion of an almost dynamic procedure for transforming one manifold to another cobordant one by "attaching" handles at the critical points of the Morse function. Following Milnor [Milnor (1959)] and [Milnor (1965a)] we look into the these technique more carefully, eventually obtaining the h-cobordism and other powerful topological theorems. Of course, since we are so centrally concerned with smoothness questions, we must note that the Morse techniques, involving smooth orbits, necessarily apply only to smooth manifolds, whereas more general cobordism and surgery techniques can be studied in the more general TOP category.

To have a physical model, consider a lump of clay as a model for a D^3. A physically "continuous" deformation of this lump can change its shape (geometry), for example into a bowl, but to change its topology we must "punch holes," or, "attach handles" transforming the bowl to a cup-with-handle. These handlebody and surgery techniques allow us to go from topologically simple manifolds such as non-compact, euclidean $\mathbb{R}^n$ or compact manifolds, such as the disk, D^n, (with boundary) and the sphere S^n, without boundary, to produce more complicated manifolds. In this process, we keep "control" of the topology in the sense of being able to calculate homology/cohomology changes at each stage.

Finally, since we are concerned with differential topology, we must be sure that the surgery transitions are smooth. This last condition is ensured by the folk theorem that "corners can be smoothed." So, even though for ease in rendition our diagrams may contain curves and surfaces with right angles, they should be thought of as smoothed.

Handle and Handlebody
An *n-dimensional k-handle* is defined as a copy of $D^k \times D^{n-k}$ (for $0 \le k \le$

n). Thus, in 2 dimensions there are only three possible handles: a 0-handle ($D^0 \times D^2 = D^2$), a 1-handle ($D^1 \times D^1$), and a 2-handle ($D^2 \times D^0 = D^2$) which is dual to the 0-handle. The *attaching of a k-handle to an n-manifold* X is accomplished by means of an embedding $\varphi \colon \partial D^k \times D^{n-k} \to \partial X$, known as *attaching map*, as represented in Figure 6.8. The case $k = 0$ is exceptional, since then $\partial D^0 = \emptyset$, and there can be no attaching map from the empty set to the ∂X. By convention, the attachment process here means adding D^n by forming a disjoint union of D^n with ∂X (which may be empty itself). Again, although this figure appears to have "unsmooth"

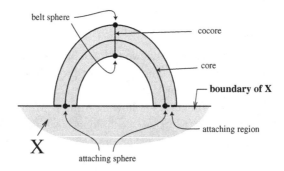

Fig. 6.8 k-handle glued along the boundary of X (figure from [Gompf and Stipsicz (1999)] p. 100, Fig. 4.1)

corners, these can be smoothed. $D^k \times 0$ is called the *core* of the handle, and $0 \times D^{n-k}$ is the *co-core*, φ is the *attaching map*, $\partial D^k \times D^{n-k}$ (or its image $\varphi(\partial D^k \times D^{n-k})$) is the *attaching region*, $\partial D^k \times 0$ is the *attaching sphere*, and $0 \times \partial D^{n-k}$ is the *belt sphere*. The number k is called the *index* of the handle. Such a procedure is the basic technique for obtaining more complicated and interesting smooth manifolds from a trivial one such as an n-ball[4]. The great advantage of such procedures is the control they provide over the homology and cohomology by use of the Mayer-Vietoris sequence after the attachment of a k-handle. For a more complete discussion of these constructions, see Lawson [Lawson (2003)], or Gompf and Stipsicz [Gompf and Stipsicz (1999)].

Example: Handle attachment to D^2
Start with a disk D^2. A 0-handle is a copy of $D^0 \times D^2$ which is D^2. Glue this handle along the boundary onto D^2 to get a disjoint union of two D^2's.

---

[4]Except for dimension 4 there is also a topological counterpart of this smooth procedure. The failure is directly related to the failure of Whitney's trick and will be explained in section 6.6.

Next, glue a 1-handle $D^1 \times D^1$ along the boundary of the disk $\partial D^2 = S^1$, using an embedding
$$S^0 \times D^1 = \{\star\} \times D^1 \cup \{\star\} \times D^1 \to \partial D^2 = S^1.$$
This produces a new manifold D_H^2 which is now homeomorphic to the region of a disk between two concentric circles. See Figure 6.9, noting that the last step is a deformation, or homotopy equivalence of the ring between the two circles into a circle. For another example, attach a 2-handle,

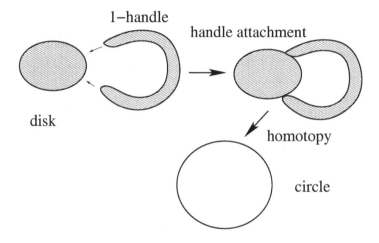

Fig. 6.9   Gluing of a 1-handle to get a circle (=1-sphere)

$D^2 \times D^0 = D^2$, to the boundary $\partial D^2 = S^1$ of a 2-disc to produce a 2-sphere S^2. (See Figure 6.10). In terms of homology, we can say that the process of attaching a k-handle results in an additional generator in the k-th homology group.

Handlebody decomposition

The above examples can be generalized, decomposing an arbitrary n-manifold X by handlebody attachment, providing a *handle decomposition*. More precisely, the handlebody decomposition of X is a sequence of submanifolds starting with D^n and ending with X where each term is obtained from the previous one by attaching a corresponding handle. A manifold X with a given handle decomposition is called a *relative handlebody*.

(1) We can arrive at D_H in the previous example by starting with D^2 and simply attaching a 1-handle. This is, of course, what we did in going through the critical point at $f(p) = 1$, for the torus in Figure 6.2.

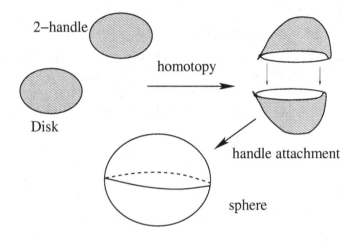

Fig. 6.10   Gluing of a 2-handle to get a sphere (=2-sphere)

(2) Let us look at the full construction of the torus, $T^2 = S^1 \times S^1$, from the critical points of the Morse function in Figure 6.2. The first critical point, of index 0, is at the transition from the empty set. So we attach a 0-handle. Using the convention above, this gives the disjoint union of D^2 with $\{x : f(x) < 0\} = \emptyset$, or simply D^2. The next two critical points are of index 1, so we attach two 1-handles. Finally, we cap it off at the top, index 2, attaching the 2-handle, D^2.

As stated in the previous section, and illustrated in simple examples, we can consider (in the case of smooth manifolds) the handlebody decomposition using a Morse function on the manifold. But in itself this information is not enough to completely define the manifold. In fact, the information that we need about the attaching map is its "deformation" (homotopy) class, given by the degree map of the boundary.

Framing

An attaching map $\varphi \colon \partial D^k \times D^{n-k} \to \partial X$ is an embedding of the boundary of the handle $\partial D^k \times D^{n-k}$ into the boundary ∂X. Obviously, there are many such embeddings, so we need to determine the appropriate classification for them. It turns out that we need the **isotopy class** of the attaching map. An isotopy of two embeddings is a family (smooth in our case) of maps parameterized by $t \in [0, 1]$. Again referring to Gompf and Stipsicz,

[Gompf and Stipsicz (1999)], especially chapter 4, consider the restriction, $\phi_0 : \partial D^k \times \{0\} \to H_0 \subset \partial X$, for $\{0\} \in D^{n-k}$. Consider the normal bundle, in $N_0(H_0) \subset T(X)$, restricted to this image. It can be shown that this bundle is trivial, and a trivialization of it, that is, a map f identifying N_0 with $\partial D^k \times \mathbb{R}^{n-k}$ then determines the full attaching map, ϕ, up to isotopy. Such an f is a **framing**. The choice of the term "framing" is appropriate since each such map can be understood as a map from $\partial D^k = S^{k-1}$ to a "frame" of $n - k$ independent vectors in $\mathbb{R}^{n-k}$.

Now pick some framing, f_0, as a standard. Then any other framing can be obtained by operating on the image of f_0 by an element of $GL(n-k)$, that is, by a map $S^{k-1} \to GL(n-k)$. Since only the isotopy class of such maps is important we have the result that, given some standard f_0, there is a bijection of the framings into $\pi_{k-1}(GL(n-k)) = \pi_{k-1}(O(n-k))$. It turns out that the dependence of this identification of framings with homotopy elements may not be entirely independent of the choice of f_0, but we will not look into this question further here.

To illustrate these technicalities, consider the example of a disk D^2 and glue a 1-handle along the boundary $\partial D^2 = S^1$. The framing of the attaching map is an element of $\pi_{1-1}(O(2-1)) = \pi_0(O(1)) = \mathbb{Z}_2$. The figure 6.11 shows both cases. It is interesting to note that the first non-finite framings

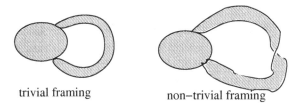

Fig. 6.11 The two possible framings for adding a 1-handle on D^2

occur for $n = 4, k = 2$ leading to $\pi_1(O(2)) = \mathbb{Z}$. That is, beginning with four dimensions we get the problem of possibly infinite framings.

Surgery

To this point we have been concerned with handle attachment, basically gluing a copy of $S^{k-1} \times D^{n-k}$ onto the *boundary* of a piece of a manifold, $M_{a-\epsilon}$, to understand the transition across the level, a, of a critical point. Another way to look at this process is *surgery theory*, involving attaching to the *interior* of a manifold. For references and comparison see both of Milnor's books, [Milnor (1963)](for handle attachment) and [Milnor (1965a)], or [Gompf and Stipsicz (1999)]. See [Lance (2000)] for a collection of papers surveying various topics in surgery theory.

Specifically, surgery involves embedding $\varphi\colon S^k \to M^n$ ($-1 \leq k \leq n$) with $\partial D^0 = S^{-1} = \emptyset$ with (normal) framing f. Then (φ, f) determines an embedding $\hat{\varphi}\colon S^k \times D^{n-k} \to M$, a procedure which is unique up to isotopy. *Surgery* proceeds by removing $\hat{\varphi}(S^k \times \text{int } D^{n-k})$ and replacing it with $D^{k+1} \times S^{n-k-1}$, using the gluing map $\hat{\varphi}|_{S^k \times S^{n-k-1}}$. In terms of Morse functions, this corresponds to passing a critical point of index k.

Example: Surgery on the torus

As an example, consider the case $n = 2, k = 1$ for the 2-manifold $S^1 \times S^1 = T^2$. Embed the cylinder $S^1 \times \text{int } (D^1) = S^1 \times (0,1)$ into the torus T^2 in an obvious way, i.e. the cylinder wraps around the "hole part" of the torus. Then we cut out this cylinder (making the former torus into a cylinder) and glue in $D^2 \times S^0$ (the disjoint union of two D^2's). This gluing "closes" the cylinder disjointly at both ends and results in a 2-sphere S^2.

For us, the important fact is

Surgery and Smoothness: *The smooth structure of the resulting manifold is only influenced by the isotopy class of the maps (φ, f).*

Another important cobordism result is that two closed, oriented manifolds M_1 and M_2, can be transformed into each other by a sequence of surgeries if and only if M_1 and M_2 are *oriented cobordant*, i.e., there is a compact, oriented manifold whose boundary is $\overline{M}_1 \cup M_2$ (and similarly for non-oriented manifolds and arbitrary surgeries)(see the section 6.1.1).

Handle manipulation: Canceling and Sliding of Handles

It turns out that the Morse function, or handlebody/surgery description of a manifold is not unique. For example, there may be more handles than necessary. A simple one dimensional example is described in Fig. 6.12, with one minimum, index 0, and a maximum, index 1. The first corresponds to

Fig. 6.12 Annihilation/creation of handles described by Morse functions (figure from [Gompf and Stipsicz (1999)] p. 110, Fig. 4.9)

the attachment of a 0-handle $D^0 \times D^1$ and the second a 1-handle $D^1 \times D^0$. The obvious deformation eliminates these two curves without changing the topology, as indicated in this figure. For a two dimensional model, where the maximum has index 2 and the minimum index 1, see the figure 5.2, page 49, of [Milnor (1965a)].

Fig. 6.13 Three dimensional example of cancellation of critical points

Milnor's book also contains a detailed proof that in general a $(k-1)$-handle and a k-handle can be canceled if the attaching sphere of the k-handle intersects the belt sphere of $k-1$-handle transversely in a single point. Such a pair of handles is called a *canceling pair*. By cancellation and other techniques it is possible to arrive at a canonical form for the Morse function presentation of a cobordism.

Another important technique which does not influence the diffeomorphism type involves *handle slides*. Given two k-handles h_1 and h_2 attached to ∂X, a *handle slide* of h_1 over h_2 is given by pushing the handle h_1 over h_2 according to the Figure 6.14. More carefully described, we isotope the

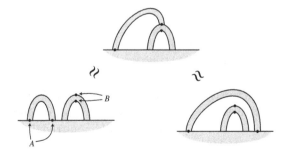

Fig. 6.14 Sliding the handle A over B (figure from [Gompf and Stipsicz (1999)] p. 109, Fig. 4.8

attaching sphere A of h_1 in $\partial(X \cup h_2)$, pushing it through the belt sphere B of h_2 (Figure 6.14). At the intermediate stage, the spheres will intersect in one point p (with $T_pA \oplus T_pB$ of codimension 1 in $T_p\partial X$). We will have a choice of directions for pushing A off B. One direction gives the original picture, and the other gives the result of the handle slide. Cerf has provided the definitive result for these techniques as summarized in the following theorem.

Theorem 6.4. *Cerf [Cerf (1970)]*
Given any two handle decompositions (ordered by increasing index) of a compact manifold X with boundary ∂X, it is possible to get from one to the other by a sequence of handle slides, creating/annihilating/cancelling handle pairs and isotopies.

Before closing this section and proceeding to apply these techniques to manifolds of specific dimensions, we want to remark that there is a more general way to introduce surgery, initiated by Wall [Wall (1970)]. The main techniques make use of purely algebraic objects known as cell complexes (fulfilling the abstract Poincaré duality) to construct topological manifolds. In this presentation, a surgery is given by an algebraic relation. Wall associated to the surgery operation so-called L-groups expressing obstructions to surgery involving an exact surgery sequence. These matters are more involved than we would like to pursue here. However, hopefully the further development of L-theory will provide tools for problems in 4 dimensions and help to unify all of manifold theory (see also [Kirby and Tylor (1998)]).

The reader may wonder why we need to consider topologically non-trivial spaces, when we are largely concerned in this book with the topologically trivial $\mathbb{R}^4$, and its possibly no-trivial smoothness. Without knowing whether an exotic $\mathbb{R}^4$ exists or not, we can know some of its properties. Consider an embedding $\mathbb{R}^4 \to M^4$ into some compact 4-manifold then there are two possible cases: 1. M^4 is a topologically non-trivial 4-manifold (i.e. different from S^4) or 2. M^4 is S^4. In this case the image of the embedding cannot be homeomorphic to $int(D^4)$ but of course must be contractible. In both cases we will need to know something about non-trivial topology. In fact, the first exotic $\mathbb{R}^4$ was constructed by finding it embedded in a non-trivial manifold and by the failure of the extension of topological surgery theory to the smooth case.

6.2 Application of Surgery to Low-dimensional Manifolds

By low-dimensional, we mean here dimensions less than five. Starting with dimension 1 and 2 it turns out that the homological methods provide complete classification. In dimension, 3 the tools of Dehn surgery along a knot/link lead to the construction of all 3-manifolds. But this procedure is not unique i.e. given two different (i.e, non-isotopic) knots or links may lead to homeomorphic 3-manifolds. In fact, Kirby and Fenn, and Rourke introduced a set of knot/link moves which leaves the topology of a 3-manifold unchanged[5]. Thurston's "geometrization program," [Thurston (1997)] has introduced a new and entirely different approach to the classification problem of 3-manifolds, relating geometry to smoothness starting with the fact that every smooth manifold can be endowed with a Riemannian metric.

For dimension 4, previous techniques are inadequate. In fact, there is no guarantee that a handlebody theory on a given topological 4-manifold exists. However, a new tool, Casson handles, has been developed for this topological case. Smoothness on a 4-manifold leads to a new version of handlebody studies, the so-called Kirby calculus. These techniques will turn out to be of central importance for our purposes, since, in fact, **if two handle bodies are transformed by the Kirby calculus then the corresponding 4-manifolds are diffeomorphic.** But first, we look at dimensions 1 and 2 which are by now classical.

6.2.1 1- and 2-manifolds: algebraic topology

For 1-manifolds, compactness divides the two possible cases. The non-compact case is simply the real line $\mathbb{R}$ whereas the compact case is the circle S^1. All other connected 1-manifolds are homeomorphic to one of these. Furthermore, every 1-manifold admits a differential structure which is unique. See the Appendix of [Milnor (1965b)] for a complete proof of these facts.

The first non-trivial case involves 2-manifolds. The classification problem for compact, connected 2-manifolds was solved in the 19th century by Klein, Poincaré et al. We will sketch an approach to the ultimate result using handlebody techniques as discussed above. Especially note that the framing of a k-handle attached to an n-manifold is determined by the assignment of a frame normal to the embedded S^{k-1} and is thus an element of $\pi_{k-1}(O(n-k))$.

---

[5]This was the original motivation for the development of Kirby calculus.

As an introduction to the theorem below, let us review handle body techniques for 2-manifolds discussed in section 6.1.1 above. Here we will be interested in arriving at closed 2-manifolds, compact with null boundary. Start with a 0-handle, equivalent to a D^2, with boundary S^1. To this we can attach either a 2-handle or one or more 1-handles. Attaching a 2-handle, $D^2 \times \{pt\}$, means identifying the two boundary circles with a diffeomorphism, $S^1 \to S^1$. If this diffeomorphism is isotopic to the identity, we obtain the standard S^2. In the second case we attach a 1-handle $D^1 \times D^1$. This involves embedding the two disjoint intervals of $\partial D^1 \times D^1$ into the $S^1 = \partial D^2$. Note that this can be done with either ± 1 framing. For $+1$, we can view this in terms of Morse functions as discussed in section 6.1.1 at the first critical point of the height function, at the bottom of the "hole." This intermediate result is a cylinder. If we now attach another 1-handle we arrive at the middle image in figure 6.5. The boundary of this space turns out to be simply S^1. An attachment of a 2-handle to this closes the space into the standard 2-torus, $T^2 = S^1 \times S^1$. However, if we attach the first handle with a twist, -1 framing, the space is non-orientable, with boundary not two disjoint S^1's, but only a single one. We can close this off by attaching a 2-handle, but arrive at a non-orientable closed 2-manifold which is actually $\mathbb{R}P^2$.

Summarizing, starting from the basic 0-handle, D^2,

- attach a 2-handle to get S^2,
- attach two 1-handles with +1 framing, then a 2-handle, get T^2,
- attach one 1-handle with framing -1, then a 2-handle, get $\mathbb{R}P^2$.

This procedure can be iterated. For the torus, if we attach $2g$ 1-handles we arrive at the connected sum of g tori. The number g turns out to be the topological genus of the space, or equivalently the dimension of the first integral homology group. The last and important point is that this procedure turns out to be exhaustive, that is, it produces all smooth closed 2-manifolds. The precise statement is the following theorem (see [tom Dieck (1991)]) using the connected sum notation #.

Theorem 6.5. *Every compact, closed, oriented 2-manifold is homeomorphic to either S^2 or the connected sum*

$$\underbrace{T^2 \# T^2 \# \ldots \# T^2}_{g}$$

of T^2 for a fixed genus g. Every compact, closed, non-oriented 2-manifold *is homeomorphic to the connected sum*

$$\underbrace{\mathbb{R}P^2 \# \mathbb{R}P^2 \# \ldots \# \mathbb{R}P^2}_{g}$$

of $\mathbb{R}P^2$ for a fixed genus g.

Every compact 2-manifold with boundary can be obtained from one of these cases by cutting out the specific number of disks D^2 from one of the connected sums.

Finally, every 2-manifold admits a unique differential structure.

Homology provides the natural tool to describe this classification. Historically, homology was developed by Poincaré to understand the generalization of Euler's polyhedra formula[6]. Consider the torus $T^2 = S^1 \times S^1$ with homology $H_n(T^2) = \mathbb{Z}$ for $n = 0, 2$ and $H_1(T^2) = \mathbb{Z} \oplus \mathbb{Z}$. For the connected sum of g copies of T^2, the first homology group changes to

$$H_1(\underbrace{T^2 \# \ldots \# T^2}_{g}) = \underbrace{\mathbb{Z} \oplus \ldots \mathbb{Z}}_{2g}$$

whereas the other homology groups are unchanged. Thus, we have a method to compute the decomposition into tori for any compact, closed, oriented 2-manifold Σ, simply calculate the first homology group of this surface. The order of this group is twice the genus g, so the space is the connected sum of g copies of T^2. Similarly for non-oriented 2-manifolds, the homology of the real projective space is $H_n(\mathbb{R}P^2) = \mathbb{Z}$ for $n = 0, 2$ and $H_1(\mathbb{R}P^2) = \mathbb{Z}_2$. Thus the number of $\mathbb{Z}_2$-factors in the first homology gives us the decomposition for non-orientable, compact, closed 2-manifolds.

Let us now consider the two dimensional definition of a tool, the **intersection form**, whose generalization to the 4-dimensional case will be of great importance later. Within a two-manifold consider two closed curves γ_1 and γ_2. As an example, look at the torus T^2. By deformations we can make γ_1 transverse to γ_2 so that the curves intersect in a finite number of points. This number is not an invariant with respect to small deformations because some of the intersection points disappear after deformations[7]. However, this number modulo 2 turns out to depend only on the homology class of γ_1 and γ_2 in $H_1(T^2, \mathbb{Z}_2)$ and thus is invariant with respect to

---

[6]This formula is the relation: points-edges+surfaces=2. Today we formalize this relation into the Euler characteristic. Every polyhedron is homeomorphic to S^2 with Euler characteristics $\chi(S^2) = 2$.

[7]The 1-handle can be glued along the boundary with 2-framings. That is the reason for using the $\mathbb{Z}_2$.

small deformations. In our example of a torus, $S^1 \times S^1$, we have two generators of the first homology represented by the curves $\gamma_1 = S^1 \times \{\text{point}\}$ and $\gamma_2 = \{\text{point}\} \times S^1$. These two curves intersect in a single point. In general, if x and y are two generators of $H_1(T^2, \mathbb{Z}_2)$, define the **intersection form** μ of T^2 as the map $\mu : H_1(T^2, \mathbb{Z}_2) \times H_1(T^2, \mathbb{Z}_2) \to \mathbb{Z}_2$ with $\mu(x, y) = $ (number of intersection points between x and y) mod 2. Obviously this form is a symmetric bilinear form on $H_1(T^2, \mathbb{Z}_2)$. With respect to the above defined basis (γ_1, γ_2) we obtain the symmetric matrix

$$H = \begin{pmatrix} 0 & 1 \\ 1 & 0 \end{pmatrix}$$

as the representation of the bilinear form, μ. Thus, γ_1 and γ_2 intersect in one point but have no self-intersections. For the real projective space $\mathbb{R}P^2$, there is a curve which intersects itself. To visualize this, we construct the real projective space by gluing a disk along the boundary to the Möbius strip. Now we consider a closed curve on the Möbius strip which is the latitude of the strip. This curve[8] then turns out to have one self-intersection point. Thus, the intersection form for the space $\mathbb{R}P^2$ is simply $\mu = (1)$.

This algebraic structure provides another classification tool for 2-manifolds and there have been extensive studies of such forms, represented by symmetric bilinear matrices over $\mathbb{Z}_2$. Such forms with non-diagonal elements equal to one, such as H above, are called type I. Others, i.e., diagonal ones, are type II (including the sphere, $\mu = 0$). In terms of the general surface decomposition described in Theorem 6.5 above, the non-orientable $\mathbb{R}P^2$ is of type I, while the orientable surfaces, sphere and torus, are of type II. In fact, we have the following theorem (see [Lawson (1985)]):

Theorem 6.6. *Two compact connected surfaces are diffeomorphic if and only if their intersection forms are abstractly equivalent. Surfaces can be separated into two classes: type I and type II. Those of type I are non-orientable and can be decomposed into a connected sum of real projective planes. Those of type II are orientable and are either spheres, or can be decomposed into a connected sum of tori.*

Later we will see that the classification of simply-connected 4-manifolds is similar to this classification but a bit more complicated.

---

[8] A tubular neighborhood of this curve is diffeomorphic to the Möbius strip.

6.2.2 3-manifolds: surgery along knots and Thurston's Geometrization Program

Now consider 3-manifolds. There are many similarities between the surgery operations of 3- and 4-manifolds. Kirby [Kirby (1978)] established his calculus for surgeries on 3-manifolds but today the second Kirby operation (handle slides, see below) is only used in the 4-manifold surgery, so this section is in some sense a "warm-up" for later considerations, especially for 4-manifolds with boundary.

Historically after the complete topological classification of surfaces via homology, the next task was the classification of 3-manifolds. Poincaré, a major figure in the invention of the concept of homology, believed at first that homology is sufficient to classify 3-manifolds. But he found a counterexample, the "Poincaré homology sphere," see, for example, pages 353, 354 in [Bredon (1993)]. The existence of such a counterexample means that a more subtle invariant is needed. He thought to have one which he denoted as the fundamental group, now called the first homotopy group. The corresponding conjecture is now known as *Poincaré conjecture* and is one of the hardest unsolved problems[9]. Milnor has recently provided an excellent review of the problem in [Milnor (2003)], which, as of this writing may well have been established as we will see later.

We may expect that the construction and classification of 3-manifolds is much more complicated than the lower dimensional cases. In fact, there are many different decompositions of a 3-manifold into manageable pieces: the prime-decomposition, the torus-decomposition and the Heegard decomposition. This last one involves a surgery procedure, the so-called Dehn surgery.

Dehn surgery

Consider the S^3 and cut out a filled torus $D^2 \times S^1$ with boundary the usual torus $T^2 = S^1 \times S^1$. Then glue in the filled torus $S^1 \times D^2$, having the same boundary as $D^2 \times S^1$, by a diffeomorphism $T^2 \to T^2$ of the boundaries. Every diffeomorphism of the torus can be decomposed by elementary transformations introduced by Dehn. In the trivial case that $T^2 \to T^2$ is the identity we obtain the decomposition of the 3-sphere S^3 by two filled tori $D^2 \times S^1$ and $S^1 \times D^2$ by the identity map of the boundaries $T^2 \to T^2$. The same is also true for a map $T^2 \to T^2$ which is isotopic to the identity. In all other cases, the isotopy class of the gluing map determines another

---

[9]The Clay Mathematics Institute (CMI) has founded a millennium prize for the solution of the seven hardest mathematical problems. One of these problems is the Poincaré conjecture. The other six problems are described in www.claymath.org.

3-manifold. A generalization of the above procedure is given by considering a knot K (i.e. an embedding $K : S^1 \hookrightarrow M$) in the 3-manifold M together with a tubular neighborhood $\nu K \approx K \times D^2$. The removing of νK and gluing in $S^1 \times D^2$ by any diffeomorphism of the boundary tori $T^2 \to T^2$ is called *Dehn surgery* or *rational surgery* (or sometimes just *surgery*, a term that we will avoid in this context). Of course this procedure depends strongly on the gluing map $T^2 \to T^2$ which will be considered now (See Rolfsen [Rolfson (1976)] for further reading). According to the *Dehn-Lickorish theorem* [Lickorish (1962)] we can arrange any such diffeomorphism by a sequence of *Dehn twists*, i.e. we cut the torus along a curve, twist one side a number of times and glue it together. Fig. 6.15 shows the case where we cut along the meridian α and make one curl. The self-diffeomorphisms of T^2

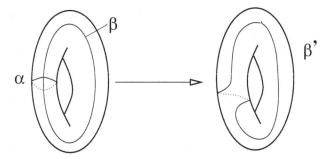

Fig. 6.15 Dehn twist along the meridian of the torus

are also derivable purely algebraically from the fact that the torus is given by $T^2 = \mathbb{R}^2/\mathbb{Z}^2$ i.e. the identification of opposite sites of a square. So every map $\mathbb{Z}^2 \to \mathbb{Z}^2$ of the lattice to itself generates a homeomorphism of the torus which is actually a diffeomorphism. These maps of the lattice form a group $SL(2,\mathbb{Z})$ and the Dehn twists are the generators of this group. Let (μ, λ) be a basis for $H_1(T^2;\mathbb{Z})$. Note that the solid torus $S^1 \times D^2$ can be built by attaching one 2-handle $D^2 \times D^1$ to D^3 along the boundary by using a map $\partial D^2 \times D^1 = S^1 \times D^1 \to \partial D^3 = S^2$. For Dehn surgery, all that is needed is the homology class α in $H_1(T^2;\mathbb{Z})$ of this circle. This class is then given uniquely by relatively prime integers, $\alpha = p\mu + q\lambda$. Reversing the orientation of K or α reverses the signs of both p and q but doesn't affect the diffeomorphism type obtained by the Dehn surgery, so all we need is the quotient p/q, and we call this a Dehn surgery with *coefficient* (or *slope*) $\frac{p}{q} \in \mathbb{Q} \cup \{\infty\}$.

Remark 6.3.
Let (θ, ϕ) be two angle coordinates (range $[0, 2\pi)$), for each S^1 factor. If $u(\alpha)$ is a smooth function equal to one near π but zero elsewhere, we represent the (p, q) twist by

$$(\theta, \phi) \to (\theta + p2\pi u(\phi), \phi + q2\pi u(\theta)).$$

Or, we can define the $(p, q) \in \mathbb{Z}^2$ twisted torus as identification space resulting from identifying $(x, y) \sim (mpx, nqy)$ for any $(m, n) \in \mathbb{Z}^2$ and $(x, y) \in \mathbb{R}^2$.

Framing and Integer Surgery along a Link

A **link** is a multi-connected generalization of a knot, a set of embedded circles, S^1's, intertwined with each other. The Dehn surgery case $q = \pm 1$ is called *integer surgery*. For a link we may have different coefficients for each component of the link. The following theorem shows that integer surgery is enough to describe 3-manifolds (for a proof see [Gompf and Stipsicz (1999)] p.159).

Theorem 6.7. *Any closed, oriented, connected 3-manifold M is realized by integer surgery on a link L in S^3.*

As we noted above this procedure is not unique, i.e. different (i.e. non-isotopic) knots and links can describe the same (i.e. homeomorphic or diffeomorphic) 3-manifold. Some of these moves will be described now.

Blow-up and Blow-down Surgery

Fenn and Rourke [Fenn and Rourke (1979)] introduced a set of moves described as **blow-ups** and **blow-downs**. We will not describe these moves in detail, but simply note that they are defined by the adding and removing of unknotted circles with framing ± 1 to the link or knot. Using these techniques Fenn and Rourke proved the following theorem.

Theorem 6.8. *Let L and L' be two framed links in S^3 describing (orientation-preserving) diffeomorphic 3-manifolds (by integer surgery). Then we can get from L to L' by blowing up and down (and isotopy). In fact, we can realize any preassigned orientation-preserving diffeomorphism in this manner.*

Historically, Kirby [Kirby (1978)] proved the analogous theorem with handle slides also allowed, but Fenn and Rourke [Fenn and Rourke (1979)] eliminated the handle slides. The latter paper also proved a corresponding theorem for non-orientable 3-manifolds, where S^3 is replaced by a twisted S^1-bundle over S^2 (classified by the elements of the group $\pi_1(S^1) = \mathbb{Z}$), handle sliding is retained and an additional move is introduced. (For a different proof, see [Lu (1992)] or [Matveev and Polyak (1994)].) This theorem has turned out to be particularly useful since the discovery in recent years of many new 3-manifold invariants (for example, [Reshetikhin and Turaev (1990, 1991)]). One can define such invariants by means of integer surgery diagrams, and then one only needs to prove invariance under blowing up and down.

We now switch to an entirely different approach to 3-manifold classifications, a geometric one, applicable, of course, only in the smooth case.

Thurston's Geometrization program

In 1982, W.P. Thurston presented a program intended to classify smooth 3-manifolds and solve the Poincaré conjecture by investigating the possible geometries on such 3-manifolds. For a survey of this topic see [Thurston (1997)]. The key ingredient of this classification ansatz is the concept of a model geometry. Again, in this section, all manifolds are assumed to be smooth.

A *model geometry* (G, X) consists of a simply connected manifold X together with a Lie group G of diffeomorphisms acting transitively on X fulfilling certain set of conditions. One of these is that there is a G-invariant Riemannian metric. For example, reducing the dimension, we can consider 2-dimensional model geometries of a 2-manifold X. From Riemannian geometry we know that any G-invariant Riemannian metric on X has constant Gaussian curvature (recall that G must be transitive). A constant scaling of the metric allows us to normalize the curvature to be 0, 1, or -1 corresponding to the Euclidean ($\mathbb{E}^2$), spherical (S^2) and hyperbolic ($\mathbb{H}^2$) space, respectively. Thus, there are precisely three two-dimensional model geometries: spherical, Euclidean and hyperbolic.

It is a surprising fact that there are also a finite number of three-dimensional model geometries. It turns out that there are 8 geometries: spherical, Euclidean, hyperbolic, mixed spherical-Euclidean, mixed hyperbolic-Euclidean and 3 exceptional cases. A *geometric structure* on a more general manifold M (not necessarily simply connected) is defined by a model geometry (G, X) where X is the universal covering space to M i.e. $M = X/\pi_1(M)$. This is equivalent to a representation $\pi_1(M) \to G$ of the fundamental group into G. Of course a geometric structure on a 3-manifold may not be unique but Thurston explored decompositions into pieces each of which admit a unique geometric structure. This decomposition proceeds by splitting M into essentially unique pieces using embedded 2-spheres and 2-tori in such a way that a model geometry can be defined on each piece. Thus,

Thurston's Geometrization conjecture can be stated:

The interior of every compact 3-manifold has a canonical decomposition into pieces (described above) which have one of the eight geometric structures.
For the all important Poincaré conjecture we have an M which is simply connected. The decomposition of Thurston's conjecture implies that M must be an irreducible 3-manifold with finite fundamental group and, in fact, that M is geometric and thus $M = S^3/\Gamma$ for some finite group Γ. But the fundamental group of M is zero (simple-connectivity) so M must be a geometric model for some group G. Since M is assumed to be closed, this together with a theorem of Hamilton, would imply that there is only one possibility: $(S^3, SO(3))$ the spherical geometry, so M would be diffeomorphic to standard S^3. For more details, refer to Thurston's book, [Thurston (1997)] and the review by Milnor, [Milnor (2003)].

Remarkable progress on the Thurston, and thus Poincaré, conjecture has been made by G. Perelman by converting this Geometrization Conjecture to one involving the Ricci flow equations (see the original papers [Perelman (2002)],[Perelman (2003b)],[Perelman (2003a)] and the overview article by Anderson [Anderson (2004)]). The technicalities of this work are daunting, but many workers in the field now (2005) believe that a positive solution to the Poincaré conjecture, if not the more general one of Thurston, is near at hand.

6.3 Higher-dimensional Manifolds

Now we skip the next dimension, 4, because it is so special, and instead go from 3 to ≥ 5, where special techniques, not applicable in dimension 4, solve the homotopy classification problem. A complete solution of this problem was found in the sixties and seventies culminating in the h- and s-cobordism theorem which include such powerful algebraic tools as algebraic K-theory and non-commutative localizations of rings. Here we concentrate on applications of the h-cobordism theorem leading to a classification of differential structures on higher-dimensional manifolds.

6.3.1 *The simply-connected h-cobordism theorem*

After the introduction and calculation of the cobordism ring in 1954, it became clear that this approach could be powerful enough to classify manifolds up to homotopy opening new doors in the description of equivalences between manifolds using sophisticated algebraic techniques. A particularly important type of cobordism, *h-cobordism* (homotopy-cobordism) was developed by Thom, Milnor, Smale and others leading to the h-Cobordism Theorem of Smale. These techniques provide important ingredients in the smooth classification program for manifolds with dimension $n \geq 5$, which we will discuss later.

The h-cobordism

Two simply connected, closed n-dimensional manifolds X_-, X_+ are *h-cobordant* if there is a cobordism W between them such that the inclusion $i_\pm : X_\pm \hookrightarrow W$ induces a homotopy equivalence between $X_\pm$ and W. An obvious trivial example is the product $W = I \times X$ as h-cobordism from X to itself.

The h-cobordism theorem

In 1961, Smale [Smale (1961)] proved the h-cobordism theorem for dimension $n > 4$ showing that two simply connected manifolds are h-cobordant if and only if both are diffeomorphic.

Theorem 6.9. (The h-Cobordism Theorem) *If W is an h-cobordism between the smooth, simply-connected, closed n-dimensional manifolds X_-, X_+ and $n \geq 5$, then W is diffeomorphic to the product $I \times X_-$. In particular, X_- is diffeomorphic to X_+.*

Note: *For our purposes, the outstanding point here is that the topological cobordism induces a diffeomorphism.*

The failure of this theorem in dimension 4 was a natural hint that there might be many exotic differential structures on certain 4-manifolds. We should note, of course, that since the h-cobordism theorem refers only to **compact** manifolds, modifications are required to understand diffeomorphisms between non-compact manifolds, especially our important $\mathbb{R}^4$ case. One approach is the so-called "engulfing" theorem of Stallings, [Stallings (1962)]. These questions are discussed in Chapter 8 which specializes to the Euclidean case.

We will now sketch an outline of the proof of this theorem and some of its consequences. For full details see [Milnor (1965a)] or [Rourke and B.J. (1972)]. An h-cobordism W between X_-, X_+ defines an $(n+1)$-dimensional manifold W with $\partial W = X_+ \amalg X_-$ such that the inclusions $i_\pm : X_\pm \hookrightarrow W$ induce homotopy-equivalences. Furthermore, there is a relative handlebody decomposition of (W, X_-) so that the interior int W is a handle body with

$0,\ldots,n+1$-handles. We are done with the proof if we can show that this relative handle body decomposition is empty, that is, there is no handle body in int W. If this is true, the handle body decomposition of W is induced from the boundary and W must be a product handle body $W = X_- \times I$. To prove that the handlebody is trivial, use techniques of handle sliding and cancellation discussed in subsection 6.1.3. First, it is shown that the 0-,1-,n- or $(n+1)$- handles can be canceled[10]. Note that this proposition holds for $n = 4$ too, that is, an h-cobordism between simply connected 4-manifolds involves only 2- and 3-handles. After this, only the k-handles with $1 < k < n-1$ are left.

Algebraic versus geometric intersection numbers

Reviewing subsection 6.1.3 we find that the cancellation of the intermediate dimension handles can be carried out if the attaching sphere of the k-handle and the belt sphere of $k-1$ handle intersect transversely in one point. Of course, in general that may not be the case. To analyze the problem consider geometric and algebraic intersection numbers. Let Y_1^n, Y_2^m be transversely intersecting oriented, smooth submanifolds of complementary dimensions in the oriented manifold X^{n+m}. The *geometric* intersection number of Y_1^n and Y_2^m is simply the cardinality of the set $Y_1^n \cap Y_2^m$. The *algebraic* intersection number of Y_1^n and Y_2^m is by definition the sum of the signs of the intersection points. The important fact is that the cancellation of the k and $k-1$ handles is only possible if the geometric intersection number is equal to 1. Recall that the dimension of the attaching sphere of the k-handle is $k-1$ while that of the belt sphere is $n+1-(k-1)-1 = n+1-k$. However, the attaching is done in the boundary, ∂W, of dimension n. In fact, it turns out that by rearranging the handle decomposition through handle slides, we can pair up the intermediate dimension handles that the attaching sphere intersects the belt sphere *algebraically* once, fulfilling this condition[11]. At this point we would have a proof of Theorem 6.9 if we can establish the equality of geometric and algebraic intersection numbers. For this, we need the Whitney trick.

Whitney trick and generalized Poincaré conjecture

To show that Y_1^n and Y_2^m can be homotopically deformed so that their algebraic intersection number (which is 1) equals their geometric intersection number, we use the *Whitney trick*.

---

[10] Here use the fact that a h-cobordism has an equal number of k and $k-1$.
[11] This step involves some algebraic considerations based on the fact that $H_*(W, X_-; \mathbb{Z}) = 0$ for the h-cobordism.

Theorem 6.10. (Whitney trick) *Let Y_1^n, Y_2^m be transversely intersecting connected, smooth submanifolds of complementary dimensions in the simply connected $(n+m)$-manifold X^{n+m}. Assume furthermore that $n+m \geq 5$, $m \geq 3$ and $n \geq 2$ (for $n=2$ $X \setminus Y_2$ has to be simply connected). If p, q are intersection points with opposite signs, then there exists an isotopy φ_t ($t \in [0, 1]$) (of id_X) such that the intersection points p, q can be canceled.*

Using this result cancel all remaining handles in W to arrive at the h-Cobordism Theorem. For more figures, see also page 349 of [Gompf and Stipsicz (1999)]. Among many important consequences of this theorem is the following:

Theorem 6.11. (Smale) *If X^n is a closed, simply connected, smooth manifold of dimension $n \geq 5$ and $H_*(X^n; \mathbb{Z}) \cong H_*(S^n; \mathbb{Z})$, then X^n is homeomorphic to S^n.*

This result is also known as the *Generalized (Topological) Poincaré Conjecture*. The case $n = 4$ is covered by the theorem of Freedman (Theorem 8.8). The $n = 3$ case is still open but the work of Perelman mentioned at the end of section 6.2 above holds great promise for establishing the general Thurston conjecture, or at least the special Poincaré case, that every 3-manifold homotopic to S^3 is also homeomorphic to S^3. The extension to the smooth case, that is, the question of whether a smooth manifold X^n that is homeomorphic to S^n is actually diffeomorphic to it, is only known to be true in a few dimensions, e.g. 1,2,3,5,6. The question is still open in dimension 4 but there is a conjectured candidate. The higher dimensional cases, $n > 6$, are known to include some exotic spheres, homeomorphic but not diffeomorphic to S^n. We will look at these problems in section 7.8.

6.3.2 The non-simply-connected s-cobordism theorem*

Barden, Mazur and Stallings [Kervaire (1965)] extended the h-cobordism theorem to the non-simply connected case proving what is now called the s-cobordism theorem. Here an additional invariant, the so-called Whitehead torsion, is needed. This Whitehead torsion serves as an obstruction to the reduction of an h-cobordism to a product, but is generally difficult to calculate. We will give a brief overview of this interesting topic. First we state the theorem:

Theorem 6.12. *Let (W, V, V') be a compact connected cobordism V to V' which is relative in the sense that $\partial W \setminus int(V \cup V') \simeq \partial V \times [0,1]$. Suppose that the inclusions $i : V \hookrightarrow W, i' : V' \hookrightarrow W$ are homotopy equivalences, $\dim W \geq 6$, and i has zero torsion in the Whitehead group in $Wh(\pi_1(W))$. Then $(W, V, V') \simeq V \times ([0,1], 0, 1)$.*

Since the Whitehead torsion depends only on the fundamental group $\pi_1(W)$ of the cobordism W, if $\pi_1(W) = 0$, so that V, V' are simply-connected then the Whitehead torsion must be trivial $Wh(\pi_1(W)) = 1$.

We now briefly survey the rather technical tools needed to define the Whitehead torsion.

Definition 6.1. Let R be a ring with unit. Let $GL(n, R)$ be the group matrices with values in the ring R. From the natural inclusions

$$\ldots \subset GL(n-1, R) \subset GL(n, R) \subset GL(n+1, R),$$

we obtain the inductive limit $GL(R) = \bigcup_n GL(n, R)$. Define an $n \times n$ matrix as **elementary** if it has 1's in the diagonal and at most one non-zero off-diagonal entry, that is, if $a \in R$ and $1 \leq i, j \leq n$, $i \neq j$. Now define a particular elementary matrix $e_{ij}(a)$ to be the $n \times n$ matrix with elements $(e_{ij}(a))_{kl}$ with $1 \leq k, l \leq n$

$$(e_{ij}(a))_{kl} = \begin{cases} 1 \text{ if } k = l \\ a \text{ if } k = i, l = j \\ 0 \text{ elsewhere.} \end{cases}$$

The set of such matrices forms a subgroup $E(n, R)$ of $GL(n, R)$ with the limit denoted by $E(R)$.

Remark 6.4.
The elementary matrices over a ring R satisfy the relations

$$\begin{aligned} e_{ij}(a)e_{ij}(b) &= e_{ij}(a+b) & a, b \in R \\ e_{ij}(a)e_{kl}(b) &= e_{kl}(b)e_{ij}(a) & j \neq k \text{ and } i \neq l \\ [e_{ij}(a), e_{jk}(b)] &= e_{ik}(ab) & i, j, k \text{ distinct} \\ [e_{ij}(a), e_{ki}(b)] &= e_{kj}(-ab) & i, j, k \text{ distinct} \end{aligned}$$

Recall the definition of the group commutator, $[g_1, g_2] = g_1 g_2 g_1^{-1} g_2^{-1}$. Furthermore, any upper-triangular or lower-triangular matrix with 1's on the diagonal belongs to $E(R)$.

Next we state *Whitehead's lemma* leading to the definition of the so-called $K_1(R)$, the first algebraic K-group.

Lemma 6.1. *For any ring R, the subgroup $[GL(R), GL(R)]$ formed by the commutators between two elements of $GL(R)$ is identical to $E(R)$, i.e.*

$$[GL(R), GL(R)] = E(R)\qquad .$$

In particular $E(R)$ is a normal subgroup and thus the quotient $GL(R)/E(R)$ is the Abelian quotient.

Now, a series of definitions:

Definition 6.2. Let R be a ring with unit. The first algebraic K-group $K_1(R)$ of the ring is defined by

$$K_1(R) = GL(R)/E(R) \quad .$$

Definition 6.3. Let G be a group. We denote by $\mathbb{Z}G$ the integral group ring made from the group i.e. any element $z \in \mathbb{Z}G$ can be represented by the formal sum

$$z = \sum_{g \in G} a_g\, g \qquad a_g \in \mathbb{Z}$$

Note the special elements, images of the units $\pm g$, $g \in G$. Finally, the Whitehead torsion is defined as follows.

Definition 6.4. If G is a group, its Whitehead torsion $Wh(G)$ is the quotient of $K_1(\mathbb{Z}G)$ by the image of $\{\pm g : g \in G\}$, i.e.,

$$Wh(G) = K_1(\mathbb{Z}G)/\{\pm g : g \in G\} \quad .$$

This invariant can be difficult to calculate in general. Here are a few examples.

- Let $\pi_1(W)$ be trivial. From the standard fact $K_1(\mathbb{Z}) = \mathbb{Z}_2$ we obtain $Wh(\pi_1(W)) = Wh(1) = 1$.
- Let $\pi_1(W) = \mathbb{Z}_2$ then one obtains after a long calculation (see [Rosenberg (1994)] for instance): $Wh(\mathbb{Z}_2) = 1$.
- For a less trivial case, consider $G = \mathbb{Z}_p$ for a prime p. According to [Rosenberg (1994)] p. 100, there is a surjective map $Wh(G) \to \mathbb{Z}^{\frac{p-3}{2}}$ showing that $Wh(G)$ must be non-trivial for some primes $p > 3$.

For further applications of Whitehead torsion in geometry and topology of manifolds and stratified spaces see [Weinberger (1994)].

6.4 Topological 4-manifolds: Casson Handles*

As mentioned above, the topological handle body theory in 4 dimensions is completely different from that in the other dimensions. In dimensions $n = 1, 2, 3$ every topological $\mathbb{R}^n$ admits a unique smoothness structure as we described for each individual dimension. For $n > 4$ Whitney's trick is used to cancel handles of dimensions $k, k+1$, proving the h-cobordism theorem and establishing that two h-cobordant manifolds must necessarily be diffeomorphic. The failure of the trick in dimension 4 (see [Kervaire and Milnor (1961)]) means that we cannot use it to guarantee a *smooth embedding* of a D^2 in a general 4-manifold. Of course such a disk can be embedded, but perhaps not smoothly, or it can be smoothly immersed, but not embedded (it may have double points). However, these weaker conditions are not sufficient to cancel the handle bodies necessary to prove four dimensional h-cobordism theorem. In 1973, Casson developed a theory of flexible handles constructing something which "looks like" an open 2-handle at least up to "proper-homotopy" rather than finding the smooth 2-handle directly. In 1981 Freedman showed that all Casson handles are homeomorphic to the open 2-handle $D^2 \times \mathbb{R}^2$ containing a topologically embedded disk which can be used as Whitney disk, topologically. Bizaca has applied Casson techniques to provide further insights into the handle body decomposition of exotic $\mathbb{R}^4$'s, as we will discuss later (see Chapter 8).

Singular points, double points and the disk problem

The four dimensional disk embedding problems focus on maps having only "double point" singularities. Specifically, a smooth map $f : X^n \to M^{2n}$ is said to have only double points if the following three conditions are satisfied: $\#|f^{-1}(f(x))| \leq 2$, there are only a finite number of points with $\#|f^{-1}(f(x))| = 2$ and at each singularity $p = f(x) = f(y)$, there is a coordinate chart $(u_1, \ldots, u_n, \nu_1, \ldots, \nu_n)$ in M^{2n} around p where the two coordinate subspaces $\mathbb{R}^n \times 0 (\nu_1 = \cdots = \nu_n = 0)$ and $0 \times \mathbb{R}^n (u_1 = \cdots = u_n = 0)$ are exactly the immersed images of f near x and y respectively. That is, the map is "self-transverse." The simplest example for such a double point is the figure "eight" in the plane. As a corollary of general position theorems in dimension 4, every map $f : D^2 \to M^4$ can be deformed by an isotopy, relative to ∂D^2, to a map which has only double points. All these questions about disk embeddings can be summarized in the long outstanding smooth disk problem: *Under what conditions (on f and W) does there exist a smooth embedding $g : D^2 \to W^4$ with $g|_{\partial D^2} = f|_{\partial D^2}$ such that framings determined by f, g are homotopic?*

The skeleton of a Casson handle

Casson's idea was that, instead of finding the smooth 2-handle directly, try to construct something which "looks like" an open 2-handle at least up to "proper-homotopy". He named the objects he found "flexible handles" although they are now generally called Casson handles. We will introduce the concept of a Casson handle in two steps. First consider immersed disks with one double point. An infinite construction to kill the singular points at each stage leads to a kind of "skeleton" of a Casson handle. In the second step this "skeleton" is enlarged through the use of the so-called "kinky handles" to produce the "real" Casson handle.

The iterative construction of the Casson handle begins with the first stage tower. Consider the isolated double points of the disk immersion $f : D^2 \to M^4$ and construct a closed curve on every double point. For each double point, choose a closed curve bounding that point. Each such curve is a generator of the fundamental group $\pi_1(f(D^2))$. The double points in the first stage are killed by the singular disks of the second stage which also produce double points. After an infinite union of stages we end up with a simply connected space X, the skeleton of the Casson handle.

Remark 6.5.
Here is a very brief summary of Casson's procedure. Suppose $f : (D, \partial D) \to (W, \partial W)$ is a proper immersion with isolated double points, with $x_1, \ldots, x_k$ the singular points. There are disjoint simple closed curves $D^1 := \{c_1, \ldots, c_k\}$ in $f(D)$ such that $x_i \in c_i$ for all i, and $\pi_1(f(D))$ is freely generated by the homotopy classes of these curves D^1 which provide a basis for $f(D)$. Now suppose each element in a basis c_i bounds a singular disk D_i^2 in W with the following two properties:

(1) disjointedness: $D_i^2 \cap D_j^2 = \emptyset$ for all $i \neq j$,
(2) attaching property: $D_i^2 \cap D^1 = c_i$.

The union $D^1 \cup \bigcup_i D_i^2$ is called the second stage tower. Similarly the i-th stage singular disks $D_1^i, \ldots, D_2^i$ are constructed by an inductive process from the $i-1$-th stage. The union $X = \bigcup_{i=0}^{\infty} \bigcup_s D_s^i$ defines the skeleton of the Casson handle. The space X is simply-connected ($\pi_1(X) = 0$) because at any stage the generators of $\pi_1(D_s^i)$ are killed by the next stage singular disks.

The pictures 6.16, 6.17 and 6.18 provide a symbolic representation of the procedure. Obviously, this is a rather technical procedure and we recommend the original work of Casson [Casson (1986)] and the review in [Gompf and Stipsicz (1999)] for more complete explanations.

Self-plumbing and kinky handle
A Casson handle is a union of specially constructed regular neighborhoods of the D_j^i's. The corresponding double points can be described by the process of self-plumbing leading to the concept of *kinky handles* [Casson (1986)]. More technically, the neighborhood of the singular disk $f(D^2)$ with double points can be illustrated as in Figure 6.19.

Remark 6.6.
Choose $2k$ points in $int(D^2) \times \{0\}$, say $\{x_i, y_i\}_{i=1}^k$. Let $D(x_i), D(y_i)$ be closed disks in

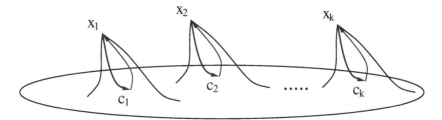

Fig. 6.16 The 1-stage tower (figure from [Freedman and Luo (1989)] p.51 Fig. 4.8)

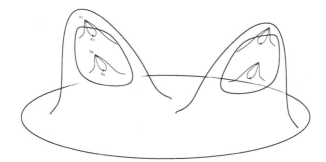

Fig. 6.17 The 2-stage tower built from the 1-stage (figure from [Freedman and Luo (1989)] p.51 Fig. 4.9(a))

Fig. 6.18 Symbol description of the second stage (figure from [Freedman and Luo (1989)] p.52 Fig. 4.9(b))

$int(D^2)$ such that $x_i \in int(D(x_i))$ and $y_i \in int(D(y_i))$; and let $\phi : D(x_i) \to D^2$, and $\psi_i : D(y_i) \to D^2$ be orientation preserving diffeomorphisms. Then the k-fold plumbing is the quotient space $D^2 \times D^2/\sim$ obtained by identifying $D(x_i) \times D^2$ with $D(y_i) \times D^2$ by sending (s,t) to $(\psi_i^{-1}(t), \phi_i(s))$ for $i = 1, 2, \ldots, k$.

The Casson handle

Replace the singular disks D_j^i by the regular neighborhoods $N(D_j^i)$ and de-

186   Exotic Smoothness and Physics

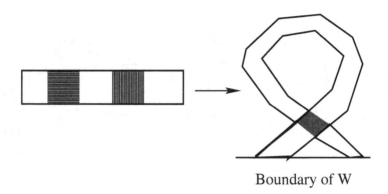

Fig. 6.19   Self-plumbing of a 2-handle

fine the infinite union $CH = \bigcup_{i,j} N(D_j^i)$ to be the Casson handle $(CH, \partial CH)$, with $\partial CH = N(f(\partial D))$. CH is simply-connected and there is a (proper) homotopy to the open 2-handle $(D^2 \times \mathbb{R}^2, \partial D^2 \times \mathbb{R}^2)$. Under some mild conditions, Casson [Casson (1986)] proved the existence of such a Casson handle. A very important result concerning Casson handles was obtained in 1981 by Freedman. It can be stated as follows:

Theorem 6.13. *Any Casson handle $(CH, \partial CH)$ is homeomorphic to the standard open handle $(D^2 \times \mathbb{R}^2, \partial D^2 \times \mathbb{R}^2)$.*

In summary, even though the immersed disks at each finite stage of the tower have double points, the limit does not, but rather this limit is an embedded handle, but only *topologically*. That is, smoothness may be lost in the limiting process required to remove the double points and arrive at an embedding.

As a corollary to this theorem one obtains disk embedding theorems for dimension 4 leading to the solution of many topological problems (see section 8.2). From the point of view of Casson handles we have a topological handle body theory solving important TOP classification problems in dimension 4 in a way similar to the higher-dimensional classification techniques. In particular, the topological h-cobordism theorem was proved by Freedman. Recall that the smooth variant of this theorem fails (see section 6.6). As a corollary Freedman obtained the complete classification of simply connected, compact 4-manifolds.

Casson handle techniques solve the TOP classification problems in dimension four, providing a TOP solution for the four dimensional Whitney trick. Of course, we should point out that these procedures only solve the TOP classification problem for manifolds which are already known to be h-cobordant. For the smooth category, Gompf [Gompf (1988)], [Gompf (1989b)], shows that there are uncountably many diffeomorphism types of Casson handles (see also [Gompf and Stipsicz (1999)] p.372 Exercise 9.4.13 (b)). Furthermore, as we will see in the following, Gompf and Bizaca used these techniques to construct the so-called small exotic $\mathbb{R}^4$'s.

6.5 Smooth 4-manifolds: Kirby Calculus

Kirby calculus is a technique for going from one decomposition of a smooth 4-manifold by handles to another. In Section 6.1.3, we introduced a complete set of moves for handle bodies, namely handle pair creation/cancellation and handle sliding, which (together with isotopies) are sufficient to go between any two relative handle presentations of a given pair $(X, \partial_- X)$ (Theorem 6.4). In this section, we will describe these moves in the context of so-called Kirby diagrams providing visual representations of the handle body decomposition.

Kirby diagram

A *Kirby diagram* is a description of a 4-dimensional handlebody with 3-dimensional diagrams for a compact closed 4-manifold X. In this case the attaching regions of the handles lie in the boundary manifold, locally $\mathbb{R}^3$. For instance, the attaching region of each 1-handle is $D^3 \coprod D^3$, which we draw as a pair of round balls in $\mathbb{R}^3$. Because of non-trivial framings, the most interesting case is given by 2-handles, represented as embeddings of thickened circles $(S^1 \times D^2)$ in $\mathbb{R}^3$ perhaps as links. For a closed, compact 4-manifold X it turns out that the complexity of a 4-dimensional handlebody primarily resides in the 2-handles. [Trace (1982)].

Handle sliding

The process of "sliding" involves smoothly moving, or sliding, the attaching region of one handle along the body of another. For a k-handle, $D^k \times D^{n-k}$, the attaching region is $S^{k-1} \times D^{n-k} \in \partial X$. We can then slide this $(n-1)$ manifold along the full n-dimensional body of a second handle and take advantage of the extra dimension to deform, link, unlink, etc., the S^{k-1}. In more detail, for a pair of handles h_1 and h_2 of the same index k attached to the manifold X the sliding can be described by an isotopy of the attaching sphere for h_1 in $\partial(X \cup h_2)$ along a disk $D^k \times p \subset \partial h_2$ ($h_2 = D^k \times D^{n-k}$), which "returns" the attaching sphere to ∂X. This can perhaps best be understood in terms of an illustration. Figure 6.20 shows a 1-handle slide on a 2-manifold. In general, the handle slide will change the attaching sphere of h_1 (that is unlinking it from that of h_2), as well as the framing[12]. It is also easy to draw 1-handle slides in Kirby diagrams for higher dimensional manifolds by pushing the attaching ball of h_1 through the 1-handle h_2. Duality provides similar descriptions for the sliding of 3-handles. Because

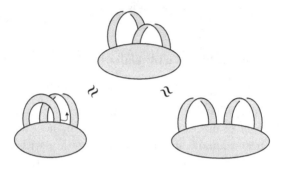

Fig. 6.20 Handle sliding of a 1-handle on a 2-manifold (figure from [Gompf and Stipsicz (1999)] p. 140, Fig. 5.1)

of the non-trivial framing, the sliding of 2-handles is more interesting and difficult. We will only touch on the highlights and again refer the reader to the [Gompf and Stipsicz (1999)], especially chapter 4 for more details and illustrative diagrams. For X a four manifold, $M = \partial X$ is three dimensional and the attachment of a 2 handle is defined by a map, $\phi : S^2 \times D^2 \to M$. The image of the circle, $\phi_{|S^1 \times 0}$ is a knot, K. This knot (and thus the handle) is **framed** by choosing some vector transverse to it, or equivalently a "neighboring" knot, K', which can be chosen to be the image $\phi_{|S^1 \times p}$ for some p near 0. Consider two 2-handles h_1 and h_2 attached along the framed knots K_1 and K_2. Define the framing of K_2 by a parallel curve K'_2, which will bound a disk $D^2 \times p \subset \partial(Y \cup h_2)$. Now we slide h_1 by an isotopy of K_1 over one such disk is the sliding of h_1 over h_2. The result is called the *band-sum* of K_1 and K'_2. Figure 6.21 visualizes this procedure. Note that there are two possible types of such operations (depending on the orientation): the *handle addition* and the *handle subtraction*.

Framing for 2-handles after the handle slide

Since the framing of a circle in a 3-space depends on the choice of a transverse vector, this choice can be identified with the homotopy class of a map: $S^1 \to O(2)$, that is, an element $n \int \pi_1(O(2)) \cong \mathbb{Z}$. Figure (6.22) represents

---
[12]In this case the framing is a well-defined element of $\pi_0(O(1)) \cong \mathbb{Z}_2$.

A Guide to the Classification of Manifolds 189

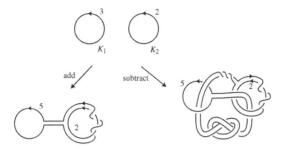

Fig. 6.21 Handle sliding of 2-handles: handle addition (left) and handle subtraction (right) (figure from [Gompf and Stipsicz (1999)] p.142, Fig. 5.5)

two such framings. The left hand one is the trivial one, n=0, while the right hand one corresponds to $n = 3$. Finally, note that handle sliding can

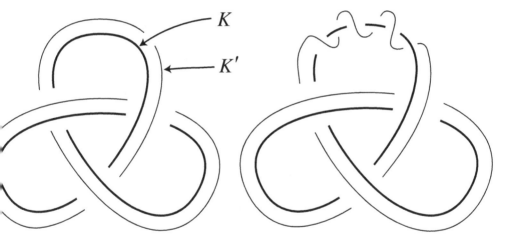

Fig. 6.22 Framing of knots (figure from [Gompf and Stipsicz (1999)] p. 118, Fig. 4.18,4.19)

change framings. If the framings of h_1 and h_2 are n_1, n_2 before the slide, then after the slide of h_1 over h_2, the resulting handle, h'_1 will have framing N_{12}, given by

$$n_{12} = n_1 + n_2 \pm 2\ell k(K_1, K_2) ,$$

where ℓk denotes the linking number of the two knots and the sign is $(+)$ for handle addition and $(-)$ for subtraction. We note that this formula can also be applied in case of 1-handles.

6.6 Why is Dimension 4 so Special?

As we noted in the subsection 6.3.1, the Whitney trick does not provide the needed smoothly embedded 2-disk to cancel the middle handles for 4-dimensional manifolds. This is the reason for the exclusion of dimension four in the proof of the h-cobordism theorem. However, not only does the displayed proof technique not work but in fact there are now counter examples to the possible extension of the statement in Theorem 6.10 to dimension four.

Why does Whitney's trick fail in dimension 4?

Fortunately, Freedman, [Freedman (1984)], has provided an easily read informal account of this matter. Consider the 5-dimensional h-cobordism W between X_1 and X_2. The interior int W of W has $n-$handles for all $n = 0, 1, 2, 3, 4, 5$. The proof techniques of Theorem 6.10 can still be applied to cancel all adjacent handle pairs, except for the $n = 2, 3$ pairs. To ensure these cancellations, we would need the Whitney trick, the embedding of a special disk, or more accurately, the embedding of an open 2-handle $D^2 \times \mathbb{R}^2$. Casson handle techniques discussed in section 6.4 show that such an embedding exists *topologically*. To visualize the situation, see Figure 6.23 showing 2 submanifolds (indicated by two curves) intersecting each

Fig. 6.23 Two intersecting submanifolds

other. The Whitney trick involves the embedding of the disk, called the Whitney disk, indicated by the shaded area on the left side of Figure 6.24. An isotopy of this disk to a point results in the desired result in the right side of Figure 6.24. But, in dimension four, the Whitney disk may have self-intersections which cannot be removed smoothly to ensure the cancellation of the 2-handle, 3-handle pairs. However, the Casson techniques lead to such cancellation in the TOP category. So, we can say that **topologically,** $W = X_1 \times [0, 1]$ so that the two end 4-manifolds, X_1, X_2, are homeomorphic but perhaps not diffeomorphic (see Kreck [Kreck (1999)]).

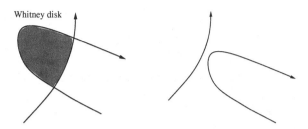

Fig. 6.24 Simple visualization of Whitney's trick

Again, *Whitney's trick fails only for smooth 4-manifolds.*
Consequences for the handle calculus
However, some results in the DIFF category for dimension four can be obtained. For example, there is a theorem of Wall:

Theorem 6.14. [Wall (1964a)] *If closed, smooth, simply connected, orientable 4-manifolds X, Y are homeomorphic, then $X \# k S^2 \times S^2$ is diffeomorphic to $Y \# k S^2 \times S^2$ for some $k \geq 0$.*

Thus, given two homeomorphic but not diffeomorphic 4-manifolds the connected sum addition of $S^2 \times S^2$ k times to both manifolds results in diffeomorphic manifolds. We can understand this theorem in the context of Casson handle theory. As Freedman shows, every Casson handle is homeomorphic to $D^2 \times \mathbb{R}^2$ which is equivalent to $S^2 \times (S^2 \setminus D^2)$. Consider now an h-cobordism between two smooth X and Y which, from the work of Casson and Freedman, can be topologically trivialized by adding k Casson handles. But this is equivalent to changing X to $X \# k S^2 \times S^2$ and Y to $Y \# k S^2 \times S^2$ to get diffeomorphic manifolds, according to Theorem 6.14. One consequence of this remarkable fact is that the exoticness of 4-manifolds is an **unstable** phenomenon, that is, the connected sum addition of rather trivial spaces destroys the exotic structure. This is in contrast to the differential structures on higher-dimensional manifolds where we have the product structure theorem of Kirby and Siebenmann (see section 7.9).

The theorem 6.14 can be extended to any compact orientable 4-manifold (possibly with nonempty boundary).

Remark 6.7.

Actually Theorem 6.14 can be extended to non-orientable manifolds by allowing connected sums with $S^2 \tilde{\times} S^2$, that is, the non-trivial S^2-bundle over S^2, as well. (See [Gompf (1991)].) However, the non-trivial product of the S^2's is necessary since there is an exotic smooth structure on $\mathbb{RP}^4$ which can never be trivialized by connected sum additions of any number of

$S^2 \times S^2$ [Cappell and Shaneson (1976)].

There are no general results concerning the minimum value of k for a given pair X, Y satisfying Theorem 6.14. There are many examples with $k = 1$, and it is possible that $k = 1$ is always sufficient. Up to now, pairs requiring $k > 1$ have not been detected due to the lack of suitable invariants.

h-cobordant 4-manifolds

To summarize, we quote the theorem Freedman [Freedman (1982)], the topological version of the h-cobordism theorem for dimension 4. Freedman used Casson handle theory to replace the smooth Whitney trick for the middle handles.

Theorem 6.15. *Simply connected, smooth 4-manifolds are h-cobordant if and only if they are homeomorphic, i.e. having isomorphic intersection forms.*

The existence of exotic (non-diffeomorphic) smooth structures on 4-manifolds shows that this result cannot be extended to the DIFF category in dimension 4 (see [Donaldson (1990)] for the first example).

Remark 6.8.

Using similar techniques it is possible prove the "converse" of Theorem 6.14 (see [Lawson (1998)]):

If $X \# kS^2 \times S^2$ is diffeomorphic to $Y \# kS^2 \times S^2$, then there exists an h-cobordism W between X and Y which is built on $I \times X$ with k 2-handles and k 3-handles.

Akbulut corks

We have seen that if W^5 is an h-cobordism between simply connected 4-manifolds X and Y, then W^5 is not necessarily the trivial **smooth** product, $I \times X$. However, W is "not far" from being trivial in the sense of the following theorem. The main statement of this theorem was found by Curtis and Hsiang. Variations and addenda were proved by Freedman and Stong, Kirby and Bižaca, and by Matveyev. (See also [Kirby (1997)].)

Theorem 6.16. [Curtis *et al.* (1997)] *Let W be a cobordism between X and Y. There is a sub-cobordism $V \subset W$ between compact 4-manifolds $A_1, A_2 \subset X, Y$ such that $W -$int V is the product cobordism (i.e. diffeomorphic as a triple to $I \times (X-$int $A_1))$, and V, A_1, A_2 are contractible.*

Furthermore we have the additional properties:

(1) $W - V$ and each $X - A_1, Y - A_2$ are simply-connected,

A Guide to the Classification of Manifolds   193

(2) $A_1 \times [0,1]$ and $A_2 \times [0,1]$ are diffeomorphic to D^5,
(3) A_1 is diffeomorphic to A_2 by a diffeomorphism which, restricted to $\partial A_1 = \partial A_2$, is an involution, i.e. a map $\tau : M \to M$ with $\tau \circ \tau = id_M$ but $\tau \neq \pm id_M$,
(4) the boundary $\partial A_1 = \partial A_2$ is a homology 3-sphere.

Please note that we have shifted from h-cobordism to simple cobordism (without the retract assumptions) for the boundary manifolds and the failure of the h-cobordism can be confined in a small region.

These V cobordisms are delicate objects; their non-triviality vanishes when a trivial h-cobordism is added. The V's are called *Akbulut corks* after the first author to describe them. In subsection 8.5.1 we will use this theorem to construct the handle body decomposition of an exotic $\mathbb{R}^4$.

6.7 Constructing 4-manifolds from Intersection Forms

In this section we will concentrate on 4-manifolds generally in the TOP category. It turns out that an integer symmetric matrix, the *intersection form* is a powerful tool for the classification of such spaces. So, certain topological results can be obtained from algebraic techniques associated with standard forms for integer matrices. We end this section and chapter with Freedman's important theorem establishing existence and uniqueness results relating unimodular symmetric integer matrices to topological 4 manifolds.

6.7.1   The intersection form

In subsection 6.2.1 we introduced a bilinear form for 2-manifolds. This form is derived from the intersection of pairs of representative elements $H_1(M^2, \mathbb{Z}_2)$. This led to the classification of 2-manifolds (surfaces) from purely algebraic properties. Now, we double the dimension and look at the "middle" homology group $H_2(M^4, \mathbb{Z})$ of a compact, oriented, topological 4-manifold M^4.

Definition 6.5. Suppose $a, b \in H_2(M^4, \mathbb{Z})$ can be represented by oriented surfaces A, B in M^4. Define

$$Q_{M^4} : H_2(M^4, \mathbb{Z}) \times H_2(M^4, \mathbb{Z}) \to \mathbb{Z}$$

by $Q_{M^4}(a,b) = A \cdot B$, where $A \cdot B$ is the (oriented) algebraic intersection number of the cycles A and B.

This bilinear form is clearly symmetric. From the work of Lee and Wilczynski [Lee and Wilczynski (1993)] every homology class in $H_2(M)$ of a topological 4-manifold M can be represented topologically by locally flat 2-spheres. In general, this does not carry over to the smooth case. In fact, in dimension four, it may be necessary to use surfaces of genus $g > 0$ to represent second homology classes smoothly. The genus depends on the complexity of the 4-manifold.

From Poincaré duality $H_2(M^4, \mathbb{Z}) \cong H^2(M^4, \partial M^4, \mathbb{Z})$ we obtain another, fully equivalent, definition of the intersection using cohomology groups and the cup product. Recall that when M^4 is oriented, it has a fundamental class $[M^4] \in H_4(M^4, \partial M^4; \mathbb{Z})$.

Definition 6.6. The symmetric bilinear form

$$Q_{M^4} \colon H^2(M^4, \partial M^4; \mathbb{Z}) \times H^2(M^4, \partial M^4; \mathbb{Z}) \to \mathbb{Z}$$

defined by $Q_{M^4}(a, b) = \langle a \cup b, [M^4] \rangle = a \cdot b \in \mathbb{Z}$ is called the *intersection form* of M^4.

Recall that the cup product $\cup$ is the generalization of the integral of the wedge product in deRham cohomology. Furthermore, the pairing $\langle \,.\,,\,.\, \rangle$ between cohomology and homology is the integer coefficient replacement of the integral in deRham, real coefficient, cohomology. If M^4 is closed (so $\partial M^4 = \emptyset$), then by Poincaré duality Q_{M^4} is *unimodular*, that is, $det\ Q_{M^4} = \pm 1$. So for this case, we can represent Q_{M^4} by a symmetric unimodular matrix with integer number entries.

Remark 6.9.
Note that for this definition of Q_{M^4} we only need the topological structure of M^4. From linearity, $Q_{M^4}(a, b) = 0$ if a or b is a torsion element. Thus Q_{M^4} descends to a pairing mod torsion, hence by choosing a basis of $H_2(M^4; \mathbb{Z})/\text{Torsion}$, we can represent Q_{M^4} by a matrix. In case of a simply connected 4-manifold there is no torsion (by the universal coefficient theorem). The matrix M of Q_{M^4} transforms by a basis transformation C as CMC^T; we say that Q_{M^4} is represented by the matrix M if M is the matrix of Q_{M^4} in an appropriate basis. Note that the determinant $det\ M$ is independent of the choice of the basis over $\mathbb{Z}$; we sometimes denote it by $det\ Q_{M^4}$.

Another way to present the intersection form is to define elements of $H_2(M^4, \mathbb{Z})$ by embedded surfaces. Suppose, for simplicity, that $H_2(M, \mathbb{Z})$ is of rank 2 and that Σ_1, Σ_2 are surfaces representing the corresponding generators $[\Sigma_1], [\Sigma_2] \in H_2(M^4, \mathbb{Z})$. Then the intersection form can be computed as

$$Q_{M^4} = \begin{pmatrix} \Sigma_1 \cdot \Sigma_1 & \Sigma_1 \cdot \Sigma_2 \\ \Sigma_2 \cdot \Sigma_1 & \Sigma_2 \cdot \Sigma_2 \end{pmatrix}$$

where $\Sigma_1 \cdot \Sigma_2$ denotes the algebraic intersection number of Σ_1 and Σ_2.

As a simple example consider the 4-manifold $S^2 \times S^2$. Using the Künneth formula (see [Bredon (1993)]) for the second homology we have

$$H_2(S^2 \times S^2, \mathbb{Z}) = H_2(S^2, \mathbb{Z}) \oplus H_2(S^2, \mathbb{Z}) = \mathbb{Z} \oplus \mathbb{Z},$$

so that $H_2(S^2 \times S^2, \mathbb{Z})$ is generated by the two generators $\Sigma_1 = S^2 \times \{\text{point}\}$ and $\Sigma_2 = \{\text{point}\} \times S^2$. Obviously there are no self-intersections, i.e. $\Sigma_i \cdot \Sigma_i = 0$ for all $i = 1, 2$ but the spheres do meet at one point, so $\Sigma_1 \cdot \Sigma_2 = \Sigma_2 \cdot \Sigma_1 = 1$. Thus the matrix representing the intersection form $Q_{S^2 \times S^2}$ is

$$Q_{S^2 \times S^2} = H = \begin{pmatrix} 0 & 1 \\ 1 & 0 \end{pmatrix}.$$

Representations of the intersection form
Note that the cohomological definition of the intersection form via the cup product implies that we can omit the torsion. Recall the first Chern class $c_1 \in H^2(M^4, \mathbb{Z})$, of a complex line bundle over M^4 is defined as the curvature form of any holomorphic connection on the bundle. Furthermore, $\int_{M^4} c_1(L) \in \mathbb{Z}$. In fact every image of the map $H^2(M, \mathbb{Z}) \to H^2(M, \mathbb{R})$ is the Chern class of some line bundle. We can now relate the cohomology pairing to the intersection number approach by noting that the Poincaré dual $PD(c_1(L))$ is a non-trivial 2-manifold[13] embedded in M^4. An important result is that every embedded 2-manifold can be represented in this manner by an appropriate complex line bundle. Thus, given two complex line bundles L_1 and L_2 with corresponding surfaces Σ_1 and Σ_2, respectively, we obtain for intersections between Σ_1 and Σ_2

$$\Sigma_1 \cdot \Sigma_2 = \int_{M^4} c_1(L_1) \wedge c_1(L_2).$$

This formula is one of the key ingredients of the gauge theory approach to smooth 4-manifold theory known as Donaldson and Seiberg-Witten theory.

---
[13] Another construction of this 2-manifold is given by the zero set of a section in this line bundle L.

Some properties of the intersection form

Let us now look at some of the relationships between algebraic properties of the intersection form and topological properties of a 4-manifold, X, and its 3-manifold boundary, ∂X. A very useful and more detailed account of these properties can be found in [Gompf and Stipsicz (1999)], §1.2.

For oriented 4-manifolds, the intersection form can be regarded as an integer valued bilinear form on the second integer homology group:

$$Q_X : H_2(X, \mathbb{Z}) \times H_2(X, \mathbb{Z}) \to \mathbb{Z}.$$

We will say more about the algebraic properties of such integer quadratic forms in 6.7.2 below. Here note that the vector space on which the form Q_X acts is over the integers. So any change of basis must be by invertible integer basis. However, it is easy to see that any invertible integer matrix must be unimodular, that is, have determinant ± 1. So under any basis change described by the invertible matrix C,

$$Q_X \to Q'_X = C^T Q_X C,$$

with $|\det C|^2 = 1$, so $\det Q_X$ is independent of the basis chosen for $H_2(X, \mathbb{Z})$.

From the standard exact sequence for relative homology we have

$$... H_3(X, \partial X) \to H_2(\partial X) \to H_2(X) \xrightarrow{\pi} H_2(X, \partial X) \to H_1(\partial X) \to H_1(X) \to ... \quad (6.3)$$

Now assume that X is compact and simply connected, so $H_1(X) = 0$. By Poincaré-Lefschetz duality, we then also have $H^3(X, \partial X) = 0$, so (6.3) reduces to

$$...0 \to H_2(\partial X) \to H_2(X) \xrightarrow{\pi} H_2(X, \partial X) \to H_1(\partial X) \to 0 \to ... \quad (6.4)$$

Thus, π is an isomorphism if and only if

$$H_2(\partial X) = H_1(\partial X) = 0. \quad (6.5)$$

But of course (6.5) is precisely the condition for ∂X to be a homology sphere (or possibly disjoint union of such). Now return to the algebra of Q_X, noting that it defines a map, $L : H_2(X) \to H^2(X)$ by

$$L(u) \in H^2(X), \text{ by } L(u)(v) = Q_X(u, v). \quad (6.6)$$

Thus, L is an isomorphism if and only if Q_X is invertible, and thus unimodular. Also from Poincaré-Lefschetz duality we have the isomorphism

$$P : H^2(X) \to H_2(X, \partial X).$$

So, PL is an isomorphism, and thus so is π if and only if Q_X is unimodular. But as we saw from (6.4), ∂X is an homology sphere if and only if π is an isomorphism.

This brief discussion illustrates some of the interdependence between the algebraic properties of the intersection form and the topology of the underlying manifold.

We should note that not all homology 3-spheres are realized as boundaries of *smooth*, contractible, non-closed 4-manifolds (otherwise that would contradict the Donaldson theorem see chapter 8 theorem 8.11). For instance the Poincaré homology 3-sphere cannot be the boundary of such smooth, contractible 4-manifold (see [Fintushel and Stern (1985)] for further examples). Also note that by the definition of the intersection form we have $Q_{\overline{X}} = -Q_X$. Given two 4-manifolds X_1 and X_2 with the same boundary a homology sphere then the connected sum along this boundary leads to a new manifold $X = X_1 \cup_{\partial X_1 = \partial X_2} X_2$ with $Q_X = Q_{X_1} \oplus Q_{X_2}$. The reverse is also true (see [Freedman and Taylor (1977)]): the splitting of the intersection form leads to the topological splitting of the 4-manifold. Other properties of the intersection form directly connected with the topology will be reviewed at the end of this chapter.

6.7.2 Classification of quadratic forms and 4-manifolds

We now digress to survey the rather specialized algebraic issues involved in the classification of unimodular integral forms. Husemoller and Milnor have provided a full, detailed treatment [Husemoller and Milnor (1973)]. For a nice summary see the introductory chapter in [Gompf and Stipsicz (1999)]. We follow this with some examples, and close the chapter with the culminating theorem of Freedman relating these apparently purely algebraic questions to the existence and uniqueness of four dimensional topological manifolds.

Basic algebraic information about integral quadratic forms

Consider symmetric, bilinear integer-valued forms Q on the (finitely generated) free Abelian group A. Three classifying pieces of informations, **rank**, **signature** and **parity** are defined in the following way:

- The *rank* $rk(Q)$ of Q is the dimension of A.
- Extend and diagonalize Q over $A \otimes \mathbb{R}$. The number of positive/negative eigenvalues of Q is denoted by $b_2^{\pm}$. The difference $b_2^+ - b_2^-$ is the *signature* $\sigma(Q)$ of Q. Note that if Q is unimodular (i.e. $\det Q = \pm 1$), then $rk(Q) = b_2^+ + b_2^-$.
- The *parity* of Q is *even* if $Q(\alpha, \alpha) \equiv 0 \pmod 2$ for every $\alpha \in A$; Q is *odd* otherwise.

In addition note,

- Q is *positive (negative) definite* if $rk(Q) = \sigma(Q)$ ($rk(Q) = -\sigma(Q)$ resp.). Q is *indefinite* otherwise.
- The *direct sum* $Q = Q_1 \oplus Q_2$ of the forms Q_1 and Q_2 (given on A_1, A_2 respectively) is defined on $A_1 \oplus A_2$ in the following way. If $a, b \in A = A_1 \oplus A_2$ splits as $a = a_1 + a_2$ ($b = b_1 + b_2$) with $a_i, b_i \in A_i$, then $Q(a,b) = Q_1(a_1, b_1) + Q_2(a_2, b_2)$. If $k > 0$ then kQ denotes the k-fold sum $\oplus_k Q$; for negative k we take kQ to be $|k|(-Q)$; finally if $k = 0$ then the form kQ equals the zero form on the trivial group (represented by the empty matrix $\emptyset$) by definition. The form on $A = \mathbb{Z} \oplus \mathbb{Z}$ represented by the matrix $\begin{bmatrix} 0 & 1 \\ 1 & 0 \end{bmatrix}$ will be denoted by H.
- An element $x \in A$ is a *characteristic element* if $Q(\alpha, x) \equiv Q(\alpha, \alpha)$ (mod 2) for all $\alpha \in A$. Note that if Q is even, then $0 \in A$ is characteristic. An element $\alpha \in A$ is *primitive* if $\alpha = d\beta$ ($\beta \in A$, $d \in \mathbb{Z}$) implies that $d = \pm 1$. For any $x \in A$ there is a primitive element α such that $x = d\alpha$; $|d|$ is the *divisibility* of x.
- An integer valued matrix, C, has an (also integer valued) inverse if and only if it is **unimodular**, that is, $\det C = \pm 1$.

Canonical forms for integral matrices, Q, are obtained by integral changes of basis, that is by

$$Q \to C^T Q C,$$

for some unimodular integral C. Thus, although every Q can be diagonalized (since they are symmetric) over the reals, they may not be diagonalizable over $\mathbb{Z}$.

Classification of integral quadratic forms

For the rest of the section, consider only unimodular forms. First, note that indefinite forms are fully classified by the three algebraic numbers, rank, signature and parity.

Theorem 6.17. *If indefinite unimodular forms Q_1, Q_2 (defined on A_1, A_2 respectively) have the same rank, signature and parity, then they are equivalent.*

For a proof, see the Introductory chapter in [Gompf and Stipsicz (1999)]. Also note

Lemma 6.2. *If $x \in A$ is characteristic, then $Q(x,x) \equiv \sigma(Q)$ (mod 8); in particular if Q is even, then the signature $\sigma(Q)$ is divisible by 8.*

A Guide to the Classification of Manifolds   199

Thus every such a unimodular form is given by a matrix consisting of parts represented by 8×8 matrices.

Remark 6.10.
For a sketch of the proof of Lemma 6.2 note that if x is characteristic in (A, Q), then $x + e$ is characteristic in $(A \oplus \mathbb{Z}, Q \oplus \langle -1 \rangle)$, where e generates the $\mathbb{Z}$ summand. By Theorem 6.17, $Q' = Q \oplus \langle -1 \rangle \cong b_2^+ \langle 1 \rangle \oplus (b_2^- + 1)\langle -1 \rangle$, and a characteristic vector has odd components in this new basis. Since the square of an odd number is congruent to 1 modulo 8, we have that $Q(x, x) - 1 = Q'(x + e, x + e) \equiv b_2^+ - (b_2^- + 1) = \sigma(Q) - 1$ (mod 8). If Q is even, then 0 is a characteristic element, which proves $8|\sigma(Q)$.

Without further justification at this point, define the matrix E_8:

$$E_8 = \begin{bmatrix} 2 & 1 & 0 & 0 & 0 & 0 & 0 & 0 \\ 1 & 2 & 1 & 0 & 0 & 0 & 0 & 0 \\ 0 & 1 & 2 & 1 & 0 & 0 & 0 & 0 \\ 0 & 0 & 1 & 2 & 1 & 0 & 0 & 0 \\ 0 & 0 & 0 & 1 & 2 & 1 & 0 & 1 \\ 0 & 0 & 0 & 0 & 1 & 2 & 1 & 0 \\ 0 & 0 & 0 & 0 & 0 & 1 & 2 & 0 \\ 0 & 0 & 0 & 0 & 1 & 0 & 0 & 2 \end{bmatrix}.$$

The perceptive reader will recognize this as the matrix corresponding to the Dynkin-diagram of the exceptional Lie-algebra. Soon we will relate this to a smooth four manifold. Straightforward algebraic calculations show that this matrix has non-integer eigenvalues, and thus cannot be diagonalized over $\mathbb{Z}$, a fact of critical importance in our later smoothness studies. As the matrix of an intersection form Q on $\mathbb{Z}^8$, E_8 gives a positive definite, even, unimodular form with $\sigma(Q) = 8$. By a slight abuse of notation, from now on E_8 will denote that bilinear form. Recall that H is used for the form corresponding to the matrix $\begin{bmatrix} 0 & 1 \\ 1 & 0 \end{bmatrix}$. Using E_8 and H as building blocks, for every pair $(\sigma, rk) \in \mathbb{Z} \times \mathbb{N}$ with $8|\sigma$, $rk \geq |\sigma|$ and $rk \equiv \sigma$ (mod 2) one can construct an intersection form $Q = aE_8 \oplus bH$ with $\sigma = \sigma(Q)$ and $rk = rk(Q)$, with $a = \frac{\sigma}{8}$ and $b = \frac{rk-|\sigma|}{2}$. In fact,

Theorem 6.18. *Suppose that Q is an indefinite, unimodular form. If Q is odd, then in an appropriate basis it is isomorphic to $b_2^+ \langle 1 \rangle \oplus b_2^- \langle -1 \rangle$; if Q is even then it is isomorphic to $\frac{\sigma(Q)}{8} E_8 \oplus \frac{rk(Q) - |\sigma(Q)|}{2} H$. QED*

On the other hand, in the definite case there is no such nice description of all unimodular forms. For a given rank there are only finitely many definite symmetric unimodular forms (see [Husemoller and Milnor (1973)]);

this number, however, can be very large. (For example, there are more than 10^{50} definite forms of rank 40.) Finally we state all results of the classification in the following table 6.1.

type	definite form	indefinite form
type I	$\pm\overset{p}{\bigoplus}(1)$ and many others e.g. $E_8 \oplus (1)$ and $\Gamma^{4(2k+1)}$	$\overset{p}{\bigoplus}(1)\overset{q}{\bigoplus}(-1)$
type II	e.g. E_8, Γ^{16}, $E_8 \oplus E_8$ ... (rank can be divided by 8)	$H \oplus ... H \oplus E_8 \oplus ... \oplus E_8$

6.7.3 Some simple manifold constructs

We now review some elementary but still powerful techniques for constructing manifolds.

Algebraic subsets of the $\mathbb{R}^n$

This was perhaps the first source of non-trivial examples of manifolds. For example, the two sphere, S^2, is the subset of all tuples $(x, y, z) \in \mathbb{R}^3$ fulfilling the equation $x^2 + y^2 + z^2 = 1$. In four dimensions we have the sphere $S^4 = \{x \in \mathbb{R}^5 \mid ||x|| = 1\}$. Since $H_2(S^4; \mathbb{Z}) = 0$, the intersection form Q_{S^4} is identically zero. Generalizations of this approach lead to one of the main problems in algebraic geometry, determining whether or not the solution set of the polynomial equation $p(x_1, x_2, \ldots, x_n) = 0$ form a manifold. If there are singularities can they be "unfolded" to get a smooth manifold? Such questions are important in catastrophe or bifurcation theory as well as number theory.

Manifolds from equivalence classes and group actions

Start with a simple space like S^n or $\mathbb{R}^n$ and define a relation on this space. For example on $\mathbb{R}^2$ define the relation:

$$(x, y) \sim_e (u, v) \leftrightarrow x + 1 = u, y + 1 = v.$$

The set of equivalence classes $\mathbb{R}^2 / \sim_e$ is a manifold denoted by T^2, the torus (or doughnut). It is easy to construct the 4-dimensional analog T^4. Somewhat more complicated examples are the *complex projective spaces*. Given the obvious free action of $\mathbb{C}^* = \mathbb{C} \setminus \{0\}$ on $\mathbb{C}^{n+1} \setminus \{0\}$ (that is $\lambda(z_0,...,z_n) = (\lambda \cdot z_0, ..., \lambda \cdot z_n)$ for $\lambda \in \mathbb{C}^*$), take the quotient $\mathbb{C}P^n = (\mathbb{C}^{n+1} \setminus \{0\})/\mathbb{C}^*$. The resulting space is the n-dimensional complex projective space

$\mathbb{C}P^n$; $\mathbb{C}P^1 = S^2$ is the *complex projective line* and $\mathbb{CP}^2$ is the *complex projective plane*. Using $\mathbb{R}$ instead of $\mathbb{C}$ one defines the real projective spaces $\mathbb{R}P^n$. If $P \in \mathbb{C}P^n$ and $(z_0, ..., z_n) \in P$, then we can uniquely identify P by the ratios, the *homogeneous coordinates* $[z_0 : z_1 : ... : z_n]$. Note that $\mathbb{C}P^n$ can be covered by the *affine coordinate* charts $\psi_i \colon \mathbb{C}^n \to \mathbb{C}P^n$ ($i = 0, ..., n$) where $\psi_i(z_1, ..., z_n) = [z_1 : ... : z_i : 1 : z_{i+1} : ...z_n]$ provides the i-th coordinate map from $\mathbb{C}^n$ into $\mathbb{C}P^n$. As in the real case, it is easy to see that any two (distinct) points of $\mathbb{CP}^2$ lie on a unique projective line ($\mathbb{C}P^1$), and every two (distinct) projective lines in $\mathbb{CP}^2$ intersect each other in exactly one point. Since $H_2(\mathbb{CP}^2; \mathbb{Z}) \cong \mathbb{Z}$, and two generic lines intersect each other in a point with positive sign, $Q_{\mathbb{CP}^2} = \langle 1 \rangle$. Subsets of $\mathbb{C}P^n$ provide the basis for an important class of complex manifolds.

Complex algebraic surfaces So far S^4 and $S^2 \times S^2$ are the only examples of manifolds we discussed with even intersection form. Now consider the zero set of a *homogeneous* polynomial p of degree d i.e. polynomials[14] in the variables $\{z_0, ..., z_n\}$ on $\mathbb{C}P^n$ that is $p(\lambda z_0, ..., \lambda z_n) = \lambda^d p(z_0, ..., z_n)$.

Definition 6.7. If p is a homogeneous polynomial of degree d then the set $V_p = \{[z] \in \mathbb{C}P^n \mid p(z) = 0\}$ is called the *hypersurface* corresponding to the polynomial p. The complex submanifolds of $\mathbb{C}P^n$ are called *complex algebraic* manifolds.

It turns out that not every complex manifold can be embedded into $\mathbb{C}P^n$, so not all complex manifolds are algebraic. However, a very important class of them are algebraic. One such class is defined as follows

$$S_d = \{[z_0 : z_1 : z_2 : z_3] \in \mathbb{C}P^3 \mid \sum z_i^d = 0\} \subset \mathbb{C}P^3,$$

where d is a positive integer. For these the following theorem follows

Theorem 6.19. S_d *is a smooth, simply connected, complex surface. If d is odd, then Q_{S_d} is equivalent to $\lambda_d \langle 1 \rangle \oplus \mu_d \langle -1 \rangle$, where $\lambda_d = \frac{1}{3}(d^3 - 6d^2 + 11d - 3)$ and $\mu_d = \frac{1}{3}(d-1)(2d^2 - 4d + 3)$; if d is even, then Q_{S_d} is equivalent to $l_d(-E_8) \oplus m_d H$, where $l_d = \frac{1}{24}d(d^2 - 4)$ and $m_d = \frac{1}{3}(d^3 - 6d^2 + 11d - 3)$.*

---

[14] If the polynomial p vanishes at a point $z \in \mathbb{C}^{n+1} \setminus \{0\}$ then it vanishes on its entire equivalence class $[z] \in \mathbb{C}P^n$.

Again, the matrix E_8 defined by:

$$E_8 = \begin{bmatrix} 2 & 1 & 0 & 0 & 0 & 0 & 0 & 0 \\ 1 & 2 & 1 & 0 & 0 & 0 & 0 & 0 \\ 0 & 1 & 2 & 1 & 0 & 0 & 0 & 0 \\ 0 & 0 & 1 & 2 & 1 & 0 & 0 & 0 \\ 0 & 0 & 0 & 1 & 2 & 1 & 0 & 1 \\ 0 & 0 & 0 & 0 & 1 & 2 & 1 & 0 \\ 0 & 0 & 0 & 0 & 0 & 1 & 2 & 0 \\ 0 & 0 & 0 & 0 & 1 & 0 & 0 & 2 \end{bmatrix} \qquad (6.7)$$

mentioned above in the classification problem of quadratic forms (see section 6.7.2). It also describes the Dynkin diagram of the exceptional Lie-algebra E_8.

The case $d = 4$ provides one of several possible constructions of a particular simply connected complex surface of major importance in the first discoveries of exotic $\mathbb{R}^4$, the *K3-surface*. Algebraic geometric methods can be used to show that different constructions lead to diffeomorphic $K3$-surfaces, so from the differential topological point of view, we can call S_4 the $K3$-surface. From the previous formula $Q_{S_4} = 2(-E_8) \oplus 3H$. We remind the reader not to confuse these S_d with spheres, S^d.

More examples of simply connected 4-manifolds come from generalizations of the S_d. Take homogeneous polynomials p_i of degree d_i in $n+1$ variables ($i = 1, ..., n-2$). Note that each p_i defines a hypersurface in $\mathbb{C}P^n$. Assume that

$$S(d_1, ..., d_{n-2}) = \{P \in \mathbb{C}P^n \mid p_i(P) = 0; \ i = 1, ..., n-2\}$$

is a smooth submanifold of complex dimension 2 in $\mathbb{C}P^n$. If so, S is called a *complete intersection* surface of multidegree $(d_1, ..., d_{n-2})$. Furthermore, it can be shown that the diffeomorphism type of $S(d_1, ..., d_{n-2})$ depends only on the multidegree $(d_1, ..., d_{n-2})$. Without loss of generality we can assume that each $d_i \geq 2$. Furthermore by the so-called Lefschetz Hyperplane Theorem, the surfaces are simply-connected i.e. $\pi_1(S(d_1, ..., d_{n-2})) = 1$. For further information about complex surfaces we refer to the standard book [Barth *et al.* (1984)].

Gluing and sewing of spaces

Here, we start also with a simple object like a n-ball D^n with boundary $\partial D^n = S^{n-1}$ and glue a k-handle along the boundary to get a non-trivial space. This general scheme using Kirby calculus can be used to represent all smooth 4-manifolds.

Note that for simply-connected, closed 4-manifold, one needs only 1- and 2-handles. By the operation of the connected sum we can glue simple pieces together to get much more complicated spaces. For completeness we define the *connected sum* of two connected, oriented n-dimensional manifolds X_1 and X_2 by

Definition 6.8. For $i = 1, 2$ let $D_i^n \subset X_i$ be an embedded disk, and let $\varphi \colon D_1^n \to D_2^n$ be an orientation-reversing diffeomorphism. The smooth manifold $(X_1 \setminus \text{int } D_1) \cup_{\varphi | \partial D_1} (X_2 \setminus \text{int } D_2)$ is called the *connected sum* $X_1 \# X_2$ of X_1 and X_2; it does not depend on the choices of D_i^n or φ (since any two orientation-preserving embeddings of a disk are smoothly isotopic). In particular, $\# mX$ denotes the manifold we get by the connected sum of m ($m \geq 0$) copies of the same manifold X. (If $m = 0$, then $\# mX = S^n$ by definition.)

The iterated application of the connected sum operation for $\mathbb{CP}^2$, $\overline{\mathbb{CP}}^2$ and $S^2 \times S^2$ gives other examples of simply connected 4-manifolds. Applying this result, one can easily prove that the intersection form of $n\mathbb{CP}^2 \# m\overline{\mathbb{CP}}^2$ is equivalent to $n\langle 1 \rangle \oplus m\langle -1 \rangle$ ($n, m \geq 0$), so these intersection forms can be realized by smooth manifolds.

Of course, in physics the idea of cutting and gluing has been an integral part of general relativity from the early days in relation to Einstein-Rosen bridges, wormholes, etc. See, for example, [Visser (1996)].

6.8 Freedman's Classification

We close this chapter with brief statement of a central *topological* theorem in the early studies of exotic $\mathbb{R}^4$'s. For the simply connected case $\pi_1(X) = 0$, the first and the third homologies and cohomologies vanish (by Poincaré duality), and $H_2(X;\mathbb{Z}) \cong H^2(X;\mathbb{Z}) \cong Hom(H_2(X;\mathbb{Z}),\mathbb{Z})$ has no torsion, so Q_X contains all the (co)homological information about X. For topological manifolds, Q_X is more or less a complete set of invariants. More precisely,

Theorem 6.20. (Freedman, [Freedman (1982); Freedman and Quinn (1990)]) *For every unimodular symmetric bilinear form Q there exists a simply connected closed topological 4-manifold X such that $Q_X \cong Q$. If Q is even, this manifold is unique (up to homeomorphism). If Q is odd, there are exactly two different homeomorphism types of manifolds with the given* *intersection form.*

Roughly speaking, simply connected topological manifolds are classified by their intersection forms. Thus in Table 6.1 every unimodular form is realized by the intersection form of one 4-manifold. Section 8.2 will contain a discussion of some of the contributions of Freedman relevant to this result. Section 8.3 will address the smooth form of these issues for four manifolds and intersection forms, and discuss their importance for exotic smoothness discoveries.

Chapter 7

Early Exotic Manifolds

7.1 Introduction

The successful classification of principal bundles and sphere bundles in the 50's, led to a plethora of interesting results, including a discovery by Milnor of topological seven spheres possessing non-standard, or **exotic** smoothness structures. This discovery was first published in his landmark paper [Milnor (1956c)]. Until this publication it was generally assumed that, at least for topologically simple spaces, topology would uniquely determine smoothness. Thus, the announcement of the first six exotic structures on a 7-sphere was a great shock for the mathematical community[1]. Fortunately, Milnor has recently [Milnor (2000)] given a brief but very informative retrospective review of this paper and its historical context.

Milnor's original paper was succeeded by further results relating topology to smoothness. Notable was the landmark theorem of Smale [Smale (1962)] now known as h-cobordism theorem, suggesting that smoothing (introducing differential structures on topological manifolds) is much simpler than expected for the higher-dimensional cases, ≥ 5. Kervaire and Milnor, [Kervaire and Milnor (1963)] were able to classify all possible differential structures on higher dimensional spheres, and Kirby and Siebenmann [Kirby and Siebenmann (1977)] used Milnor's notion of microbundle to classify both the smoothness and piecewise linear structures on manifolds of dimension greater than 4.

We begin our study of these early exotic structures by reviewing Milnor's original work, defining the first exotic 7-spheres as 3-sphere bundles over the 4-sphere. We compare this construction to the compactified Eu-

---

[1] In 1962, Milnor won the Fields medal, the most important award in mathematics, for this result.

clidean Yang-Mills models from quantum theory. Next, we briefly mention some consequences of exotic smoothness for **geometry** and **spectra** of the differential operators. After a review of the classifying work of Kervaire, Milnor and Siebenmann we close this chapter with the proof that the higher-dimensional Euclidean spaces ($\mathbb{R}^n$ with $n \geq 5$) have **no** exotic differential structures.

7.2 Some Physical Background: Yang-Mills

In quantum field theory it has become common to consider spacetime models changed to remove some of the mathematically inconvenient features that apparently occur in physically "reasonable" models. In particular the indefinite Minkowski metric is replaced by the positive definite Euclidean one and spacetime is compactified. We will not attempt to provide justification for these changes from the physics viewpoint, but simply point out that their use is widespread in quantum field theory. So, let us begin with the mathematically standard one-point compactification of the basic spacetime model $\mathbb{R}^4 \cup \{\infty\} = S^4$. Many studies in quantum field theory thus start with S^4 as the spacetime model. Furthermore, to make sense out of the path integral formulation of quantum field theory, it is customary to replace the indefinite Minkowski metric with the Euclidean one. Of special interest to us here are the **Yang-Mills models** proposing force fields constructed as connection fields of an $SU(2)$ gauge symmetry. Formally then we look at $SU(2)$ principal bundles over this S^4,

$$\begin{array}{c} S^3 = SU(2) \to M^7 \\ \downarrow \pi \\ S^4 \end{array} \qquad (7.1)$$

This particular symmetry group was inspired by the "isotopic" spin symmetry recognized in low energy nuclear physics of the 50's and 60's, motivated by the apparently identical action of the nuclear (**strong**) force on neutrons as on protons. Thus, at that time the "elementary" particle in the nucleus was the **nucleon,** which could exist in two-dimensional isotopic spin state, $\mathbb{C}^2$, spanned by the two orthogonal base states, neutron and proton. Of course, this isotopic spin group was soon replaced by $SU(3) \supset SU(2)$, the original quark model of Gell-Mann et al., and then a plethora of generalizations, but the Yang-Mills formalism is important as the first non-Abelian physical gauge theory.

As briefly mentioned in an earlier discussion, the physics of force fields starts with a connection on such a principal bundle and imposes field equations through a variational principle involving some form of curvature invariants. For this four dimensional case, the second Chern class, the trace of the "square" of the $SU(2)$ curvature, $Tr(\Omega \wedge \Omega)$, provides the natural Lagrangian density. This interplay between the topology, geometry and index theory of differential equations has proved to be very productive for contemporary mathematics, especially in the work of Donaldson and others leading to exotic $\mathbb{R}^4$'s as we discuss in chapters 5 and 8. For now, we simply recall the importance of $SU(2)$ bundles over S^4 in certain physical models.

7.3 Mathematical Background: Sphere Bundles

Let us start with a general discussion of bundles over S^4. The basic one is of course the bundle of frames, an $SO(4)$ principal bundle, a non-trivial bundle. The standard two hemisphere chart decomposition of the base space sphere S^4 is $S^4 = H_+ \cup H_-$ with $H_+ \cap H_- = \mathbb{R}^1 \times S^3$. Since this latter space is homotopically equivalent to S^3, a principal bundle is defined by the transition map $g_{+-} : H_+ \cap H_- \to SO(4)$ with homotopy class in $\pi_3(SO(4))$. First, recall that Spin(4), the two-fold covering of $SO(4)$, is homeomorphic to $SU(2) \times SU(2)$. Since $SU(2) = S^3$ and Spin(4) and $SO(4)$ have the same π_n for $n > 1$, we find that every element of $\pi_3(SO(4)) = \mathbb{Z} \oplus \mathbb{Z}$ can be described by a pair of maps $S^3 \to S^3$ classified by $\pi_3(S^3) = \mathbb{Z}$. For Milnor's argument, we will need a reduction of the bundle group, $SO(4) \to \text{Spin}(4) \to SU(2)$ given by a relation based on two numbers $(i,j) \in \pi_3(SO(4))$ characterizing the topological class of the bundle.

In physics, using a natural generalization of the Yang-Mills formalism, we can interpret the sections of such a bundle as particles with left- and right-handed isospin. Milnor's discovery of exotic smoothness on some of these bundles then leads us to speculate on possible physical significance of such "exotic" Yang-Mills models.

Before proceeding with the details of Milnor's results, let us review the classical Hopf bundle presentation of S^7, starting with the projective representation of spheres. If u represents a Euclidean vector in $\mathbb{R}^n$, and $(t,u) \in \mathbf{R}^{n+1}$, with $t \in \mathbb{R}^1$, then S^n is the locus: $t^2 + |u|^2 = 1$. Now define two maps, "projections"

$$p_{\pm}(u) = (\pm\frac{1-|u|^2}{1+|u|^2}, \frac{2u}{1+|u|^2}) \in S^n. \tag{7.2}$$

Clearly,

$$p_+(u) = p_-(u/|u|^2).$$

The hemispheres $H_\pm$ discussed in the $n = 4$ case above are clearly the images of $p_\pm$. Now use this formalism to define the base space, S^4, replacing $\mathbb{R}^n$ by $\mathbb{H}$, the space of quaternions, so that u now has multiplicative algebraic properties. Finally, let v represent a unit quaternion, thus a general element of S^3. We can now explicitly describe S^7 as a Hopf bundle using the functions in (7.2) as projection maps. The two coordinate patches are $H_\pm$, and the local presentations are, on $H_+ \times S^3$,

$$\phi_+(p_+(u), v) = \frac{(v, uv)}{\sqrt{1+|u|^2}} \in S^7 \subset \mathbb{R}^8, \tag{7.3}$$

and, on $H_- \times S^3$,

$$\phi_-(p_-(u'), v') = \frac{(\bar{u}'v', v')}{\sqrt{1+|u'|^2}} \in S^7 \subset \mathbb{R}^8, \tag{7.4}$$

where u and u' correspond to the same point in S^4, so $p_+(u) = p_-(u')$, or $u' = u/|u|^2$. The fact that the right side of these two equations represent the same point in S^7 then is equivalent to

$$v' = uv/|u| = g_{+-}(u) \cdot v, \tag{7.5}$$

and the transition functions are

$$g_{+-}(u) = L_{u/|u|} \in SO(4)), \quad u \in H_+ \cap H_-, \tag{7.6}$$

as required.

7.4 Milnor's Exotic Bundles

Milnor's original description uses a generalization of the Hopf bundle formulation of S^3 bundles over S^4 described by the quaternion algebra above. Here we follow the spirit of Milnor's argument, but modify the order somewhat. We begin with a summary of the main points:

(1) A closed smooth seven manifold, M_k^7, is defined as the total space of an S^3 bundle over S^4, for each odd integer k. This bundle is denoted by ξ_{hj}, where $h + j = 1, h - j = k = 2h - 1 =$ odd.

(2) Using Morse functions, we find that M_k^7 is **homeomorphic** to S^7.
(3) M_k^7 is the boundary of a compact topological manifold, B^8.[2]
(4) Milnor defines an integer (mod 7), $\lambda(M^7)$, where $M^7 = \partial B^8$, using the topology of B^8. However, its value is independent of the actual choice of B^8, with $\partial B^8 = M^7$.
(5) For the total space of the bundle ξ_{hj}, $h-j = k$, $\lambda(M_k^7) = k^2 - 1 \mod 7$.
(6) If $\lambda(M^7) \neq 0$, B^8 **cannot** be homeomorphic to the Euclidean eight-ball, or disk.
(7) If M_k^7 were diffeomorphic to standard S^7, it would be the boundary of the standard eight-ball.
(8) Therefore if $k^2 - 1 \neq 0 \mod 7$, then M_k^7 is homeomorphic, but **not diffeomorphic**, to the standard S^7. Together with (2) this establishes the fact that such a M_k^7 is an **exotic sphere**.

Milnor's striking result can then be summarized by his Theorem 3, [Milnor (1956c)], the item (8) above, which we quote here

Theorem 7.1. *For $k^2 \not\equiv 1 \mod 7$ M_k^7 is homeomorphic but not diffeomorphic to S^7.*

Now we fill in a few of the technical details. For each pair of integers, (h, j) where $h + j = 1$, we define a generalization of the transition function, (7.6), by $g_{hj}(u) \cdot v = u^h v u^j / |u|^3$. Since h and j are not independent, we can characterize the pair by $k = h - j$. Denote the sphere bundle corresponding to this transition function by ξ_{hj}. Explicitly, take two copies of $\mathbb{R}^4 \times S^3$ and identify the subsets $(\mathbb{R}^4 - \{0\}) \times S^3$ under the diffeomorphism

$$(u, v) \longrightarrow (u', v') = \left(\frac{u}{||u||^2}, \frac{u^h v u^j}{||u||} \right) \quad (7.7)$$

using quaternion multiplication. This presents M_k^7 as a smooth manifold, point (1).

Referring back to the **standard** S^7 Hopf construction, (7.3), (7.4), we see that the first coordinate in the enveloping $\mathbb{R}^8$ is

$$t = \mathfrak{Re}(\frac{v}{\sqrt{1 + |u|^2}}), \text{ in } H_+, \quad (7.8)$$

and

$$t = \mathfrak{Re}(\frac{\bar{u}'v'}{\sqrt{1 + |u'|^2}}), \text{ in } H_-. \quad (7.9)$$

---
[2] Please note that B^8 will not necessarily be a ball.
[3] The standard Hopf bundle presentation of the standard S^7 above clearly corresponds to $h = 1$.

Using this as a "height" function, with precisely two critical points (at the poles), Morse theory confirms that this is a topological S^7. We discuss this in more detail in the 7.5. Now, Milnor uses the same function, but with the non-standard transition function defined by (h,j). Again, it has only two non-degenerate critical points, from which Morse theory establishes that M_k^7 is **topological** S^7, that is, point (2).

From the definition and calculation of the cobordism ring due to Thom [Thom (1954)] we know that every closed 7-manifold is the boundary of a compact 8-manifold, establishing point (3).

The next step is the most technical. First, since each fiber, S^3 of ξ_{hj} is the boundary of a standard ball, B^4, the bundle can be embedded in a 4-cell bundle. The total space of this bundle is one of the B^8 in point (3). Milnor begins with some basic mathematical facts about oriented closed seven-manifolds, M^7, which are boundaries, $M^7 = \partial B^8$. Assume Milnor's **hypothesis ***,

$$* : \quad H^3(M^7) = H^4(M^7) = 0. \tag{7.10}$$

From the exact sequences of algebraic topology this implies that

$$i^* : H^4(B^8, M^7) \to H^4(B^8), \tag{7.11}$$

is an isomorphism. Next, let $\mu \in H_7(M^7)$ be an orientation. There then exists $\nu \in H_8(B^8, M^7)$, $\mu = \partial \nu$, which defines the orientation of M^7 in B^8. Because $8 = 2 \times 4$ we can then define a quadratic form over the cohomology group, $H^4(B^8, M^7)$ by

$$|\alpha|^2 = <\nu, \alpha^2>, \quad \alpha \in H^4(B^8, M^7). \tag{7.12}$$

In deRham cohomology, α would be a four-form and $\alpha^2 = \alpha \wedge \alpha$. From physics we are accustomed to defining the **signature** of a quadratic form as the sum of ± 1, corresponding to the signs of the diagonal elements. Milnor uses the term **index** for the signature of the form defined in (7.12) and labels it $\tau(B^8)$.

Now let $p_1(B^8) \in H^4(B^8)$ be the first Pontrjagin class [4] of the tangent bundle to B^8. From the isomorphism in (7.11), let $\beta = (i^*)^{-1}(p_1) \in H^4(B^8, M^7)$ and the function q is defined by

$$q(B^8) = <\nu, \beta^2>. \tag{7.13}$$

We now state Milnor's first important theorem:

THEOREM 1: *The residue class of $2q(B^8) - \tau(B^8)$ modulo 7 does not depend on the choice of the manifold B^8.*

---
[4] See the discussion in section 5.3.

That is, this number depends only on $\partial B^8 = M^7$ and is denoted by $\lambda(M^7)$. We will not repeat Milnor's proof of this theorem which calls on the Hirzebruch index theorem and other results. This establishes point (4).

Next, in two Lemmas, Milnor is able to compute $2q - \tau$ for the total space of the bundle, ξ_{hj}, to be $8k^2 - 1$, so that

$$\lambda(M_k^7) = 8k^2 - 1 \equiv k^2 - 1 \mod 7, \quad (7.14)$$

establishing point (5).

In particular, if $H^4(B^8, \mathbb{R}) = 0$, then the quadratic forms defining $\tau(B^8)$ and $q(B^8)$, and thus $\lambda(M^7)$ are identically zero. This is point (6). On the other hand, if $M^7 = S^7$, the "standard" sphere which by definition can be smoothly embedded in $\mathbb{R}^8$ as the boundary of the standard ball. Since this ball is topologically trivial point (6) then results in point (7). Finally, point (8) summarizes these points and establishes Milnor's main theorem.

Of course, what this establishes is simply that these M_k^7 are not diffeomorphic to the standard S^7. However, at this point we do not know whether any of them are diffeomorphic to each other, or precisely how many exotic spheres there are. The answer to these questions is found in the work of Kervaire and Milnor [Kervaire and Milnor (1963)], which we review in section 7.8.

Finally we mention that Brieskorn has been able to construct certain exotic spheres by algebraic geometric means. The Brieskorn spheres are defined by

$$\Sigma(a_1, \ldots, a_n) = \{(z_1, \ldots, z_n) \in \mathbb{C}^n \mid z_1^{a_1} + \cdots + z_n^{a_n} = 0\} \cap S^{2n-1} \quad (7.15)$$

where $(a_1, \ldots, a_n)$ is an arbitrary n-tuple of integers $a_i > 1$. For $n = 3$ these manifolds are homology 3-spheres and $\Sigma(2, 3, 5)$ represents the Poincaré sphere. In higher dimensions $n > 3$ with $n \equiv 1 \mod 4$ these manifolds are homeomorphic to S^{2n-3} but in general not diffeomorphic. In particular, all Milnor spheres, M_k^7 can be obtained from this construction for $n = 5$. We refer the reader to the original paper [Brieskorn (1970)].

7.5 Coordinate Patch Presentation

In physical applications, explicit coordinate patch presentations are often required. Of all of the exotic manifolds the Milnor spheres seem most likely to be explicitly coordinatized. In this section we will look at this problem by reducing Milnor's bundle presentation of M_k^7 to one explicitly involving

a gluing of the S^6 boundaries of two disks. Consider the standard seven sphere as a subset of $\mathbb{R}^8$, defined by
$$(t, \mathbf{w}), \text{ with } t^2 + |\mathbf{w}|^2 = 1, \ \mathbf{w} \in \mathbb{R}^7. \tag{7.16}$$
Define $D^7_\pm$ as the two polar caps, up to the equator. Thus, $D^7_\pm$ are defined by the conditions $\pm t \geq 0$. Clearly the embedding identifies each with the standard D^7. The corresponding boundaries are two (orientation reversed) copies of the equatorial six-sphere. This provides a model for the standard, smooth
$$S^7 = D^7 \cup_1 D^7, \tag{7.17}$$
two standard disks patched together by the identity map on their bounding spheres. Alternatively, the boundaries of the two D^7's can be joined by a collar, $[0,1] \times S^6$,
$$S^7 = D^7 \cup_1 \left([0,1] \times S^6\right) \cup_1 D^7. \tag{7.18}$$
Return to the Hopf bundle presentation of Milnor's M^7_k, with the height function, t, defined in (7.8) and (7.9). It is easy to see that the critical points of t occur at $u = 0, v = \pm 1$, both in $u' \neq 0$, and that these are non-degenerate and of positive index. First, we follow the argument of Reeb[Reeb (1952)] as outlined in Hirsch[Hirsch (1976)], pages154,155. Use the height function, t, to divide M^7_k into three pieces,
$$M^7_k = t^{-1}[1,a] \cup t^{-1}[a,b] \cup t^{-1}[b,-1], \text{ with } -1 < a < b < 1, \tag{7.19}$$
where the Morse function argument, discussed in more detail in Chapter 6.1.1, shows that the first part is a disk, D^7_+, the second part, C, is diffeomorphic to $[0,1] \times S^6$, and the last is another D^7_-. Now we look at the relationship of the two sets defined in (7.18) and (7.19). Let F be the diffeomorphism of $[0,1] \times S^6$ onto C, with $F[0,x] = x$, the identity on the boundary of the top cap of S^7 in (7.18), and $F[1,x] = f(x)$, a diffeomorphism of the boundary of the bottom cap with S^6. The only remaining task is to extend this map from the boundary of the bottom cap to the full disk. For this, we can use "Alexander's trick,"[5] defining it on the full disk D^7 from the boundary sphere map, f, by
$$F(\vec{x}) = \begin{cases} |\vec{x}| f(\vec{x}/|\vec{x}|), & \vec{x} \neq 0, \\ 0, & \vec{x} = 0, \end{cases} \tag{7.20}$$
where we use the usual vector notation within D^7. Now, the important fact is that the *homeomorphism,* (7.20), cannot be a *diffeomorphism,* else M^7_k

---
[5]An alternative procedure is given by Milnor[Milnor (1965a)], pages 109,110.

would be diffeomorphic to S^7. Clearly any breakdown of the smoothness is "concentrated" at the origin. On the other hand, if f were (smoothly) isotopic to the identity, it could be isotoped to this identity using $|x|$ as a re-scaled parameter before it gets the zero. In this case, F, defined in (7.20) would actually be a diffeomorphism. So, defining

$$\Sigma_f = D^7_+ \cup_f D^7_-$$

for some diffeomorphism, $f : S^6 \to S^6$, then Milnor has shown:

Fact: *The space Σ_f is diffeomorphic to the standard smooth S^6 if and only if f is smoothly isotopic to the identity.*

The very enlightening historical review by Milnor, [Milnor (2000)], of his path in the late 1950's to the exotic seven-spheres closes with a review of the critical relationship between the smoothness structures of seven-spheres and the smooth isotopy classes of maps between the boundaries of the two coordinate patches, $S^6 \to S^6$, or equivalently, $\pi_6(S^6)$.

We close this section by noting very interesting recent constructive results obtained by Durán et al., [Durán (2001)],[Durán et al. (2004)],[Abresch et al. (2005)]. These papers provide an explicit statement, using algebraic quaternion methods, of just such an S^6 diffeomorphism not isotopic to the identity which constructs Milnor's M_3^7, with $(h,j) = (2,-1)$. These purely algebraic quaternion techniques appear to offer insights to further exotic constructions. They also describe natural geometries, as we discuss in the next section.

7.6 Geometrical Consequences

Of course, an essential component of physical models is geometry and the possibilities of interesting, non-standard, geometric characteristics for exotic manifolds is one of the main lures of exotic smoothness. Certain well known relationships between curvature and homology hold for any smooth manifold and thus can be used in studying exotic ones. For example, the basic theorem of Gauss-Bonnet, and its generalization by Chern and other developments are described in some detail in [Gallot et al. (1980)], especially in their chapter III. Milnor's book on Morse theory, [Milnor (1963)], also contains an excellent survey of the topic. For the case of Lorentzian signature metrics the indefinite character of the norm requires some changes. In fact, Beem et al have devoted an entire book to the subject, [Beem et al. (1996)].

In many cases the presentation of a geometry, i.e., metric and/or connection, is given in terms of coordinate patches, which are not explicitly available to us for exotic spheres. However, Gromoll and Meyer [Gromoll and Meyer (1974)] have been able to use a group theoretical representation of the Milnor spheres M_k^7 to obtain explicit geometries on these spaces. In particular, they construct a metric with nonnegative sectional curvature. Today we know (see [Grove and Ziller (1999)]) that 15 of all 27 exotic spheres in dimension 7 admit metrics of non-negative sectional curvature[6]. Work of Hitchin [Hitchin (1974)] demonstrates the existence metrics on exotic spheres with non-positive scalar curvature. Another class of problems relates exotic spheres with positive Ricci curvature. The paper [Wraith (1998)] shows that every exotic sphere has a metric of positive Ricci curvature. For additional recent work see [Abresch et al. (2005)].

We now present a brief review of the work of Gromoll and Meyer. Let $Sp(n)$ denote the symplectic group of $n \times n$ quaternion matrices. Identify the field of quaternions with the $\mathbb{R}^4$. Consider the action σ of $Sp(1) \times Sp(1) \simeq SU(2) \times SU(2)$ on $Sp(2)$ given by

$$\sigma : (Sp(1) \times Sp(1)) \times Sp(2) \longrightarrow Sp(2)$$
$$(q_1 \times q_2, Q) \longmapsto \begin{pmatrix} q_1 & 0 \\ 0 & q_1 \end{pmatrix} Q \begin{pmatrix} \bar{q}_2 & 0 \\ 0 & 1 \end{pmatrix}$$

where $\bar{q}_2$ denotes the conjugate of q_2. The action σ is free, i.e. if $\sigma(q_1 \times q_2, Q) = Q$ for all Q, then $q_1 \times q_2$ must be the unit element $e \times e$ (or unit quaternion 1). Then we can define the equivalence relation:

$$Q_1 \sim Q_2 \iff \exists q_1, q_2 \in Sp(1) \quad \sigma(q_1 \times q_2, Q_1) = Q_2$$

for all $Q_1, Q_2 \in Sp(2)$. The quotient manifold $Sp(2)/\sim$ (sometimes denoted by $Sp(2)/Sp(1) \times Sp(1)$ is diffeomorphic to S^4 (see [Gromoll and Meyer (1974)]). The diagonal $\triangle = \{(q,q) \in Sp(1) \times Sp(1)\} \subset Sp(1) \times Sp(1)$ also acts freely on $Sp(2)$. The quotient manifold $\Sigma^7 = Sp(2)/\triangle$ is a 7-manifold admitting the structure of an S^3-bundle over S^4, providing an alternative route to Milnor's construction of M_k^7 [Milnor (1956c)]. The following diagram reflects this bundle structure.

$$Sp(1) \times Sp(1)/\triangle \simeq Sp(1) = S^3$$
$$\downarrow$$
$$Sp(2)/\triangle = \Sigma^7$$
$$\downarrow$$
$$Sp(2)/Sp(1) \times Sp(1) \simeq S^4$$

---

[6]These are exactly all Milnor spheres i.e. each sphere can be written as an S^3 bundle over S^4.

In Milnor's notation this bundle corresponds to $\xi_{2,-1}$ as shown by Gromoll and Meyer [Gromoll and Meyer (1974)]. Thus Σ^7 is diffeomorphic to the exotic sphere M_3^7.

On the geometric side, Gromoll and Meyer were able to provide an explicit expression for the sectional curvature of embedded surfaces in terms of their $Sp(2)$ group presentation of the sphere. Among other results is a positive definite Ricci curvature. For further results we have the work of Kreck and Stolz[Kreck and Stolz (1988)], showing that there are 7-dimensional manifolds with the maximum number of 28 exotic smooth structures each of which admits an Einstein metric with positive scalar curvature.

7.7 Eells-Kuiper Smoothness Invariant

The work of Eells and Kuiper [Eells and Kuiper (1962)], and others, has led to some important relations between the spectral properties of differential operators on certain topological spheres and their smoothness structures. The important theorem is

Theorem 7.2. *Suppose $k = 7, 11$ and M_1, M_2 are two topological k-spheres with codimension one metrics. If M_1, M_2 are isospectral then they are diffeomorphic.*

Here the phrase "codimension one metric" means that the manifolds are endowed with a metric geometry from an immersion in an $\mathbb{R}^{k+1}$ of one higher dimension. "Isospectral" means that the Hodge operators (generalized Laplacians) of these metrics have the same eigenvalue spectrum. This result comes from relating the Atiyah-Patodi-Singer spectral invariant [Atiyah et al. (1973)] to a differential topological invariant defined by Eells and Kuiper. Although we will not explore this result further at this time, we mention it here because it provides another tool to investigate our fundamental question of whether or not two manifolds are diffeomorphic.

7.8 Higher-dimensional Exotic Manifolds(Spheres)

In this section we will review the work of Kervaire and Milnor [Kervaire and Milnor (1963)] which together with the work of Smale [Smale (1961)] gives an (almost) complete classification of possible non-diffeomorphic smoothness structures that can be put on topological spheres by using the h-cobordism theorem stated in section 6.3.1. The number of such structures for each dimension up to 18 is given in Table 7.1. The argument proceeds as follows:

(1) First it is shown that every manifold homotopic to a sphere, called a "homotopy sphere," is actually homeomorphic to it, for dimension $n \geq 6$ using the (topological) h-cobordism theorem.
(2) A group action using connected sums $\#$ is established on the set of homotopy spheres of dimension n, defining the group Θ_n.
(3) Define the group of all homotopy spheres bP_{n+1} that bound parallelizable manifolds and show that Θ_n/bP_{n+1} is finite.
(4) The group bP_{n+1} is finite too. The order of the group is divided into two cases with respect to n and can be calculated.
(5) With much extra work, the order of Θ_n can be calculated. The factor Θ_n/bP_{n+1} is determined by the stable homotopy groups of spheres.
(6) The smooth h-cobordism and Alexander's trick shows that there is precisely one smoothness structure for each element of Θ_n.
(7) Finally, this implies that the number of smooth manifolds of dimension n which are *homeomorphic* to the standard sphere, but not *diffeomorphic* to any other is the number of elements in Θ_n.

Starting with (1), let us denote a general homotopy sphere by the symbol Σ^n. First note that every homotopy sphere can be decomposed $\Sigma^n \simeq D^n \cup_f D^n$, a union of two balls glued by the diffeomorphism $f : S^{n-1} \to S^{n-1}$. To see this, cut out two small disks from Σ^n, viewed as the "polar caps" of the homotopy sphere. What remains is a manifold W^n with the homotopy type of the cylinder and two boundary components each homeomorphic to S^{n-1}. Since $n - 1 \geq 5$ and S^{n-1} is simply connected, the hypotheses of the h-cobordism theorem 6.9 are satisfied and there is a diffeomorphism from W to $S^{n-1} \times [0, 1]$ which is the identity on the boundary component corresponding to the south polar cap, and is the diffeomorphism f on the boundary of the other cap. This provides the decomposition of the homotopy sphere by $\Sigma^n \simeq D^n \cup_f D^n$.

The diffeomorphism equivalence class of this homotopy sphere depends only on the isotopy class of f, since an isotopy of the f's gives an h-cobordism of the corresponding homotopy spheres and we can apply the h-cobordism theorem again. Conversely, if there is an orientation-preserving diffeomorphism from $D^n \cup_f D^n$ to the standard sphere, it is not hard to see that there must be an isotopy from f along the cylinder to the identity. Thus the smooth homotopy spheres are classified by the isotopy classes of the equator, that is, by the elements $\pi_0(DIFF(S^{n-1}))$. In the topological case, any self-homeomorphism f of S^{n-1} extends via "Alexander's trick" described in (7.20) above to a self-homeomorphism of D^n itself. This yields a homeomorphism from $D^n \cup_f D^n$ to S^n, proving the generalized Poincaré conjecture. *Thus every homotopy sphere ($n \geq 5$) is topologically equivalent to the standard sphere*[Smale (1961)]. However, Alexander's trick is only in the TOP category, so we do not yet know about smooth equivalence of Σ^n to S^n. This discussion leads to a classification of the smooth structures on the sphere S^n for $n \geq 6$ in terms of homotopy spheres.

The fundamental work of Kervaire and Milnor [Kervaire and Milnor (1963)] led to the result that there is only a *finite* number of exotic spheres in higher dimensions. The key ingredient in this approach is the relationship between the exotic spheres and the (stable) homotopy groups of spheres. Thus a problem in differential topology is converted to a problem in algebraic topology for higher dimensions (≥ 5). The proof can be described in steps (2)-(4) indicated above.

(2) provides an abelian group structure for the homotopy spheres Θ_n. Let M_1 and M_2 be manifolds with the connected sum $M_1 \# M_2$. This defines a group operation on the set of h-cobordism classes of manifolds endowing this set with an Abelian group structure. Next, the subset of homotopy spheres defines a subgroup Θ_n. We note that a simply connected manifold M is h-cobordant to the sphere if and only if M bounds a contractible manifold (Lemmas 2.3 and 2.4 in [Kervaire and Milnor (1963)]). If M is a homotopy sphere then $M \# \overline{M}$ bounds a contractible manifold and is itself h-cobordant to the sphere S^n. This gives the group structure on Θ_n. The sphere S^n is the identity element in this group.

As preparation for (3), we introduce the concept of S-parallelizability to characterize the elements of the group Θ_n further. Recall that a manifold is defined to be *parallelizable* if its tangent vector bundle is trivial. That is, vector fields can be described in terms of one fixed global frame.

Definition 7.1. Let M be a (smooth) manifold with tangent bundle[7] $\tau(M)$. Denote by ϵ^1 the trivial line bundle over M. M is said to be S-parallelizable if the Whitney sum $\tau(M) \oplus \epsilon^1$ is a trivial bundle.

In some sense these manifolds are "almost" parallelizable. As an example we know that S^2 is not parallelizable (If a billiard ball had hair, it couldn't be combed flat all over without a "whirl"). Thus the tangent bundle $\tau(S^2)$

---

[7] According to Milnor [Milnor (1964)] it is possible to introduce a tangent bundle, known as micro tangent bundle, also in case of a topological manifold.

is non-trivial. However we can always embed $f : S^2 \to \mathbb{R}^3$, from a standard result (see e.g. page 31 [Milnor and Stasheff (1974)])

$$f^*\tau(\mathbb{R}^3) = \tau(S^2) \oplus \nu(S^2).$$

The triviality of $f^*\tau(\mathbb{R}^3)$ and $\nu(S^2)$ then establishes that S^2 is S-parallelizable:

$$f^*\tau(\mathbb{R}^3) = \epsilon^3 = \tau(S^2) \oplus \nu(S^2) = \tau(S^2) \oplus \epsilon^1.$$

Clearly this proof extends to all spheres S^n. But Kervaire and Milnor go further to show:

Theorem 7.3. *Every homotopy sphere is S-parallelizable.*

Of course this theorem contains our example. The next remark gives a short outline of the proof which also summarizes the arguments of Kervaire-Milnor.

Remark 7.1.
The proof of this theorem can be obtained by considering the obstruction $o_n(\Sigma)$ to the triviality of $\tau(\Sigma) \oplus \epsilon^1$ for a homotopy n-sphere Σ. By the standard methods of obstruction theory (see the next point below) this element $o_n(\Sigma)$ is contained in the group $H^n(\Sigma, \pi_{n-1}(SO(n+1))) = \pi_{n-1}(SO(n+1))$. Now it is enough to switch to the stable groups $\pi_{n-1}(SO)$ which are periodic of period 8 according to Bott. Then for $n = 3, 5, 6, 7 \mod 8$ the group $\pi_{n-1}(SO) = 0$ vanishes and thus the obstruction vanishes also. The case $n = 0, 4 \mod 8$ is related to the signature theorem of Hirzebruch, but because of $H^{2k}(\Sigma) = 0$ the signature and thus the obstruction vanish. For the last case $n = 1, 2 \mod 8$ Kervaire and Milnor used the fact that the J-homomorphism,

$$J_n : \pi_n(SO(k)) \to \pi_{n+k}(S^k)$$

is actually a monomorphism in the stable range $k > n$. consequently, $o_n(\Sigma) = 0$ in the stable range.

To complete step (3) consider the following subgroup bP_{n+1} of Θ_n and show that Θ_n/bP_{n+1} is finite.

Definition 7.2. A homotopy n-sphere M ($\in \Theta_n$) represents an element of bP_{n+1} if and only if M is the boundary of a parallelizable manifold.

Of course, it must first be established that this defines bP_{n+1} as a subgroup. Assuming this, we have the following important theorem:

Theorem 7.4. *The quotient group Θ_n/bP_{n+1} is finite.*

Thus it remains to consider the subgroups bP_{n+1}.

Remark 7.2.
We briefly outline the proof of this result. First choose an embedding $i : M \to S^{n+k}$ of a closed S-parallelizable manifold M of dimension n with $k > n + 1$. The Pontrjagin-Thom

construction defines a map $p(M, \phi) : S^{n+k} \to S^k$ depending on the framing ϕ. For a varying frame we define the class $p(M)$ to be $\{p(M,\phi)\}$ which is a subset of elements of the stable homotopy group $\Pi_n = \pi_{n+k}(S^k)$. The kernel of the map

$$p' : \Theta_n \to \Pi_n/p(S^n)$$

forms a group which is indeed bP_{n+1}. Thus Θ_n/bP_{n+1} is isomorphic to a subgroup of $\Pi_n/p(S^n)$ and according to Serre [Serre (1951, 1953)] Π_n is finite and thus Θ_n/bP_{n+1} is also finite. Note that these quotient groups are actually isomorphic if $n \neq 2^{j+1} - 2$.

For step (4) we must consider the two possible cases (bP_{2k+1}) and (bP_{2k}). The key theorem for the first case is

Theorem 7.5. *If a homotopy sphere of dimension $2k$ bounds an S-parallelizable manifold M, then it also bounds a contractible manifold M_1.*

To prove this, Kervaire and Milnor produce M_1 using surgery operations on M leaving the boundary invariant. These operations can be considered as killing the homotopy groups of M. Then by definition of bP_{n+1} we obtain $bP_{2k+1} = 0$. The non-trivial case bP_{2k} is much more complicated. First note that bP_{4k+2} has at most two elements, i.e. $bP_{4k+2} = \mathbb{Z}_2$. Thus we have to consider only the case $k = 2m$. The first value $k = 2$ is still unresolved because Kervaire and Milnor surgery technique excludes this case. In particular, a homology class $H_2(M^4)$ of a 4-manifold need not be representable by a smoothly embedded sphere.

- *Again, we have the special circumstance arising in the critical dimension four.*

However, for $k = 2m \neq 2$. Kervaire and Milnor did establish:

Theorem 7.6. *Let M be a (framed) S-parallelizable manifold of dimension $4m > 4$, bounded by the $(4m-1)$-sphere. Consider the collection M_0 of such manifolds M. The corresponding signatures $\sigma(M_0) \in \mathbb{Z}$ form a group under addition. If σ_m is the generator of the group, then bP_{4m} has order $\frac{\sigma_m}{8}$.*

To understand this theorem consider the surgery operation on M as a cobordism between two manifolds M and M' of dimension $4m$ having the same boundary. An important invariant of this process is the signature of M. The next problem is to calculate the signature σ_m of the generator. [Kervaire and Milnor (1958)] and [Adams (1963)] then obtain the closed formula:

$$\sigma_m = a_m 2^{2m+1}(2^{2m-1} - 1)\text{numerator}(B_m/4m)$$

where B_m is the mth Bernoulli number and $a_m = \frac{3-(-1)^m}{2}$. This result was obtained with a combination of the signature theorem of Hirzebruch and some results of Adams about the J-homomorphism. This completes the third step in calculating bP_{n+1}. The last step (step (5) above) involves a great deal of work involving some very strong theorems about the J-homomorphism etc. We only state the result for the full group Θ_{4m-1}:

$$\text{order } [\Theta_{4m-1}] = a_m \text{order } [\pi_{4m-1+k}(S^k)]\, 2^{2m-4}(2^{2m-1} - 1)B_m/m,$$

where $k > 4m$. In particular:

m	1	2	3	4	5
dim	3	7	11	15	19
Order of Θ_{4m-1}	1?	28	992	16256	130816

Thus, collecting these results we have a classification of the homotopy spheres given by table 7.1 leading to the classification of exotic spheres

n	1	2	3	4	5	6	7	8	9
Order of Θ_n	1	1	1?	1	1	1	28	2	8
n	10	11	12	13	14	15	16	17	18
Order of Θ_n	6	992	1	3	2	16256	2	16	16

for dimension > 4. To show this last statement we have to fill in the steps (6) and (7) above. So again, the set Θ_n is the set of homotopy n-spheres. Equivalently, an h-cobordism between two homotopy n-spheres induces also a homotopy-equivalence. Thus Θ_n is also the set of h-cobordism classes of a homotopy n-sphere. By the h-cobordism theorem, two h-cobordant homotopy n-spheres are diffeomorphic. Then, two different elements $a, b \in \Theta_n$ represent two non-diffeomorphic homotopy n-spheres. So, as shown in step (1) every homotopy n-sphere Σ can be decomposed as $D^n \cup_f D^n$ via some diffeomorphism $f : S^{n-1} \to S^{n-1}$. Thus, if the homotopy n-sphere Σ' is constructed by gluing two disk D^n along a diffeomorphism f', we may try to construct a diffeomorphism $\Sigma \to \Sigma'$ by beginning with the identity map of one disk in each sphere. This map induces a diffeomorphism $f' \circ f^{-1}$ of the boundaries of the other disks, which extends across those disks by using Alexander's trick. Such an extension is a homeomorphism which is smooth perhaps except at the origin, i.e. the map f is extended to $\tilde{f} : D^n \to D^n$ with maps the origins $\tilde{f}(0) = 0$ of the disks continuously to each other. Thus, if $f' \circ f^{-1}$ is homotopic to the identity then Σ and Σ' are h-cobordant or diffeomorphic by the h-cobordism theorem. This completes the answer to step (6). Finally we have to show that the elements of Θ_n are all homeomorphic but non-diffeomorphic n-spheres. The arguments above complete step (7) but there is more. Every homotopy n-sphere is an element in Θ_n and two diffeomorphic homotopy n-spheres define equal elements in Θ_n (see step (6) above). The connected sum $\Sigma \# \Sigma'$ between two homotopy n-spheres Σ, Σ' is the group operation in Θ_n which is commutative. According to the steps (3),(4) the group Θ_n is finite and must be a direct sum of cyclic groups. As an example consider the group Θ_7 which is isomorphic to $\mathbb{Z}_{28}$. Thus, there is a homotopy 7-sphere Σ^7 (which is M^7_{-1} in Milnor's notation, see above) which generates all other 28 spheres by the connected sum operation. All homotopy 7-spheres

$$\Sigma, \Sigma \# \Sigma, \ldots, \underbrace{\Sigma \# \Sigma \# \cdots \# \Sigma}_{27}$$

represent all exotic 7-spheres and the homotopy 7-sphere

$$\underbrace{\Sigma \# \Sigma \# \cdots \# \Sigma}_{28} = S^7$$

is diffeomorphic to the standard 7-sphere, i.e. the identity element e in Θ_7. Let a be the generator of Θ_7 represented by Σ and we obtain the relation $a + a + \cdots + a = 28a = e$ where the connected sum operation $\#$ is mapped to the $+$-operation and the identity element is e. But that is by definition the cyclic group $\mathbb{Z}_{28}$. Now the step (7) is complete.

Additionally, this solves also the (topological) Poincaré Conjecture in dimension > 4, stating that every topological n-manifold homotopic to the n-sphere is also homeomorphic to it but not in general diffeomorphic as the table above shows. For more details, see the review paper by Lance [Lance (2000)].

7.9 Classification of Manifold Structures

Kirby and Siebenmann [Kirby and Siebenmann (1977)] have been able to extend the previous results to a more general class of metrizable manifolds of dimension greater than 5. Consider metrizable, topological manifolds M of dimension ≥ 6 with $\partial M \neq \emptyset$ and ≥ 5 with $\partial M = \emptyset$. Recall that the manifold structure is defined by local coordinate patches with transition functions in their overlaps which are *continuous, piecewise linear, or smooth*. The categories of such manifolds, with corresponding morphisms, are described by the terms TOP, PL, and DIFF respectively. We will use the notation for which the generic case is CAT = {TOP, PL, DIFF} with the subset CAT_0 = {PL, DIFF} $\subset$ CAT. For general CAT we cannot define a tangent bundle (in the smooth sense) but we can define a substitute for such a bundle introduced by Milnor [Milnor (1964)] and called a *microbundle*. The classification of manifold structures is facilitated using the classification of microbundles and obstruction theory. We now provide a brief review of this technique.

In particular, the main questions are

- Can we "extend" a TOP-manifold to become a PL- or DIFF-manifold? That is, can the coordinate patch overlaps be made into PL or smooth functions?
- What are the conditions for the existence of such an extension?
- Is such an extension unique?

Now assume (≥ 5). The logical flow of arguments proceeds as follows:

(1) Define the structure "microbundle" to encode any CAT structure on the (topological) manifold M as a TOP generalization of the DIFF notion of tangent bundle.
(2) Define the stable CAT structure on M and use the stable concordance of CAT structures as equivalence relation.
(3) The Product Structure Theorem of Kirby and Siebenmann relates a CAT structure on $M \times \mathbb{R}^s$ ($s > 2$) to the CAT structure on M.
(4) The reduction of the TOP structure to CAT_0 is equivalent to the microbundle reduction of TOP to CAT_0. The equivalence classes of CAT_0 structures are isotopy classes which are identical to these microbundle reductions.
(5) Obstruction theory leads to necessary and sufficient conditions for the extension of a TOP to a CAT_0 structure (the lifting problem).
(6) For the cases PL and DIFF we calculate this obstruction.

In step (1) we seek a TOP version of the DIFF tangent bundle. In the smooth case we have smoothly consistent local trivializations of M into copies of $\mathbb{R}^n$, over which the notion of tangent vector is well-defined since locally the tangent vector fiber is also $\mathbb{R}^n$. Milnor considered a TOP generalization for which the local DIFF copies of $\mathbb{R}^n$ are replaced by TOP ones.

Early Exotic Manifolds 223

In [Milnor (1964)] Milnor recounts his thoughts on this as follows:
"...Suppose that one tries to construct something like a 'tangent bundle' for a manifold M which has no differentiable structure. Each point $x \in M$ has neighborhoods which are homeomorphic to Euclidean space. It would be plausible to choose one such neighborhood U_x for each x, and call $(x) \times U_x$ the 'fiber' over x. Unfortunately however, it seems difficult to choose such a neighborhood U_x simultaneously for each $x \in M$, in such a way that U_x varies continuously with x. Furthermore even if such a choice were possible, it is not clear that the resulting object would be a topological invariant of M. To get around these difficulties we consider a new type of bundle, in which the fiber is only a 'germ' of a topological space. ..."
This led him to the following:

Definition 7.3. A TOP-n-microbundle ξ over the topological space M is a diagram $\xi : M \xrightarrow{i} E \xrightarrow{p} M$ of spaces and maps such that $p \circ i = id_M$ and, for each $x \in M$, there exist neighborhoods U of x in M and V of $i(x)$ in E and a homeomorphism $h : V \to U \times \mathbb{R}^n$, so that

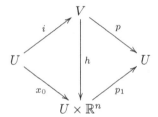

commutes, where $x_0(x) = (x, 0)$ and $p_1(x, r) = x$.

For the illustration we consider two important examples:

- The diagram $\epsilon^n : M \xrightarrow{i} M \times \mathbb{R}^n \xrightarrow{p} M$ defines via the canonical maps the trivial microbundle ϵ^n.
- The diagram $\tau(M) : M \xrightarrow{diag} M \times M \xrightarrow{p} M$ with the diagonal map $diag(x) = (x, x) \in M \times M$ for all $x \in M$ defines the (topological) microtangent bundle of M. Of course this bundle agrees with the usual tangent bundle if M carrying a DIFF structure. (see [Milnor (1964)] for details of the equivalence proof).

Definition 7.4. A micro-isomorphism $\xi_1 \to \xi_2$ of n-microbundles over M consists of a neighborhood U_1 of $i_1(M)$ in the total space $E(\xi_1)$ of ξ_1 and a topological embedding h onto a neighborhood U_2 of $i_2(M)$ in $E(\xi_2)$, making this diagram

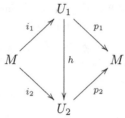

commutative. If h is an inclusion of a subspace it is called a micro-identity and ξ_1, ξ_2 are called micro-identical $\xi_1 \doteq \xi_2$.

Next we try to detect the equivalence classes under this relation. Milnor used these tools to introduce a CAT_0(=DIFF or PL) structure on a TOP manifold M. The starting point of the construction is the observation that any TOP M can be embedded in sufficiently high dimension Euclidean space. When M is so embedded in $\mathbb{R}^n$ as a retract[8] of a neighborhood N by the retraction $r : N \to M$, then the pull-back $r^*\tau(M)$ over N of the tangent microbundle $\tau(M)$ of M contains a copy of $M \times \mathbb{R}^n$ as an open neighborhood of the embedded M. Thus the pull-back tangent microbundle $r^*\tau(M)$ reflects the deformation of the TOP structure after the embedding. There results a mapping from concordance classes of CAT_0 microbundle structures on $r^*\tau(M)$ or on the stable bundle $r^*\tau(M) \oplus \epsilon^s$, $s \geq 0$, to isotopy classes of CAT_0 structures on M, which means the class of mappings $h_t : M \times [0,1] \to M$ such that h_0 is the identity map $id|_M$ and h_1 is a map $M \to M$ leaving the CAT_0 structure fixed. Call this mapping the *smoothing rule*. As Milnor [Milnor (1964)] shows there is a homotopy theorem of the form:

Theorem 7.7. *Any CAT microbundle ξ over a product $[0,1] \times M$ (M being Hausdorff paracompact) admits a CAT micro-isomorphism $f : \xi \to [0,1] \times \xi$, where $0 \times \xi$ is just a copy of $\xi|_{0 \times M}$.*

Before we can state the whole classification theorem we have to introduce the notions of stability and concordance, step (2) above. Introduced by Freudental 1937 for the homotopy groups of spheres, today it plays a key role in the unification of algebraic and analytical techniques in topology. For a simple definition we will cite J. Adams [Adams (1974)]:

---

[8]A subspace A of a space M is a retract if there is a map $f : M \to A$ called retraction that fixes every point of A.

" ... We say that some phenomenon is **stable**, if it can occur in any dimension, or in any sufficiently large dimension, provided perhaps that the dimension is sufficiently large..."
The idea of concordance is defined by a one-parameter deformation of a structure on an n-manifold to another structure. This process can be described by building a new structure on the $(n+1)$-manifold by a process similar to isotopy. An important example is cobordism.

Definition 7.5. A **stable CAT structure** on ξ means a CAT structure η on $\xi \oplus \epsilon^s$, $s \geq 0$, where $\epsilon^s : M \to M \times \mathbb{R}^s \to M$ is the trivial microbundle. If η' is a CAT structure on $\xi \oplus \epsilon^t$, $t \geq s$, then write $\eta \doteq \eta'$ if $\eta \oplus \epsilon^{t-s} \doteq \eta'$.
Let ξ be a TOP microbundle over a CAT object M. A **concordance** between CAT structures ξ_0, ξ_1 on ξ is a CAT structure γ on $[0, 1] \times \xi$ such that the restriction $\gamma|_{i \times M}$ is CAT micro identical to $i \times \xi_i$ for $i = 0, 1$. We write $\xi_0 \cong \xi_1$.
A **stable concordance** of stable CAT structures ξ_0, ξ_1 on ξ is defined by $\xi_0 \doteq \zeta_0 \cong \zeta_1 \doteq \xi_1$ for stabilizations ζ_0 of ξ_0 and ζ_1 of ξ_1.

The relation between concordance and isotopy classes of CAT_0 structures leading to a stable classification of manifold structures contains as a key ingredient the *Product Structure Theorem* of Kirby and Siebenmann, step (3) above:

Theorem 7.8. *Let Θ be a CAT manifold structure on $M \times \mathbb{R}^s$, $s \geq 1$. There exists a concordant CAT structure $\Sigma \times \mathbb{R}^s$ on $M \times \mathbb{R}^s$ obtained from a CAT structure Σ on M by taking the product with with $\mathbb{R}^s$.*

The main step in the proof of this theorem includes the usage of the s-cobordism theorem 6.12, the version of the h-cobordism theorem for non-simply connected spaces, which only works in higher dimensions ≥ 5.

Kirby and Siebenmann's classification theorem
Because of the close relationship between the concordance classes and isotopy classes of CAT_0 structures we can focus on the corresponding concordance classes of microbundles. According to Milnor's homotopy theorem for microbundles, concordance classes and homotopy classes of microbundles are the same. So, given a fixed TOP structure, in order to determine compatible CAT_0 structures, we can look at the (stable) classification of CAT_0 microbundles over a TOP-manifold. This is equivalent to considering the reduction of the TOP to the CAT_0 (=PL or DIFF) structure leading to a microbundle reduction. Briefly, the question is analogous to that of the structure group reduction in vector bundle theory. Let G be

the structure group of a vector bundle over a manifold M and $H \subset G$ a possible subgroup acting on G by left multiplication. Then this bundle admits a structure group reduction to H if and only if the bundle over M with fiber G/H has a global section. Because of the close relationship between micro bundles and CAT_0 structures, microbundle reductions can be studied as reductions from CAT to CAT_0. The number of such possible reductions should be equivalent to the number of equivalence classes of CAT_0 structures. As in the case of vector bundles any micro bundle is determined up to homotopy so that isotopy classes of CAT_0 structures are the equivalence classes of CAT_0 structures. These facts are summarized in the *Classification theorem of Kirby and Siebenmann*, step (4) above:

Theorem 7.9. *Let M be a (metrizable) topological manifold[9] of dimension ≥ 6 (or ≥ 5 if the boundary ∂M is empty). There is a one-to-one correspondence between isotopy classes of CAT_0 (=PL or DIFF) manifold structures on M and vertical homotopy classes of sections of a fibration over M expressing the classes of microbundle reductions from TOP to CAT_0.*

Thus, what can be done to the tangent bundle, $\tau(M)$, can also be done to M. Or, another way to state this is that putting a refined structure on the tangent bundle (or even the stable tangent bundle) of a topological manifold produces such a refined structure on the manifold compatible with the given refinement on the tangent bundle.

According to Milnor[Milnor (1964)], the theory of microbundle reductions is equivalent to the theory of principal bundle reductions but with respect to the structure groups TOP, PL or DIFF.

Recall earlier discussions on determining the homotopy equivalence classes of bundles in sections 5.2 and 5.3. Let P be a G-principal fiber bundle over M. Any two such bundles over homotopy-equivalent base manifolds M_1 and M_2 have isomorphic classes, $[M_1, BG] = [M_2, BG]$. The formation of a universal bundle containing all information about all G-principal bundles is the basic tool. Consider the infinite joint $EG = G \star G \star \cdots$ defined by

$$G \star G = G \times [0,1] \times G/\sim$$

with respect to the equivalence relation

$$(g_1, 0, g_2) \sim (g_1', 0, g_2) \quad \text{and} \quad (g_1, 1, g_2) \sim (g_1, 1, g_2').$$

This space EG is contractible and admits a natural G-action. The orbit space $BG = EG/G$ is known as *universal classifying space of the group*

---
[9]The non-compact case is included in the consideration!

G. Thus, we have a projection $\pi : EG \to BG$ which makes EG to a G-principal bundle over BG. Because of the homotopy classification of G-principal bundles, the set of homotopy classes

$$[M, BG] = \{[f] | f : M \to BG\}$$

are in one-to-one correspondence to the isomorphism classes of G-principal bundles over M. Now we consider any subgroup $H \subset G$ with a restricted action of H on EG defining the space $BH = EG/H$. Thus BG is naturally a quotient of BH, indeed we have a bundle $\mathcal{B} : BH \to BG$ with fiber G/H. Fix a class $c : M \to BG$ and consider the diagram:

$$\begin{array}{c} M \xrightarrow{c_l} BH \\ {}_c\searrow \downarrow \mathcal{B} \\ BG \end{array}$$

where $c = \mathcal{B} \circ c_l$. This defines the lift c_l of the map. Now it is obvious that the existence of a map $s : BG \to BH$ with $id_{BG} = \mathcal{B} \circ s$ (in other words, a section) leads to the construction of the lift $c_l = s \circ c$. Thus the set of lifts and the set of sections are equivalent. When does such a section (or lift) exist? The answer is given by obstruction theory: the obstruction is an element of the (Čech) cohomology group $H^{n+1}(M, \pi_n(G/H))$ and the number of elements of $H^n(M, \pi_n(G/H))$ is the number of inequivalent sections (or liftings).

Remark 7.3.

We now sketch some highlights of obstruction theory, assuming for simplicity that BG and BH are CW-complexes. For more details, see Steenrod[Steenrod (1999)], Part III. Because $\pi_1(BG) = \pi_0(G) = 0$, the group $\pi_1(BG)$ acts trivially on the fiber G/H of the bundle $\mathcal{B} : BH \to BG$ and we assume $\pi_1(G/H) = 0$. Suppose $\sigma : BG^{(n-1)} \to BH$ is a section over the $(n-1)$ skeleton. For each n-cell e^n of BG we have a trivialization of $\mathcal{B}^{-1}(e^n) \to e^n$ as $e^n \times G/H$. Because $\pi_1(BG) = 0$, the trivialization can be uniquely defined. Then a section σ on $BG^{(n-1)}$ induces a section of $\mathcal{B}^{-1}(e^n) \to e^n$ over $\partial e^n \simeq S^{n-1}$. The product structure allows us to define some arbitrary map $\tilde{\sigma} : \partial e^n \to e^n \times G/H$. So $\pi_2 \circ \tilde{\sigma} : \partial e^n = S^{n-1} \to G/H$ with $\pi_2(x, y) = y$. π_2 then defines an element of $\pi_{n-1}(G/H)$. Using this construction for each n-cell e^n we obtain the obstruction cocycle $\tilde{\sigma}_n : C_n(BG) \to \pi_{n-1}(G/H)$. Two such cocycles define the same section if and only if they are cohomologous. Therefore, the class $\sigma_n \in H^n(BG, \pi_{n-1}(G/H))$ is the obstruction to extending a section defined on BG^{n-1} over BG^n, keeping it fixed on BG^{n-2}.

In our case $G = TOP$ and $H = CAT_0$ for the stable version[10] of the theorem, step (5) above.

Theorem 7.10. *Let M be a (metrizable) topological manifold of dimension ≥ 6 (or ≥ 5 if the boundary ∂M is empty). There is a one-to-one correspondence between isotopy classes of CAT_0 (=PL or DIFF) manifold structures on M and the number of elements of the groups $H^n(M, \pi_n(TOP/CAT_0))$ if and only if all obstructions in $H^{p+1}(M, \pi_p(TOP/CAT_0))$ vanish for $0 \leq p \leq n$..*

Furthermore Kirby and Siebenmann proved the following results on the homotopy groups $\pi_n(TOP/CAT_0)$:

Theorem 7.11. *It is sufficient and necessary to consider the stable version of $\pi_n(TOP/CAT_0)$ instead of the unstable version $\pi_n(TOP_n/(CAT_0)_n)$. Let $\Theta_m^{CAT_0}$ be the Abelian group of oriented isomorphism classes of oriented CAT_0 m-manifolds homotopy equivalent to S^m then for $m \geq 5$ there is an isomorphism:*

$$\pi_m(TOP/CAT_0) \to \Theta_m^{CAT_0}.$$

In the last step, (6) above, we give the concrete results for the cases PL and DIFF.

Theorem 7.12. *Let M be a (metrizable) topological manifold of dimension ≥ 6 (or ≥ 5 if the boundary ∂M is empty). For the homotopy groups $\pi_k(TOP/PL)$ we obtain:*

$$\pi_k(TOP/PL) = \begin{cases} \mathbb{Z}_2 & k = 3 \\ 0 & k \neq 3 \end{cases}.$$

If the Kirby-Siebenmann class in $H^4(M, \mathbb{Z}_2)$ vanishes then there is a one-to-one correspondence between isotopy classes of PL manifold structures on M and elements of the group $H^3(M, \mathbb{Z}_2)$. The homotopy groups of $\pi_k(TOP/DIFF)$ are known as the Kervaire-Milnor groups. Thus $\pi_k(PL/DIFF)$ vanishes for $k < 7$.

Consider two simple examples: S^n and $\mathbb{R}^n$ for $n \geq 5$. The n-sphere has the non-vanishing cohomology groups $H^0(S^n, \mathbb{Z}) = H^n(S^n, \mathbb{Z}) = \mathbb{Z}$. Thus for $n \geq 5$ we obtain the fact that a topological n-sphere always

---

[10] Let TOP_n be the group of homeomorphism between $\mathbb{R}^n$. By TOP we mean the inductive limit $TOP = \cup_n TOP_n$ and call the corresponding space the stable limit of TOP_n.

admits a PL structure (because $H^4(S^n, \mathbb{Z}_2)) = 0$) which is unique (because $H^3(S^n, \mathbb{Z}_2)) = 0$). The fact that every topological n-sphere can be given a DIFF structure has been long known. If we use the fact that $H^{n+1}(S^n, \pi_n(TOP/DIFF)) = 0$ because of dimensional reasons then we obtain for the number of distinct differential structure the long-known result $H^n(S^n, \pi_n(TOP/DIFF)) = \Theta_n^{DIFF}$. There is also another way to look at this. The tangent bundle $\tau(S^n)$ of the sphere is stably trivial. Thus liftings of maps $S^n \to BTOP$ from $BTOP$ to $BCAT_0$ trivially exist in the fibration $TOP/CAT_0 \to BCAT_0 \to BTOP$. So, we can enumerate liftings by $[S^n, TOP/CAT_0] = \pi_n(TOP/CAT_0)$ as stated above.

The second example is given by the non-compact space $\mathbb{R}^n$ with $n \geq 5$ for which the classification theorem also applies. Because of the contractibility of this space, all cohomology groups except of $H^0(\mathbb{R}^n, \mathbb{Z}) = \mathbb{Z}$ vanish. Thus the topological Euclidean space admits a unique PL and DIFF structure. *There is no exotic $\mathbb{R}^n$ for $n \geq 5$.*

Chapter 8

The First Results in Dimension Four

In this chapter we review certain results for manifolds of dimension 4. We begin with a brief look at the more general problem of smoothing the prototypical manifold, $\mathbb{R}^n$. Then we proceed to the dimension 4 case, starting with the result of Freedman relating the intersection form to existence and uniqueness of TOP 4-manifolds. Later Donaldson was able to find a restriction on the set of possible intersection forms for smooth simply-connected 4-manifolds. The first exotic [1] $\mathbb{R}^4$ was found using these techniques and we review this process in some detail. In contrast to the compact case, there are uncountably many (non-diffeomorphic) exotic smooth structures for non-compact manifolds, and we review Gompf's techniques for representing a set of exotic $\mathbb{R}^4$'s as a two-parameter family and finding a universal exotic $\mathbb{R}^4$ in which all other exotic $\mathbb{R}^4$'s can be embedded. Further classification and construction results of Taylor and Gompf are also reviewed. Finally, we remind the reader that Gompf and Stipsicz have dedicated an entire book to the tools and results for 4-manifolds, [Gompf and Stipsicz (1999)].

8.1 The Smoothing of the Euclidean Space

We begin by reviewing some history of the questions surrounding existence and uniqueness of a smooth structure for the topologically trivial $\mathbb{R}^n$, starting from perspectives of the early 1960's, before the work of Kirby, Siebenmann et al. In this period it was clear that this problem depends critically on the value of n in spite of the topological triviality of each $\mathbb{R}^n$. In fact, the tools available at that time were able to resolve the question for all $n \neq 4$, but were not sufficient to settle the issue for $n = 4$. The com-

---
[1] The notation, $\mathbb{R}^4_\Theta$, is often used for an exotic $\mathbb{R}^4$, that is, a manifold homeomorphic to the smooth product of four real lines, but not diffeomorphic to it.

mon wisdom of the time was perhaps summarized by the following quote from a paper of Stallings, [Stallings (1962)]. This paper, important to the following discussion, starts with the remarkable statement:
"*Euclidean space R^n ought to have a unique piecewise-linear structure and a unique differentiable structure.*" (emphasis in the original)
What is notable here is the use of the phrase (almost a moral imperative) "ought to have," which some 20 years later was proven to be a false expectation.

Let us note two important definitions:

- An n-manifold is simplicially **triangulable** if it is homeomorphic to a (locally finite simplicial) complex.
- Two PL complexes K_1, K_2 are **combinatorially (PL) equivalent** if there are simplicial subdivisions K_1' and K_2' of K_1 and K_2, respectively, such that K_1' and K_2' are isomorphic, i.e. there is a one-to-one correspondence between the simplices of K_1' and those of K_2', preserving incidence relations.

One important link between topology and smoothness was obtained by Whitehead, [Whitehead (1940)], who showed that any differentiable manifold has a (smooth) piecewise-linear triangulation which is unique with respect to the differentiable structure. Munkres [Munkres (1960)], Corollary 6.6, established

Theorem 8.1. *Two differentiable manifolds homeomorphic to $\mathbb{R}^n$ are diffeomorphic if and only if their smoothly canonical piecewise-linear triangulations are combinatorially equivalent.*

Thus, on $\mathbb{R}^n$, the smoothness structures are categorized by the canonical PL ones.

Remark 8.1.

The general smoothing results were obtained by Munkres using obstruction theory ([Munkres (1960)] Theorem 6.2). Let M and N be combinatorially equivalent differentiable manifolds. Consider the abelian group Γ^n of orientation-preserving diffeomorphisms of S^{n-1} modulo those diffeomorphisms which are extendable to diffeomorphisms of the full n-disk D^n ([Munkres (1960)] Proposition 1.8). If $H^q(M; \Gamma^q) = 0$ for all $q < n - k$, there is a homeomorphism $g : M \to N$ which is a diffeomorphism mod the k-skeleton of a smooth triangulation of M ([Munkres (1960)] Theorem 6.2). Later this theory was extended by Kirby and Siebenmann [Kirby and Siebenmann (1977)] stated in the section 7.9 earlier. For the case of a manifold homeomorphic to $\mathbb{R}^n$, and thus homologically trivial, this argument justifies the "if" part of the preceding Theorem.

Thus, $\mathbb{R}^n$ admits more than one non-diffeomorphic differential structure if and only if it admits more than one PL triangulation not combinatorially equivalent to the trivial one. Therefore, smoothness questions for $\mathbb{R}^n$ reduce to those of PL triangulations on $\mathbb{R}^n$. Of course, there are really two questions:

(1) does a PL triangulation of $\mathbb{R}^n$ always exist, and,
(2) if it exists, is it unique under combinatorial equivalence.

The first question is easily resolved since there is always a triangulation of $\mathbb{R}^n$ induced from the trivial topology. The second question is much more difficult, and is, in fact, one of the central concerns of this book. It reduces to the question of whether or not the set of homeomorphisms is the same as the set PL morphisms (combinatorial maps). That is, is there a homeomorphism of $\mathbb{R}^n$ changing the standard triangulation to a non-combinatorially equivalent one? As noted above, the tools available in the early 1960's were not sufficient to answer this for $n = 4$. In fact, the approaches were qualitatively different for the two ranges, $n < 4$ and $n > 4$.

First, look at $n < 3$. In subsection 6.2.1 we reviewed the known classification of 1- and 2-manifolds. For $n = 1$ there are only 2 possible connected smooth manifolds without boundaries: S^1 (closed,compact case), $\mathbb{R}$ (non-compact case). All other 1-manifolds can be constructed from these (see the appendix of [Milnor (1965b)] for a proof). This also answers the question for PL structures for $n = 1$. The 2-manifold case was completely solved by Radon [Rado (1925)] by using methods from complex analysis. Thus,

Theorem 8.2. *The Euclidean space $\mathbb{R}^n$ has a unique PL and DIFF structure for $n < 3$.*

Moise[Moise (1952)] was able to resolve the $n = 3$ case, using some basic theorems in 3-dimensional topology: Dehn's Lemma, the Loop and the Sphere Theorem proved previously by Papakyriakopoulos[2] [Papakyriakopoulos (1943)]. Most important for the proof of Moise was the Sphere Theorem which states that if a 3-manifold M is orientable and $\pi_2(M) \neq 0$, then there is an embedded S^2 in M which is not contractible. Moise's completion of the proof is an involved tour-de-force using these results, so here we merely state his result:

---
[2]This 154-page-article is written in Greek.

Theorem 8.3. *The Euclidean space $\mathbb{R}^n$ has a unique PL and DIFF structure for $n = 3$.*

Now consider $n > 4$. Stallings [Stallings (1962)] was able to show that any contractible PL manifold, 1-connected at infinity, is in fact PL homeomorphic to $\mathbb{R}^n$ with the standard PL if $n > 4$. Thus, since $\mathbb{R}^n$ meets these conditions,

Theorem 8.4. *The Euclidean space $\mathbb{R}^n$ has a unique PL and DIFF structure for $n > 4$.*

The condition that M be 1-connected at infinity means that it is not compact, and for every compact subset $C \subset M$ there is a compact D with $C \subset D \subset M$ so that $M \setminus D$ is 1-connected. Given this condition, the main effort of the paper is to prove the engulfing theorem ([Stallings (1962)] theorem 3.1). From this, and other procedures which are valid only if $n > 4$, Stallings arrived at the result summarized in the preceding theorem. It should be noted that part of the proof involves deformation of 2-dimensional polyhedrons reminiscent of the Whitney trick central to the h-cobordism theorem so important in establishing DIFF equivalence for compact manifolds. However, such movements of 2-manifolds in n-manifolds could be shown to work **only when** $n > 4$. In fact, with the retrospective advantage of knowing the existence of exotically smooth (and PL) $\mathbb{R}^4$, we can see that all of the strategies above which were effective for $n < 4$ and $n > 4$ could not, in fact, be extended to this exceptional case, $n = 4$, which will concern us in the remainder of this chapter.

8.2 Freedman's Work on the Topology of 4-manifolds

The counterexample to Whitney's trick by Kervaire and Milnor [Kervaire and Milnor (1961)] shows the importance of considering the embedding of a disk into a 4-manifold. In section 6.4 we described this disk embedding problem to motivate the introduction of the Casson handle. If we have such an embedding then we can perform Whitney's trick and we are done (see section 6.3 for the explanation). The reader can find a very easily read, non-technical, overview of this process and the peculiarities of dimension 4 in a review by Freedman, [Freedman (1984)]. We recommend this as an excellent introduction to this subject.

For our purposes, the main difference between dimension 4 and the higher-dimensional case is the difference between the topological and the smooth h-cobordism theorem and the failure of the smooth h-cobordism theorem in dimension 4. By using this failure, it is possible to construct exotic (or fake) $\mathbb{R}^4$'s (see [Casson (1986); De Michelis and Freedman (1992); Bižaca and Gompf (1996)]). In some cases, there is also a non-compact version of the h-cobordism theorem (used in the previous section) called the engulfing theorem which also makes use of the Whitney trick. The landmark paper [Freedman (1982)] of Freedman presented the theorem (theorem 1.1 on page 361):

Theorem 8.5. *Any Casson handle CH is homeomorphic as a pair to the standard open 2-handle $(D^2 \times int D^2, \partial D^2 \times int D^2)$.*

The next important result requires a technical condition, $\pi_1(M)$ is NDL (Null Disk Lemma), on the fundamental group of the manifold. Rather than going into this rather involved topic, we will refer the reader to Freedman [Freedman (1983)] and Freedman and Teichner [Freedman and Teichner (1995)] for details. The basic questions refer to the word problem for finitely generated groups, and how they grow (polynomial, subexponential, etc.) as functions of the number of generators. These questions are important in group theory itself, as well as in the theory of computation. See also [Milnor (1968)]. For our purposes a sufficient condition is that π_1 be a finite group, or the integers.

At any rate, Freedman was able to show, [Freedman (1983); Freedman and Quinn (1990)]

Theorem 8.6. *Let $j : (D^2 \times D^2, \partial D^2 \times D^2) \to (M^4, \partial M^4)$ be a local homeomorphism of a 2-handle into a topological manifold which is an embedding near $j(\partial D^2 \times D^2)$. Suppose that there exists another local homeomorphism of a 2-sphere × disk, $\alpha : S^2 \times D^2 \to M^4$ and that the intersections are transverse and satisfy:*

- *the algebraic self-intersection number[3] of $j(D^2 \times 0) = 0$*
- *the algebraic self-intersection number of $\alpha(S^2 \times 0) = 0$*
- *the algebraic intersection number of $(j(D^2 \times 0), \alpha(S^2 \times 0)) = 1$.*

Assume $\pi_1(M)$ is NDL. Then j is topologically regularly homotopic (relative to a neighborhood of $\partial D^2 \times D^2$) to a topological embedding $i : (D^2 \times D^2, \partial D^2 \times D^2) \hookrightarrow (M^4, \partial M^4)$.

---

[3] See Wall [Wall (1970)] for the definition and interpretation of these "numbers" in the non-simply connected case, i.e. $\pi_1 \neq 0$, to have values in the module $\mathbb{Z}(\pi_1(M))$.

This can be regarded as Freedman's topological version of the Whitney trick, which does not work smoothly in dimension 4. As discussed in Chapter 6, the failure of the smooth Whitney trick was one of the basic reasons for the anomalous smoothings in dimension four.

Theorem 8.7. Topological h-cobordism theorem in dimension 4
Let (W, V, V') be a smooth 1-connected compact 5-dimensional h-cobordism. Then W is topologically a product, i.e. $W \simeq_{TOP} V \times [0, 1]$.

We note that theorem 10.3 in [Freedman (1982)] also covers the non-compact case for simply-connected 4-manifolds. A direct consequence of the topological h-cobordism theorem is the well-known classification of 1-connected, compact 4-manifolds first stated at the end of section 6.7.2.

Theorem 8.8. (Freedman, [Freedman (1982); Freedman and Quinn (1990)]) *For every unimodular symmetric bilinear form Q there exists a simply connected closed topological 4-manifold X such that $Q_X \cong Q$. If Q is even, this manifold is unique (up to homeomorphism). If Q is odd, there are exactly two different homeomorphism types of manifolds with the given intersection form.*

For our purposes an important result of this is the existence of a closed, 1-connected, compact, topological 4-manifold corresponding to the form E_8 (see section 6.7.2 for a discussion). But of course, as a topological theorem, this result does not tell us whether or not this manifold also admits a smooth structure. In fact, Rohlin's theorem (see the next section below) establishes one of the major steps in the study of exotic smoothness, the amazing conclusion that this manifold does **not** admit any smooth structure. Because of the fact that PL=DIFF in dimension 4, this manifold also admits no combinatorial structure.

The extension of this result to the multiply-connected case is accomplished by restricting the fundamental groups to be NDL, as in the disk embedding theorem 8.6. In fact, Freedman [Freedman (1983)] proved as a corollary to 8.6 (theorem 1 and corollary 1 in [Freedman (1983)]) the topological s-cobordism theorem (see section 6.3.2 for dimensions > 4).

Theorem 8.9. Topological s-cobordism theorem in dimension 4 *Let (W, V, V') be a compact topological 5-dimensional s-cobordism. If $\pi_1(W)$ is NDL then W is homeomorphic to $W \simeq_{TOP} V \times [0, 1]$.*

From this there follows a classification of 4-manifolds with NDL fundamental groups by changing the intersection form from coefficients in $\mathbb{Z}$ to a form with coefficients in the module $\mathbb{Z}(\pi_1(M))$. However, there remains the non-trivial problem of determining the homotopy-type of a 4-manifold with NDL fundamental group. The most general form of this problem has not been solved.

8.3 Applications of Donaldson Theory

We can summarize the issues associated with establishing a smooth structure on a given topological 4-manifold as follows:

- **Question 1** Existence: Can a topological manifold defined by an intersection form carry a smooth structure?
- **Question 2** Uniqueness: If so, how many non-diffeomorphic smooth manifolds can be supplied to this given topological manifold?

The following list of theorems illustrates some of what we know about the answers for **Question 1** and **Question 2**. For simplicity, assume that X is a simply connected, closed, oriented, smooth 4-manifold.

Theorem 8.10. (Rohlin [Rohlin (1952)]) *If X is smooth and Q_X is even, then $16|\sigma(X)$.*

This and the next theorem imply the very important fact that topological manifolds corresponding to E_8 cannot carry any smooth structure.

Theorem 8.11. (Donaldson[Donaldson (1983, 1986)]) *If X is a smooth simply connected closed 4-manifold and Q_X is negative definite, then Q_X is equivalent to $\langle -1 \rangle$.*

Note that X must be closed since a non-empty boundary would be a 3-manifold, which is known to have a PL structure. This PL structure could extend to the full X, and PL=DIFF in this dimension. This theorem takes care of manifolds with positive definite intersection forms as well.

For indefinite even intersection forms the following estimate has been proved:

Theorem 8.12. (Furata [Furata (2001)]) *If X is a smooth, simply-connected, closed, and oriented 4-manifold with even intersection form equivalent to $2kE_8 \oplus lH$, then $l \geq 2|k| + 1$.*

The theorem partly solves a strong conjecture known as the $\frac{11}{8}$-*conjecture* (Matsumoto [Matsumoto (1982)]). Consider the K3 manifold K with $l =$

$3, k = 1$ (see next section for a definition). The relation between the second Betti number $b_2(K) = 22$ and the signature $\sigma(K) = 16$ is given by

$$b_2(K) = \frac{11}{8}|\sigma(K)|.$$

Matsumoto conjectured the relation

$$b_2(M) \geq \frac{11}{8}|\sigma(M)|,$$

for **all** smooth, simply-connected 4-manifolds M with intersection form $2kE_8 \oplus lH$. The Matsumoto conjecture (11/8 conjecture) is that the condition at the end of the previous theorem is actually $l \geq 3k$. With the help of Seiberg-Witten theory, Furata [Furata (2001)] was able to prove the weaker relation

$$b_2(M) \geq \frac{5}{4}|\sigma(M)| + 2$$

know as the $\frac{10}{8}$-*conjecture*. Thus the only remaining question for answering **Question 1** in the simply connected case lies in the difference between Furata result and the $\frac{11}{8}$-conjecture.

To review the argumentation to answer **Question 1**,

(1) From Poincaré duality it is easy to show that the Q for any closed 4-manifold is unimodular, and thus either definite or indefinite.
(2) If it is definite, Donaldson's theorem, 8.11, shows that it must be $n < \pm 1 >$.
(3) If it is indefinite, theorem 6.18 shows that it is either a sum of multiples of $< \pm 1 >$ or sums of E_8 and H with coefficients to which Furata theorem and the $\frac{11}{8}$ conjecture apply.

At the end of Chapter 6 in this book, or §1.3 of [Gompf and Stipsicz (1999)], explicit examples are provided of smooth manifolds having intersection forms described in (1),(2),(3) above, thus proving their existence.

In contrast to the result of Kirby and Siebenmann [Kirby and Siebenmann (1977)] in dimension ≥ 5, there are no finiteness results on the number of non-diffeomorphic smooth structures on a topological 4-manifold, but the following is known:

Theorem 8.13. [Friedman and Morgan (1994)] *The simply-connected, topological manifolds corresponding to the intersection form $2n(-E_8) \oplus (4n-1)H$ ($n \geq 1$) and $(2k-1)\langle 1 \rangle \oplus N\langle -1 \rangle$ ($k \geq 1$, $N \geq 10k - 1$) each carry infinitely many distinct (non-diffeomorphic) smooth structures.*

In Chapter 9 we will review constructions by Fintushel and Stern [Fintushel and Stern (1996)] of a large family of smooth structures on 4-manifolds.

Theorem 8.14. [Fintushel and Stern (1996)] *Let X be a simply-connected 4-manifold with $b_2(X) > 1$ which contains an embedded torus F with no self-intersections and the complement of F in X is still simply connected. Then there are infinitely many distinct smooth structures on X (labeled by the number of knots).*

Throughout the previous part of this section we assumed that the 4-manifolds were simply connected. This assumption can be relaxed in some cases, but the general case (arbitrary fundamental group) is too difficult, since:

Theorem 8.15. *For every finitely presented group G there is a smooth, closed, oriented 4-manifold X with $\pi_1(X) \cong G$.*

The invariants we have discussed until now depend only on the homotopy type (which is the same as the homeomorphism type according to Freedman) of the manifold. All these invariants are given by algebraic topology (fundamental group, cohomology ring $H^*(X;\mathbb{Z})$). For smooth structures we need finer invariants defined by topological invariants of the moduli space of an appropriate gauge theory. The corresponding invariants of Donaldson and Seiberg-Witten are defined in chapter 9 and provide new tools to distinguish homeomorphic but non-diffeomorphic 4-manifolds.

8.4 The First Constructions of Exotic $\mathbb{R}^4$

In this section we present a brief review of the first two constructions of exotic $\mathbb{R}^4$. A critical component was provided by Donaldson theory. Consider a closed, simply-connected, smooth 4-manifold with negative or positive definite intersection form which is diagonalizable over $\mathbb{Z}$. Next, define a surgery such that the final intersection form cannot be diagonalizable over $\mathbb{Z}$. By Donaldson's theorem, this manifold admits no smooth structure. The excised part generates the exotic $\mathbb{R}^4$ by considering the embedding of the complement of one point. We now sketch the argument, with a few details.

The K3 surface

Consider the following set:

$$K3 = \{[z_1, z_2, z_3, z_4] \in \mathbb{C}P^3 \mid z_1^4 + z_2^4 + z_3^4 + z_4^4 = 0\ \},$$

which is a complex 2-manifold, and a real 4-manifold. This manifold is called the Kummer or K3 surface[4] and was an important starting point in constructing an early example of a non-smoothable topological space. For a detailed description of this manifold refer to section 6.7.2. The intersection form of $K3$ is

$$Q_{K3} = E_8 \oplus E_8 \oplus (\oplus_3 \begin{pmatrix} 0 & 1 \\ 1 & 0 \end{pmatrix}) := 2E_8 \oplus 3H$$

where the last expression contains three copies of the matrix H also known as the hyperbolic form. The proof that $K3$ actually has this intersection form can be seen in Milnor[Milnor (1958)].

According to Freedman (theorem 8.8, also [Freedman (1982)]), there exists a unique TOP manifold with given Q_{K3} built from connected sums of standard manifolds, $|E_8|\#|E_8|\#(\#_3 S^2 \times S^2)$. However, the intersection form, E_8 is not diagonalizable over $\mathbb{Z}^5$, so, from Donaldson's theorem (our theorem 8.11 and [Donaldson (1983)]), this manifold and also $|E_8|\#|E_8| = |E_8 \oplus E_8|$ **cannot be realized by a smooth manifold.** Nevertheless, the surprising result is that there is a manifold, W homeomorphic $\mathbb{R}^4$ which is diffeomorphic to standard $\mathbb{R}^4$ if and only if it $|E_8|$ is smoothable. This will establish W as the desired **fake** $\mathbb{R}^4$. The proof depends on the fact that although K3 is smooth, it contains a topological part with intersection form $E_8 \oplus E_8$ that cannot be smoothed by the previous argument.

Consider the manifold $X = \#_3 S^2 \times S^2 \setminus \text{Int } D^4$. A corollary to Freedman's theorem is:

Corollary 8.1. *A subset of $K3$ representing the homology of $3H$ in the intersection form of $K3$ can be presented as a topologically collared embedding of X in $K3$, i.e. a topological embedding $i : X \hookrightarrow K3$ with a neighborhood of $i(\partial X)$ which is a product, $C_i = \partial(i(X)) \times \mathbb{R}$. Furthermore, there is another collared embedding of $j(X)$, back into $\#_3 S^2 \times S^2$, with collar $C_j = \partial(j(X)) \times \mathbb{R}$ such that $U = i(X) \cup C_i$ and $V = j(X) \cup C_j$ equipped with the induced smooth structures are diffeomorphic.*

The diffeomorphism $\phi : U \to V$ can be chosen such that $\phi \circ i = j$, i.e. the

---

[4]Named after the three great mathematicians Kummer, Kronecker and Kodaira.

[5]This fact can be simply proven by calculating the eigenvalues of E_8. If a form has at least one non-integer eigenvalue then it cannot be diagonalizable over the integers. See section 6.7.2.

following diagram commutes.

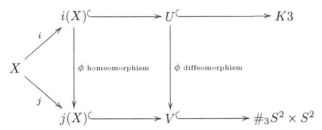

It should be noted that Freedman in his review [Freedman (1984)] used the term "bizarre" to describe the embedding, j, of X, which is a subset of $S^2 \times S^2$, back into that set in such a way the map ϕ (of the collared neighborhoods) is smooth. In other words, we only know that i is a homeomorphism, but adding the collars, C_i, C_j, results in a diffeomorphism between U and V.

Now an application of the 5-dimensional topological h-cobordism theorem shows that $W = \#_3 S^2 \times S^2 \setminus j(X)$ is homeomorphic to $\mathbb{R}^4$. Since $j(X)$ is the homeomorphic image of $X = \#_3 S^2 \times S^2 - \text{Int}D^4$, it seems reasonable that $W = \#_3 S^2 \times S^2 - j(X)$ would be homeomorphic to $\text{Int}D^4 = \mathbb{R}^4$. However, we will review a more careful argument.

8.4.1 The first exotic $\mathbb{R}^4$

In [Freedman (1982)], Freedman gave the following criteria for the existence of this homeomorphism.

Theorem 8.16. *Any non-compact 4-manifold W without boundary which is simply-connected, satisfies $H_2(W, \mathbb{Z}) = 0$, and has a single end homeomorphic to $S^3 \times [0, \infty)$ is homeomorphic to $\mathbb{R}^4$.*

Thus it is only necessary to check these three conditions to show that $W \simeq_{homeo} \mathbb{R}^4$.

Remark 8.2.
Since $\#_3 S^2 \times S^2$ is simply-connected and is the union of two open subspaces W, V with intersection $W \cap V \subset C_j (\simeq_{homeo} S^3 \times \mathbb{R})$, and since $V, W \cap V$ are simply-connected, Van Kampen's theorem shows that W is also simply-connected. Applying the Mayer-Vietoris sequence in homology to the same subspaces $(W, V, W \cap V)$ one proves that $H_2(W, \mathbb{Z}) = 0$ since $j : X \to \#_3 S^2 \times S^3$ represents the homology of $\#_3 S^2 \times S^3$ (see the corollary above). Lastly, because the embedding j is collared, W has a single end homeomorphic to $S^3 \times [0, \infty)$.

In the next step we check whether W is diffeomorphic to the standard $\mathbb{R}^4{}_{st}$ (see Figure 8.1). The idea is to construct a compact subset $D_0^4 \subset W$

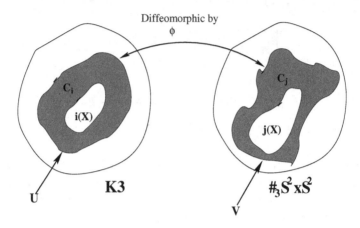

Fig. 8.1 Kirby's example of an exotic $\mathbb{R}^4$

which cannot be contained in any smoothly embedded S^3. Clearly, such a property is not compatible with the standard smoothness on $\mathbb{R}^4$. The line of argument we present follows that of the review of Freedman[Freedman (1984)].

Since $j(X) \subset V$, W intersects an open subset of C_j. Since W is homeomorphic to $\mathbb{R}^4$ it contains an infinitely expanding family of D^4's each having a topological S^3 as boundary. Eventually one of these, say D_0^4, will have a boundary, S_0^3 contained in C_j. Suppose S_0^3 could in fact be **smoothly** embedded in W. Then $S_i^3 \equiv \phi^{-1}(S_0^3)$ would be a smooth subset of $C_i \subset K' = K3 - i(X)$ and in fact separate the two ends of C_i and thus $K3$ itself. That is, we can write

$$K3 = K' \cup_{S_i^3} H'$$

with H' containing $i(X)$ and S_i^3 smoothly embedded. Clearly the smoothly embedded S_i^3 can then be identified as the boundary of a smooth disk, $S_i^3 = \partial D^4$, resulting in

$$K'' = K' \cup_{S_i^3} D^4, \qquad (8.1)$$

defining K'' as a closed smooth manifold with $Q_{K''} = 2E_8$, which is impossible from Donaldson's result, theorem 8.11. This implies that the assumption that the ever increasing disks in W will have smoothly embedded boundaries is false. So W cannot have the standard smoothness of $\mathbb{R}^4$.

- Thus there is no diffeomorphism between W and $\mathbb{R}^4{}_{st}$. The space W is topological, but not smooth $\mathbb{R}^4$, and is thus referred to as exotic $\mathbb{R}^4$.

8.5 The Infinite Proliferation of Exotic $\mathbb{R}^4$

In this section we will review the construction of families of (non-diffeomorphic) exotic $\mathbb{R}^4$'s labeled by one and then two real numbers. This proves that there are uncountably many exotic smooth structures on $\mathbb{R}^4$. Furthermore, it is possible to divide the set of exotic $\mathbb{R}^4$'s into two classes: large and small exotic according to whether or not they can be embedded smoothly in compact subsets of standard $\mathbb{R}^4$. The work of Gompf, Taubes and others led to these results.

The previous subsection introduced the two embeddings i and j of $X = \#_3 S^2 \times S^2 \setminus \text{Int } D^4$ into K and $\#_3 S^2 \times S^2$ to obtain the first exotic $\mathbb{R}^4$. The argument included the collared neighborhoods $U = i(X) \cup C_i$ and $V = j(X) \cup C_j$ diffeomorphic with respect to the induced smooth structures. Because, C_i is a product neighborhood of $i(\partial X)$, C_i is homeomorphic to $(0,1) \times S^3$. With respect to this homeomorphism, the image of $\{t\} \times S^3$ ($t \in (0,1)$) in C_i will be denoted by S_t. Hence $i(\partial X)$ can be expressed near the 0-end as $\bigcup_{t \in (0,\epsilon)} S_t$ with ϵ small. Define $U_r = i(X) \cup \bigcup_{t \in (0,r)} S_t$, $V_r = \phi(U_r)$, and $\mathbb{R}^4{}_r = \#_3 S^2 \times S^2 \setminus V_r$. The argument in the previous section can be used to show that $\mathbb{R}^4{}_r$ is homeomorphic to $\mathbb{R}^4$ for all $r \in (0,1)$. Taubes [Taubes (1987)] extended Donaldson's work to special non-compact manifolds (with one periodic end). His work and that of Gompf led to [Gompf (1985); Taubes (1987)]:

Theorem 8.17. *Let $\mathbb{R}^4{}_r$ and $\mathbb{R}^4{}_s$, $0 < r < s < 1$ be two manifolds homeomorphic to the $\mathbb{R}^4$ defined by the previous process. Then as long as $r \neq s$, the induced smoothness structures inherited from $\#_3 S^2 \times S^2$ make $\mathbb{R}^4_r$ and $\mathbb{R}^4_s$ non-diffeomorphic smooth manifolds. Thus, there are uncountably infinite many non-diffeomorphic exotic structures on topological $\mathbb{R}^4$.*

Gompf's end sum construction

Gompf [Gompf (1985)] introduced an important tool for finding new exotic $\mathbb{R}^4$ from others.

Definition 8.1. Let R, R' be two topological $\mathbb{R}^4$'s. The end-sum $R \natural R'$ is defined as follows: Let $\gamma : [0, \infty) \to R$ and $\gamma' : [0, \infty) \to R'$ be smooth properly embedded rays with tubular neighborhoods $\nu \subset R$ and $\nu' \subset R'$, respectively. For convenience, identify the two semi-infinite intervals with $[0, 1/2)$, and $(1/2, 1]$ leading to diffeomorphisms, $\phi : \nu \to [0, 1/2) \times \mathbb{R}^3$ and $\phi' : \nu' \to (1/2, 1] \times \mathbb{R}^3$. Then define

$$R \natural R' = R \cup_\phi I \times \mathbb{R}^3 \cup_{\phi'} R'$$

as the **end sum** of R and R'.

With a little checking, it is easy to see that this construction leads to $R \natural R'$ as another topological $\mathbb{R}^4$. However, if R, R' are themselves exotic, then so will $R \natural R'$ and in fact, it will be a "new" exotic manifold, since it will not be diffeomorphic to either R or R'. Gompf used this technique to construct a class of exotic $\mathbb{R}^4$'s none of which can be embedded smoothly in the standard $\mathbb{R}^4$.

2-parameter family of exotic $\mathbb{R}^4$

With the help of this sum, it is possible to construct a 2-parameter family of exotic $\mathbb{R}^4$'s. Gompf's original construction uses a special manifold in which the exotic $\mathbb{R}^4$ can be embedded. By an iteration process he obtained a class of countably infinite exotic $\mathbb{R}^4$. In the same paper [Gompf (1985)] the result of Taubes was used to produce a 2-parameter family of uncountably many exotic $\mathbb{R}^4$. The difference between this family and Taubes construction is given by the fact that this class of exotic $\mathbb{R}^4$'s changes by an orientation-reversing diffeomorphism. That is, R_s and $\overline{R}_s$ are non-diffeomorphic $\mathbb{R}^4$'s.

Theorem 8.18. *Let R_s be the one-parameter family in theorem 8.17. Then the family $\{R_{s,t} = R_s \natural \overline{R}_t \,|\, 0 < s, t < \infty\}$ defines a 2-parameter family of exotic $\mathbb{R}^4$'s such that $R_{s,t}$ embeds (orientation-preserving) in $R_{s',t'}$ if and only if $s \leq s'$ and $t \leq t'$.*

Of course, the space $R_{0,0}$ is the standard $\mathbb{R}^4$. A natural question is: what happens for the limit $R_{\infty,\infty}$?

Freedman and Taylor's universal exotic $\mathbb{R}^4$

Freedman and Taylor [Freedman and Taylor (1986)] constructed a certain smoothing of the half space $\frac{1}{2}\mathbb{R}^4 = \{(x_1, x_2, x_3, x_4)|\, x_4 \geq 0\}$ denoted by H with the properties:

- H contains all other smoothings of $\frac{1}{2}\mathbb{R}^4$
- H is unique with respect to the previous property
- the interior of H denoted by U can be naturally identified with a smoothing of $\mathbb{R}^4$

- U contains every smoothing of $\mathbb{R}^4$ embedded within it.

Thus, U is called a universal $\mathbb{R}^4$.
Gompf conjectured in [Gompf (1985)] that the space $R_{\infty,\infty}$ is diffeomorphic to the universal space U. That would imply that the moduli space of equivalence classes of exotic smoothings of $\mathbb{R}^4$ forms a space which is equivalent to a Riemann sphere S^2 rather than to an $\mathbb{R}^2$.

8.5.1 The existence of two classes

In the previous subsection we reviewed how Gompf defined the 2-parameter family determining the cardinality of the set $\mathfrak{R}$ of exotic $\mathbb{R}^4$'s. In the following we review another construction of exotic $\mathbb{R}^4$'s from the failure of the smooth h-cobordism theorem. Such $\mathbb{R}^4$'s are called *ribbon* $\mathbb{R}^4$'s. In contrast to the previously defined exotic $\mathbb{R}^4$'s these ribbon $\mathbb{R}^4$'s can be embedded into the standard smooth 4-sphere S^4 (or into the standard $\mathbb{R}^4$). This leads to the division of the set $\mathfrak{R}$ into two classes defined via the existence of an embedding into S^4. In the first class (non-existence of embedding) the exoticness of the $\mathbb{R}^4$ is "located" at the end of the space which means in some sense the neighborhood of the infinity. Then, the second class (existence of the embedding) can be characterized by the property that the exoticness is in some sense "localized" in the interior of $\mathbb{R}^4$. Thus, this is a candidate of an exotic $\mathbb{R}^4$ which can be used as coordinate patch. A description of these two classes follows.

The failure of the smooth h-cobordism theorem and ribbon $\mathbb{R}^4$
In 1975 Casson (Lecture 3 in [Casson (1986)]) described a smooth 5-dimensional h-cobordism between compact 4-manifolds and showed that they "differed" by two proper homotopy $\mathbb{R}^4$'s (see below). Freedman knew, as an application of his proper h-cobordism theorem, that the proper homotopy $\mathbb{R}^4$'s were $\mathbb{R}^4$. After hearing of Donaldson's work in March 1983, Freedman realized that there should be exotic $\mathbb{R}^4$'s and, to find one, he produced the second part of the construction below involving the smooth embedding of the proper homotopy $\mathbb{R}^4$'s in S^4. Unfortunately, it was necessary to have a compact counterexample to the smooth h-cobordism conjecture, and Donaldson did not provide this until 1985 [Donaldson (1987)]. The idea of the construction is simply given by the fact formulated in theorem 6.16 that every such smooth h-cobordism between non-diffeomorphic 4-manifolds can be written as a product cobordism except for a compact contractible sub-h-cobordism V, the Akbulut cork. An open subset $U \subset V$

homeomorphic to $[0,1] \times \mathbb{R}^4$ is the corresponding sub-h-cobordism between two exotic $\mathbb{R}^4$'s. These exotic $\mathbb{R}^4$'s are called ribbon $\mathbb{R}^4$. They have the important property of being diffeomorphic to open subsets of the standard $\mathbb{R}^4$. That stands in contrast to the previous defined examples of Kirby, Gompf and Taubes.

Akbulut corks and exotic $\mathbb{R}^4$'s

To be more precise, consider a pair (X_+, X_-) of homeomorphic, smooth, closed, simply-connected 4-manifolds. The transformation from X_- to X_+ visualized by a h-cobordism can be described by the following construction.

Theorem 8.19. *Let W be a smooth h-cobordism between closed, simply connected 4-manifolds X_- and X_+. Then there is an open subset $U \subset W$ homeomorphic to $[0,1] \times \mathbb{R}^4$ with a compact subset $K \subset U$ such that the pair $(W \setminus K, U \setminus K)$ is diffeomorphic to a product $[0,1] \times (X_- \setminus K, U \cap X_- \setminus K)$. The subsets $R_\pm = U \cap X_\pm$ (homeomorphic to $\mathbb{R}^4$) are diffeomorphic to open subsets of $\mathbb{R}^4$. If X_- and X_+ are not diffeomorphic, then there is no smooth 4-ball in $R_\pm$ containing the compact set $Y_\pm = K \cap R_\pm$, so both $R_\pm$ are exotic $\mathbb{R}^4$'s.*

Thus, remove a certain contractible, smooth, compact 4-manifold $Y_- \subset X_-$ (called an Akbulut cork) from X_-, and re-glue it by an involution of ∂Y_-, i.e. a diffeomorphism $\tau : \partial Y_- \to \partial Y_-$ with $\tau \circ \tau = Id$ and $\tau(p) \neq \pm p$ for all $p \in \partial Y_-$. This argument was modified above so that it works for a contractible *open* subset $R_- \subset X_-$ with similar properties, such that R_- will be an exotic $\mathbb{R}^4$ if X_+ is not diffeomorphic to X_-. In the next section we will see how this results in the construction of handlebodies of exotic $\mathbb{R}^4$.

Structures on the set $\mathfrak{R}$ of smoothings of $\mathbb{R}^4$

Let $\mathfrak{R}$ be the set of smoothings of $\mathbb{R}^4$ up to orientations preserving diffeomorphisms. The end-sum $R_1 \natural R_2$ between two elements $R_1, R_2 \in \mathfrak{R}$ gives $\mathfrak{R}$ the structure of a monoid $(\mathfrak{R}, \natural)$. Furthermore following [Gompf (1985)], define the following partial order $\leq$ on $\mathcal{R}$.

Definition 8.2. *Let $R_1, R_2 \in \mathfrak{R}$ and write $R_1 \leq R_2$ if every compact 4-manifold R_1 smoothly embeds (preserving orientation) in R_2. If $R_1 \leq R_2 \leq R_1$ then R_1 and R_2 are defined as compactly equivalent. The set of compact equivalence classes in $\mathfrak{R}$ is denoted by $\mathfrak{R}_\sim$.*

It can be proven that the monoid structure of $\mathfrak{R}$ descends to $\mathfrak{R}_\sim$.

Definition 8.3. Call an exotic $\mathbb{R}^4$ *large* if it contains a 4-dimensional compact submanifold that cannot be smoothly embedded in standard $\mathbb{R}^4$. Otherwise call the exotic $\mathbb{R}^4$ *small*.

Note that a $R_1 \not\subset \mathbb{R}^4$ is compactly equivalent to $\mathbb{R}^4$ if and only if it is a small exotic $\mathbb{R}^4$. Thus, the class of exotic small $\mathbb{R}^4$ maps to a point in $\mathfrak{R}_\sim$, the equivalence class of $\mathbb{R}^4$. At the same time, this equivalence class is the unique minimal element with respect to the relation $\leq$. The unique maximal element U is the universal $\mathbb{R}^4$ of Freedman and Taylor [Freedman and Taylor (1986)]. In [Gompf (1989a)], Gompf introduced a metrizable topology with countable basis into $\mathfrak{R}_\sim$ although the usefulness of this topology is not presently clear.

Taylor's invariant
Taylor's invariant is conjectured to distinguish between large and small $\mathbb{R}^4$'s. It is defined by

Definition 8.4. For $R \in \mathfrak{R}$, define $\gamma(R) \in \{0, 1, 2, \ldots, \infty\}$ to be

$$\gamma(R) = \sup_K \{\min_X \{\frac{1}{2} b_2(X)\}\},$$

where K ranges over compact 4-manifolds embedding in R and X ranges over closed, spin 4-manifolds with signature 0 in which K smoothly embeds.

Clearly, if $R_1 \leq R_2$ then $\gamma(R_1) \leq \gamma(R_2)$, so γ is well-defined on compact equivalence classes (see Definition 8.2) and gives an order-preserving function $\gamma : \mathfrak{R}_\sim \to \{0, 1, 2, \ldots, \infty\}$. For R to be a small $\mathbb{R}^4$ then $\gamma(R) = 0$ but it is not clear that $\gamma(R) > 0$ for all large exotic $\mathbb{R}^4$. A striking application by Taylor is that for R any exotic $\mathbb{R}^4$ with $\gamma(R) > 0$, any handle decomposition of R must have infinitely many 3-handles, in contrast to the examples of small exotic $\mathbb{R}^4$'s considered in the next section, which are built without 3-handles.

Remark 8.3.
Following are some properties of this invariant $\gamma(R)$:
- If $R \in \mathfrak{R}$ has a handle decomposition with only finitely many 3-handles, then $\gamma(R) = 0$.
- If R_1 and R_2 have diffeomorphic ends then $\gamma(R_1) = \gamma(R_2)$.
- Any end sum satisfies $\sup_n \{\gamma(R_n)\} \leq \gamma(\natural_n R_n) \leq \sum_n \gamma(R_n)$. In particular, for R_∞ satisfies $\gamma(R_\infty) = \sup_n \{\gamma(R_n)\}$.
- For infinitely many n including $n = \infty$ there are exotic $\mathbb{R}^4$'s $L_n \subset \mathbb{C}P^2$ with $\gamma(L_n) = n$.
- If $\gamma(R) = \infty$ then R does not embed in any compact spin 4-manifold.

Using this invariant, Taylor [Taylor (1998)] obtained results related to the isometry group of some exotic $\mathbb{R}^4$'s. A smooth manifold has *few symmetries* provided that, for every choice of a smooth (C^1 or better) metric, the isometry group for that metric is finite. Let E^4 be a smooth manifold homeomorphic to $\mathbb{R}^4$. Recall, an embedding of a manifold M into W is *(topologically) flat* if $M \times [0,1]$ embeds into W. A flat embedding $S^3 \subset E^4$ is a *barrier* S^3 provided that, given any open set $U \subset E^4$ containing S^3 and any smooth embedding $e \colon U \to E^4$, then $e(S^3) \cap S^3 \neq \emptyset$. Note that the flatness of the embedding $S^3 \hookrightarrow E^4$ is essential, because it determines how S^3 lies inside of E^4. Also, the embedding e must be required to be smooth, or otherwise, the whole construction is trivial. If there is an E^4 which is not diffeomorphic to the standard $\mathbb{R}^4$, then there is an exotic region inside of E^4 which is not separated from the rest by a smooth(!) embedding of some S^3. In that case there is no barrier S^3 surrounding the exotic region.

Given a barrier $S^3 \subset E^4$, the *inside* is the component of $E^4 - S^3$ whose closure is compact. Note that it is a smoothing of $\mathbb{R}^4$.

Theorem 8.20. *Let E^4 be a smoothing of $\mathbb{R}^4$ with a barrier S^3 whose inside does not smoothly embed in any integral homology 4-sphere. Then E has few symmetries.*

In [Taylor (1997)], Taylor constructed many examples of smooth $\mathbb{R}^4$'s. There are examples for which the isometry group is not trivial: e.g. the end-connected sum of E with itself. Furthermore, any E^4 which embeds in the standard $\mathbb{R}^4$ has no barriers. Based on these results, Sładowski [Sładkowski (1999)] showed that an empty exotic 4-space (like E^4) act as a non-trivial gravitational system, see section 10.3.

8.6 Explicit Descriptions of Exotic $\mathbb{R}^4$'s

In this section we look at the problem of constructing a coordinate patch representation of an exotic $\mathbb{R}^4$. From the mathematical point of view, this representation is equivalent to describing the handle body decomposition. In [Bižaca and Gompf (1996)], such a handle body is constructed from the existence of an Akbulut cork (see theorem 6.16). Of course the handle body, and thus the coordinate patch presentation, is infinite.

Start with the construction of the Akbulut cork. The procedure in [Bižaca and Gompf (1996)] has two main steps (we follow the argumentation in [Gompf and Stipsicz (1999)] very closely.):

- construct an Akbulut cork A_1 in a special manifold $E(n) \# \mathbb{C}P^2$
- construct an h-cobordism V of A_1 to itself relative to the product struc-

ture $[0,1] \times \partial A_1$
- construct from that a non-trivial (=non-product) h-cobordism W_n between $E(n)\#\mathbb{C}P^2$ and $X_n \approx \#(2n-1)\mathbb{C}P^2\#10n\overline{\mathbb{C}P}^2$ such that the pair (W_n, V) satisfies the conclusion of Theorem 8.19
- apply Theorem 8.19 to each of the examples W_n
- construct the exotic $\mathbb{R}^4$, R, as the interior of some subset of W_n

Since all of the handles at each stage have well known coordinate patch representations, this results in a coordinate patch representation, albeit infinite, of an exotic $\mathbb{R}^4$.

- begin with the handle $S^1 \times D^3$
- glue one 2-handle along the boundary $\partial(S^1 \times D^3) = S^1 \times S^2$ according to Link in Figure 8.2 (the circle with the dot represents $S^1 \times D^3$) to get the Akbulut cork V
- consider an infinite sequence of 2-handles $D^2 \times D^2$ with one self-intersection (self-plumbed handle) homeomorphic to $D^2 \times \mathbb{R}^2$, or a Casson handle CH
- glue the Casson handle to the boundary of the Akbulut cork V according to Figure 8.3
- the interior of the resulting manifold is an exotic $\mathbb{R}^4$

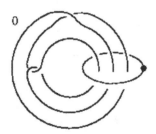

Fig. 8.2   Handle decomposition of the Akbulut cork

Note that the gluing of the Casson handle to the Akbulut cork is absolutely necessary. According to Freedman [Freedman (1982)], the interior of every Casson handle is diffeomorphic to the standard $\mathbb{R}^4$. But after the gluing to the Akbulut cork, the interior is modified to be an exotic $\mathbb{R}^4$. Thus, for the simplest Casson handle the *exotic* $\mathbb{R}^4$ R [Bižaca and Gompf (1996)] is obtained as the interior of Figure 8.3, displaying the formalized coordinate

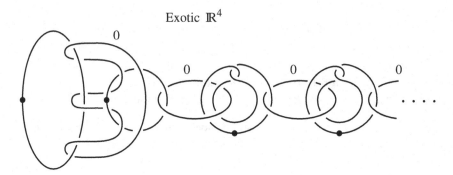

Fig. 8.3 Simplest handle decomposition of an (small) exotic $\mathbb{R}^4$ (figure from [Gompf and Stipsicz (1999)] p. 207, Fig. 6.16)

description of an exotic $\mathbb{R}^4$ by an infinite handle body containing only 2-handles. This special exotic $\mathbb{R}^4$ admits a diffeomorphism to some open subset of the standard $\mathbb{R}^4$.

8.7 Other Non-compact 4-manifolds

Before we proceed to the modern approaches Seiberg-Witten theory, consider other non-compact manifolds different from the Euclidian space $\mathbb{R}^4$. For all detailed theorems refer to [Gompf and Stipsicz (1999)] (see Corollary 9.4.25 in [Gompf and Stipsicz (1999)],p. 378, Theorem 9.4.24 is the corresponding more general, technical theorem). Recall that every non-compact 4-manifold can be chosen to admit a smooth structure. Let X be a connected, compact, topological 4-manifold X, so the non-compact manifold $X \setminus \{*\}$ is always smoothable. There are various results of this kind, e.g., given a 3-manifold $M = \partial Y$ then the non-compact 4-manifold $Y \setminus M$ admits uncountably infinite many smooth structures. The following list of examples can be extracted from the general result in [Gompf and Stipsicz (1999)]. Assuming that X is a connected topological 4-manifold and M a compact 3-manifold, then there is the following partial list of spaces with uncountably many smoothings:

- $S^3 \times \mathbb{R}$
- $X \setminus \{*\}$
- $X \setminus M$ if M is a rational homology sphere except for some linking forms (see [Edmonds (1999)] for the exceptions)

- $M \times \mathbb{R}$ if M admits a smooth embedding into $\#_n \overline{\mathbb{C}P}^2$ for some n

So far, these are the strongest results.

Chapter 9
Seiberg-Witten Theory: The Modern Approach

This chapter is dedicated to a review of the Seiberg-Witten invariants, derived from the Seiberg-Witten equations introduced in Chapter 5. These invariants have provided additional powerful tools for understanding the differential topological problems of low dimensional manifolds. Unfortunately the topic involves many technical issues which we will not be able to cover in any detail. A more thorough introduction can be found in [Akbulut (1996); Morgan (1996)] and a briefer one with a view of the topological applications in [Gompf and Stipsicz (1999)].

We begin by presenting the Seiberg-Witten field equations which replace the Yang-Mills equations of Donaldson theory. These equations provide a generalization in the sense that they require the existence only of a $Spin_C$ structure, whereas Yang-Mills require a Spin structure. From the moduli space of (perturbative) solutions to the SW equations, the SW invariants are defined. These play a role generalizing Donaldson invariants defined in section 5.7 for the Yang-Mills moduli spaces. In the SW case, the invariants are maps from the set of $Spin_C$ structures to the integers. Most importantly, these invariants characterize the smoothness of the base manifold, at least partly. That is, if two manifolds are diffeomorphic then they must have the same SW invariants, but the converse is not necessarily true in general, as shown by counter examples of Fintushel and Stern. After defining these invariants we look briefly at surgery on manifolds, the so-called logarithmic and knot transformations which change the SW invariants, and thus the diffeomorphism class, without changing the topology. We end the chapter with some brief comments on cohomotopy extensions of SW work.

9.1 The Construction of the Moduli Space

The moduli space of Seiberg-Witten (SW) formalism is formed from solutions to the SW equations to first perturbation order, as defined below. In order to state these equations, we need to review some facts about bundles with Spin_C structures introduced in 5.4. SW theory extends Donaldson theory by requiring the existence only of a Spin_C structure rather than a full Spin one. This is a weaker condition on the bundle. In fact, let M be a smooth, simply connected, closed, oriented 4-manifold with $b_2^+(M)$ odd and $b_2^+(M) > 1$. As pointed out in section 5.4, any such manifold can be the base space for at least one bundle with a Spin_C structure, although it may not support a Spin structure. Furthermore, the set of such Spin_C bundles is characterized by the second Stiefel-Whitney class, $w_2(M)$, which is an element of $H^2(M;\mathbb{Z})$ mod 2. So,

$$\mathcal{C}_M = \{K \in H^2(M;\mathbb{Z}) \mid K \equiv w_2(M) \ (\mod 2)\} \tag{9.1}$$

is the set of the so-called characteristic elements defining the family of inequivalent Spin_C structures over M. Recall that the Spin_C-group decomposes as

$$\text{Spin}_C(4) = \{B\} = \left\{ \begin{pmatrix} \lambda A_+ & 0 \\ 0 & \lambda A_- \end{pmatrix} : A_\pm \in SU(2), \lambda \in U(1) \right\}. \tag{9.2}$$

Note that this group acts on $\mathbb{C}^4$. (9.2) results in a homomorphism $\alpha : \text{Spin}_C(4) \to U(1)$ with $\alpha(B) = \det(\lambda A_+) = \det(\lambda A_-) = \lambda^2$. Let W be a Spin_C spinor bundle, with fiber $\mathbb{C}^4$, and cross sections Spin_C spinors. The fiber map, α, gives rise to a map, ρ, from the bundle W onto a line bundle, L, whose fiber elements are the λ^2 in (9.2). This is called the determinant line bundle. The first Chern class of L defines $w_2(M)$, as discussed in Proposition 2.4.16 of [Gompf and Stipsicz (1999)]. In fact, it is this characteristic class that identifies the Spin_C bundle. Using the matrix decomposition in equation (9.2), we have the natural decomposition $W = W^+ \oplus W^-$, where the groups in each factor are copies of $U(2)$ and the fibers are copies of $\mathbb{C}^2$.

Remark 9.1.

Although the group $\text{Spin}(n)$ is originally defined over the reals in terms of a Clifford algebra generated by an orthonormal tangent vector basis, $e_1, ..., e_n$, there are familiar ways of providing complex representations in the interesting cases $n = 2$ and $n = 4$. This provides the basis for equation (9.2). Note that multiplication by a single vector, regarded as a Clifford algebra element of degree one, maps $W^\pm \to W^\mp$.

This same splitting leads to a (local) bundle splitting with $W^{\pm} = S^{\pm} \otimes \sqrt{L}$.

Remark 9.2.
The complex line bundle $\sqrt{L}$ is something like the "square root" of a complex line bundle L, corresponding locally to extracting the $U(1)$ fiber element, λ^2, into one of its two square roots. The ± 1 ambiguity may not be resolvable globally, so it may not be possible to define $\sqrt{L}$ globally as a bundle, so in fact Moore [Moore (2001)] describes it as a *virtual* bundle.

Using the notation above, every element $K \in \mathcal{C}_M$ leads to a unique definition of a Spin$_C$ bundle, W, and in the following we will speak of K as Spin$_C$-structure. Let $\mathcal{A}_L$ denote the space of $U(1)$-connections on L. By choosing $A \in \mathcal{A}_L$ and coupling it to the Levi-Civita connection ∇ on M one gets a covariant differentiation $\nabla_A \colon \Gamma(M; W^+) \to \Gamma(M; W^+ \otimes T^*M)$. Note, the bundle W is a hermitian vector bundle of complex rank 4. We introduce the notation $\gamma(v), v \in TM$ to denote the vector v acting as a generator of the Clifford algebra. We can define a map $\gamma : TM \to End(W)$ satisfying $\gamma(v) + (\gamma(v))^* = 0$, $\gamma(v)(\gamma(v))^* = -||v||^2 Id$ and $\gamma(v)\gamma(w) = -\gamma(w)\gamma(v)$ for all $v, w \in TM$ with $||w - v|| > 0$. Then the Spin$_C$ connection ∇_A on W is defined by:

$$(\nabla_A)_v(\gamma(w)\Phi) = \gamma(w)(\nabla_A)_v \Phi + \gamma(\nabla_v w)\Phi$$

for all sections $\Phi \in \Gamma(M; W)$ and $v, w \in \Gamma(M; TM)$. This connection maps sections of $W^{\pm}$ to $W^{\pm}$. Note that the map $\gamma : TM \to End(W)$ extends Clifford multiplication to $C : W^+ \otimes TM \to W^-$ by $C(v, \Phi) = \gamma(v)\Phi$.

Definition 9.1. The composition of the Clifford multiplication C and ∇_A gives an operator

$$\displaystyle{\not{\partial}}_A = C \circ \nabla_A \colon \Gamma(M; W^+) \to \Gamma(M; W^-),$$

which is the *Dirac operator* of the Spin$_C$ structure K associated to the connection $A \in \mathcal{A}_L$.

Let $\{e_k\}$ be a base of the tangent space $T_x M$, then the Dirac operator is defined locally by:

$$\displaystyle{\not{\partial}}_A \Phi = \sum_k \gamma(e_k)(\nabla_A)_{e_k} \Phi$$

where $\Phi \in \Gamma(M; W^+)$ is a section on W^+. Note, that $\gamma(e_k)$ can be realized by the γ-matrices of Dirac. In arriving at equation (9.2) we used an Hermitian structure which extends to W fiber wise as $\langle\,,\,\rangle : W_x \times W_x \to \mathbb{C}$ with $\langle v, w\rangle = (\langle w, v\rangle)^*$. We should also recall that the very definition of

Clifford algebra and Spin or Spin_C assumed a Riemannian metric on M. Also note that on a 4-dimensional oriented, Riemannian manifold M the Hodge $*_g$-operator (given by the metric g) is defined and maps 2-forms to a 2-forms $*_g \colon \Omega^2(M) \to \Omega^2(M)$. Since $*^2 = 1$, the eigenvalues of $*$ are ± 1, and the space of 2-forms divides into corresponding eigenspaces, $\Omega^2(M) = \Omega^+(M) \oplus \Omega^-(M)$. These subspaces span the *self-dual* and *anti-self-dual* 2-forms.

As above, let A be a $U(1)$ connection of the determinant line bundle L with curvature F_A in L. A Spin_C connection ∇_A leads to Dirac operator $\displaystyle{\not}\partial_A$. The Lie algebra of $U(1)$ is isomorphic to $i\mathbb{R}$ and thus the curvature F_A given by a Lie-algebra valued 2-form is an element of $i\Omega^2(M)$ i.e. a purely imaginary 2-form on M. Let F_A^+ be the self-dual part of F_A i.e. $*_g F_A = F_A$. Then we can write down the Seiberg-Witten equations:

$$\displaystyle{\not}\partial_A \Phi = 0$$

$$F_A^+ = \frac{1}{4}\langle \gamma(e_i)\gamma(e_j)\Phi, \Phi \rangle e^i \wedge e^j \equiv \sigma(\Phi). \qquad (9.3)$$

These are an equivalent expanded form of the equations obtained from supersymmetry considerations in Chapter 5, (5.45) where $\{e^j\}$ is the dual basis for T^*M and $\Phi \in \Gamma(M; W^+)$ is a section of W^+. Refer to the remark in Chapter 5 following the equations (5.45).

The space of solutions belongs to $\Gamma(M; W^+) \times \mathcal{A}_L$ and the space of maps $Map(M, S^1)$ (the *gauge group*) acts on both $\Gamma(M; W^+)$ (by pointwise multiplication) and on $\mathcal{A}_L$ (as the automorphism group of the principal $U(1)$-bundle $P_L \to M$). In other words, the gauge group acts on the product $\Gamma(M; W^+) \times \mathcal{A}_L$. The resulting quotient space

$$\mathcal{B}_K = \frac{\Gamma(M; W^+) \times \mathcal{A}_L}{Map(M, S^1)}$$

is called the *configuration space*. Let $\mathcal{B}_K^*$ denote $\{[A, \Phi] \in \mathcal{B}_K \mid \Phi \text{ is not identically } 0\}$. We state without proof that the reducible solutions to (9.3) are given by $\Phi = 0$. Thus, the *moduli space* $\mathcal{M}_K(g)$ of irreducible solutions of the Seiberg-Witten equations is defined by

$$\mathcal{M}_K(g) = \{[A, \Phi] \in \mathcal{B}_K^* \mid [A, \Phi] \text{ satisfy } (9.3)\}$$

In general, it is not known whether $\mathcal{M}_K(g)$ can be provided a structure as a smooth manifold, but if we take a generic perturbation $\delta \in \Omega^+(M)$, the solution set $\mathcal{M}_K^\delta(g)$ of the equations $\displaystyle{\not}\partial_A \Phi = 0$, $F_A^+ + i\delta = i\sigma(\Phi)$ will be a smooth manifold. In this context a perturbation is a shift of the 2-form $F_A^+ \in \Omega^+(M)$ by a 2-form $\delta \in \Omega^+(M)$ which is small with respect to the norm $\int_M \delta \wedge \delta = ||\delta||^2$.

The next theorem summarizes the most important properties of the moduli space $\mathcal{M}_K^\delta(g)$. These properties ensure us that the very important Definition 9.2 stated below does indeed provide a diffeomorphism invariant of M.

Theorem 9.1. *Assume that M is a simply connected, oriented, closed 4-manifold with $b_2^+(M)$ odd. Fix a $Spin_C$ structure $K \in \mathcal{C}_M$. For a generic metric g and perturbation $\delta \in \Omega^+(M)$, the moduli space $\mathcal{M}_K^\delta(g)$ will be a smooth, compact manifold of dimension $d = \frac{1}{4}(K^2 - (3\sigma(M) + 2\chi(M))$. An orientation of $H^0(M;\mathbb{R}) \oplus H^+(M;\mathbb{R})$ determines an orientation for $\mathcal{M}_K^\delta(g)$. If $b_2^+(M) > 1$, the homology class $[\mathcal{M}_K^\delta(g)] \in H_d(\mathcal{B}_K;\mathbb{Z})$ is independent of the choice of g and δ. The space $\mathcal{M}_K^\delta(g)$ is homotopy equivalent to $\mathbb{C}P^\infty$.*

This is the main theorem of Seiberg-Witten theory (see [Akbulut (1996); Morgan (1996)] for a proof) determining the structure of the moduli space and is the Seiberg-Witten analog to that of the Donaldson theory in Chapter 5, Theorem 5.13.

9.2 Seiberg-Witten Invariants

We will now summarize some of the important properties of the Seiberg-Witten moduli space, leading to the definition of the Seiberg-Witten invariants. These are maps from the set of $Spin_C$ structures to integers which carry smoothness information on the underlying base space. For a more detailed discussion see [Salamon (1995)], [Akbulut (1996)] or [Morgan (1996)].

Let M be a smooth, simply connected, closed, oriented 4-manifold with $b_2^+(M)$ odd and $b_2^+(M) > 1$. The set $\mathcal{C}_M$ of the characteristic elements is identical to the set of $Spin_C$ structures on M. The moduli space $\mathcal{M}_K^\delta(g)$ of the solutions of the (perturbed) monopole equation can be defined. However, as stated in Theorem 9.1, the top homology group of this space is independent of the choice for metric g on M and perturbation $\delta \in \Omega^+(M)$, leading to the following observations:

- The cohomology ring $H^*(\mathcal{M}_K^\delta(g);\mathbb{Z}) = \mathbb{Z}[\mu]$ is induced by one generator $\mu \in H^2(\mathcal{M}_K^\delta(g);\mathbb{Z})$.
- Since $K \in \mathcal{C}_M$ and $b_2^+(M)$ is odd, $dim\ \mathcal{M}_K^\delta(g) = 2m$ is even.
- Since $\mathcal{M}_K^\delta(g)$ is a closed and oriented manifold, it defines a homology class $[\mathcal{M}_K^\delta(g)] \in H_{2m}(\mathcal{B}_K^*;\mathbb{Z})$. This class is independent of the choice

of metric, g, and perturbation, δ.
- The natural pairing $\langle\ ,\ \rangle$ between homology and cohomology is the generalization of the integral in the deRham theory.

These observations lead to the following definition.

Definition 9.2. The *Seiberg-Witten invariant* $SW_M \colon \mathcal{C}_M \to \mathbb{Z}$ is defined by the formula

$$SW_M(K) = \langle \mu^m, [\mathcal{M}_K^\delta(g)]\rangle,$$

where $dim\ \mathcal{M}_K^\delta(g) = 2m$. If $dim\ \mathcal{M}_K^\delta(g) < 0$ then by definition $SW_M(K) = 0$.

The next theorem is of central importance and shows that the function SW_M is in fact a *diffeomorphism invariant*.

Theorem 9.2. *The Seiberg-Witten function* $SW_M \colon \mathcal{C}_M \to \mathbb{Z}$ *is an invariant of the smooth 4-manifold* M. SW_M *does not depend on the chosen metric* g *and perturbation* δ. *For any orientation preserving diffeomorphism* $f \colon M \to M'$ *we have* $SW_{M'}(K) = \pm SW_M(f^*K)$.

Our discussion has been entirely superficial, but the reader can find more details in the standard literature such as [Akbulut (1996); Morgan (1996)].

The Seiberg-Witten *basic classes* (Spin$_C$ structures) are defined over M.

Definition 9.3. The cohomology class $K \in \mathcal{C}_M \subset H^2(M;\mathbb{Z})$ is a *Seiberg-Witten basic class* of M if $SW_M(K) \neq 0$. The set of basic classes of M will be denoted by $\mathcal{B}as_M \subset H^2(M;\mathbb{Z})$.

Note that $SW_M(-K) = (-1)^\epsilon SW_M(K)$, where $\epsilon = \frac{1+b_2^+(M)}{2}$ so $K \in \mathcal{B}as_M$ if and only if $-K \in \mathcal{B}as_M$. Further important properties can be divided into vanishing and non-vanishing cases.

Vanishing results:
Assume that M is a smooth, closed, simply connected, oriented 4-manifold with $b_2^+(M) > 1$ and odd.

(1) If $M = M_1 \# M_2$ and $b_2^+(M_i) > 0$ ($i = 1, 2$), then $SW_M \equiv 0$.
(2) If M admits a metric with positive scalar curvature (and, as always $b_2^+(M) > 1$ odd), then $SW_M \equiv 0$.
(3) If $\Sigma \subset M$ is an embedded sphere with $0 \neq [\Sigma] \in H_2(M;\mathbb{Z})$ and $[\Sigma]^2 \geq 0$, then $SW_M \equiv 0$.

For the next class, recall that a 2-form ω is a *symplectic form* on M if it is non-degenerate ($\omega \wedge \omega > 0$) and closed $d\omega = 0$.

Non-vanishing results:
Assume that M is a smooth, closed, simply connected, oriented 4-manifold with $b_2^+(M) > 1$ and odd.

(1) If S is a simply connected complex surface (hence $b_2^+(S)$ is odd) and $b_2^+(S) > 1$, then $SW_S(\pm c_1(S)) \neq 0$.
(2) More generally, if (M, ω) is a simply connected symplectic manifold and $b_2^+(M) > 1$, then $SW_M(\pm c_1(M, \omega)) = \pm 1$.

We close with the remark that manifolds with a zero-dimensional SW moduli space, referred to as manifolds of simple type, have been of central interest. We now turn to a brief look at surgical techniques important in applications of the SW invariants.

9.3 Gluing Formulas

We now begin a review of certain surgery techniques that have turned out to be useful for applications of the SW invariants. This section is based mainly on the papers [Morgan et al. (1996, 1997)] of Taubes, Szabo and Morgan.

A 4-dimensional extension of the Dehn surgery operations described in section 6.2.2 leads to the following definition:

Definition 9.4. Suppose that M is a smooth closed oriented 4-manifold, and that M contains a smoothly embedded 2-torus $T^2 \hookrightarrow M$ with trivial self-intersection number. Similarly to Dehn-surgery on knots in 3-manifolds, a *generalized logarithmic transformation* of M along T^2 is defined by deleting a tubular neighborhood $N(T^2)$ of T^2 from M and gluing it back via a diffeomorphism

$$\phi : \partial(D^2 \times T^2) \to \partial(M \setminus N(T^2)),$$

defining a new manifold denoted by $M(\phi)$.

In more detail, the diffeomorphism $\phi : \partial(D^2 \times T^2) = T^3 \to T^3 = \partial(M \setminus N(T^2))$ can be defined by an element in $SL(3, \mathbb{Z})$, that is, by a sequence of Dehn twists. Fixing a basis $a, b, c \in H_1(T^3, \mathbb{Z})$ we can describe such a Dehn twist by a curve $\gamma = pa + qb + rc$ (again, refer to section 6.2.2), with the numbers p, q, r relatively prime. Using ϕ_γ for the diffeomorphism defined by this curve, we get the resulting homology map $(\phi_\gamma)_* : H_1(\partial(D^2 \times T^2); \mathbb{Z}) \to H_1(\partial(M \setminus N(T^2)); \mathbb{Z})$. From this we can recover the curve γ as $\gamma = (\phi_\gamma)_*([\partial(D^2 \times pt)])$. γ is in fact an element of the kernel of the inclusion $i_* : H_1(\partial(M \setminus N(T^2)); \mathbb{Z}) \to H_1(M \setminus N(T^2); \mathbb{Z})$, that is, $i_*(\gamma) = 0$. Using the definitions

$$M_\gamma := M(\phi_\gamma) := N \cup_{\phi_\gamma} D^2 \times T^2 \qquad N := M \setminus N(T^2),$$

leads to an important relation between the logarithmic transformation defined by γ and the SW invariants as summarized by

Theorem 9.3. *Suppose that $b_2^+(N) > 1$, and that $\gamma \in \mathrm{Ker}\, i_*$. Let $\mathcal{B}_N$, $\mathcal{B}_{M_\gamma}$ denote the set of basic classes of N and M_γ. Define the formal series on $H_2(M_\gamma, \mathbb{R})$ by*

$$SW^*_{M_\gamma} = \sum_{K \in \mathcal{B}_{M_\gamma}} SW_{M_\gamma}(K)(\exp K).$$

Then

$$SW^*_{M_\gamma} = \left(\sum_{L \in \mathcal{B}_N} SW_N(L)(\exp(\rho(L))) \right) (\exp(T_\gamma) - \exp(-T_\gamma))^{-1},$$

where $\rho : H^2(N, \partial N, \mathbb{Z}) \to H^2(M_\gamma, \mathbb{Z})$ is given by the Mayer-Vietoris sequence, and where T_γ is the Poincaré dual of the so-called core $\{pt\} \times T^2 \hookrightarrow M_\gamma$.

The proof of this result uses cylindrical end Seiberg-Witten moduli spaces over N. Other product formulas along T^3 for the generalized fiber sum operation can be found in [Morgan et al. (1997)] using different techniques.

Next consider a set of surgery operations changing *only* the differential structure to a non-diffeomorphic one without changing the topological type.

9.4 Changing of Smooth Structures by Surgery along Knots and Links

Now we describe certain operations on a given smooth 4-manifold which preserve the underlying topological structure but alter its smooth structure. Let M be a simply connected smooth 4-manifold which contains a smoothly embedded torus T of self-intersection 0. Given a knot K in S^3, we replace a tubular neighborhood of T with $S^1 \times (S^3 \setminus K)$ to obtain the *knot surgery manifold* M_K.

Remark 9.3.
More formally, this procedure is accomplished by performing 0-framed surgery on K to obtain the 3-manifold M_K. The meridian m of K can be viewed as a circle in M_K; so in $S^1 \times M_K$ we have the smooth torus $T_m = S^1 \times m$ of self-intersection 0. Since a neighborhood of m has a canonical framing in M_K, a neighborhood of the torus T_m in $S^1 \times M_K$ has a canonical identification with $T_m \times D^2$. The knot surgery manifold M_K is given by the fiber sum

$$M_K = M \#_{T=T_m} S^1 \times M_K = (M \setminus T \times D^2) \cup (S^1 \times M_K \setminus T_m \times D^2)$$

where the two pieces are glued together so as to preserve the homology class [pt $\times \partial D^2$]. This latter condition does not, in general, completely determine the isotopy type of the gluing, and M_K is taken to be any manifold constructed in this fashion.

Because $S^1 \times (S^3 \setminus K)$ has the same homology as a tubular neighborhood of T in M (and because the gluing preserves [pt $\times \partial D^2$]) the homology and intersection form of M_K will agree with that of M. If it is also assumed that $M \setminus T$ is simply connected, then $\pi_1(M_K) = 1$ and M_K will be homeomorphic to M.

The Seiberg-Witten invariants can be used to distinguish the diffeomorphism types of the resulting M_K. Let $\{\pm\beta_1, \ldots, \pm\beta_n\}$ be the set of nonzero basic classes for M and consider variables $t_\beta = \exp(\beta)$ for each $\beta \in H^2(M; \mathbb{Z})$ which satisfy the relations $t_{\alpha+\beta} = t_\alpha t_\beta$. A lengthy calculation which we skip shows that Seiberg-Witten invariant of M can be written as the Laurent polynomial

$$\mathcal{SW}_M = \mathrm{SW}_M(0) + \sum_{j=1}^n \mathrm{SW}_M(\beta_j) \cdot (t_{\beta_j} + (-1)^{(e+\mathrm{sign})(M)/4} t_{\beta_j}^{-1}).$$

The next theorem provides important information about the effect of knot surgery. First, a smoothly embedded torus representing a nontrivial homology class $[T]$ is said to be *c-embedded* if it contains two simple closed curves which generate $\pi_1(T)$ and which bound vanishing cycles in M. Note that a c-embedded torus has self-intersection 0. Next, the Alexander polynomial, $\Delta_K(t)$, of a knot K is a Laurent polynomial in the variable t characterized by the following axioms:

(1) The Alexander polynomial of an unknot S^1 is normalized to be $\Delta_{S^1}(t) = 1$.
(2) $\Delta_K(t)$ fulfills the so-called skein-relation

$$\Delta_{K_+}(t) = \Delta_{K_-}(t) + (t^{1/2} - t^{-1/2}) \cdot \Delta_{K_0}(t) \qquad (9.4)$$

where K_+ is an oriented knot or link, K_- is the result of changing a single oriented positive (right-handed) crossing in K_+ to a negative (left-handed) crossing, and K_0 is the result of resolving the crossing as shown in Figure 9.1.

Fig. 9.1 Crossings used in the skein relation

The theorem is

Theorem 9.4. *([Fintushel and Stern (1998b)]) Let M be a simply connected oriented smooth 4-manifold with $b^+ > 1$. Suppose that M contains a c-embedded torus T with $\pi_1(M \setminus T) = 1$, and let K be any knot in S^3. Then the knot surgery manifold M_K is homeomorphic to M and has Seiberg-Witten invariant*

$$\mathcal{SW}_{M_K} = \mathcal{SW}_M \cdot \Delta_K(t),$$

where $\Delta_K(t)$ is the symmetrized Alexander polynomial of K and $t = \exp(2[T])$.

Since the K3-surface S_4 (see subsection 6.7.3) has a smooth c-embedded torus T the theorem 9.4 applies to it. Also, since $\mathcal{SW}_{E(2)} = 1$ every knot surgery along K changes the Seiberg-Witten invariant to $\mathcal{SW}_{E(2)_K} = \Delta_K(t)$. A theorem of Seifert states that any Laurent polynomial of the form $P(t) = a_0 + \sum_{j=1}^{n} a_j(t^j + t^{-j})$ with coefficient sum $P(1) = \pm 1$ is the Alexander polynomial of some knot in S^3. For each such polynomial $P(t)$, there is a manifold homeomorphic to the K3-surface with $\mathcal{SW} = P(t)$. If K_1 and K_2 have the same Alexander polynomial, Seiberg-Witten invariants are not able to distinguish M_{K_1} from M_{K_2}. For example, take M to be the K3-surface, see section 6.7.3. Then M_K has a self-intersection 0 homology class σ satisfying $\sigma \cdot [T] = 1$ which is represented by an embedded surface of genus $g(K) + 1$ where $g(K)$ is the genus of K. Can these classes be used to distinguish M_{K_1} from M_{K_2} when $g(K_1) \neq g(K_2)$? This is expressed in

Conjecture of Fintushel and Stern:
For $M = S_4$, the manifolds M_{K_1} and M_{K_2} are diffeomorphic if and only if K_1 and K_2 are equivalent knots.

However, this conjecture is wrong as shown by Akbulut [Akbulut (1999)], who constructed the handle body description of the knot manifold M_K from M. Let $-K$ be mirror of the knot K then M_K is diffeomorphic to M_{-K} contradicting the conjecture.

We conclude this section by pointing out that the knot surgery construction can be generalized to manifolds with $b^+ = 1$ and to links in S^3 of more than one component in a fairly obvious way. Glue the complements of c-embedded tori in 4-manifolds to the product of S^1 with the link complement. See [Fintushel and Stern (1998b)] for details. For example, if to each boundary component of $S^1 \times (S^3 \setminus N(L))$ we glue S_4 minus the neighborhood of a smooth c-embedded torus, we obtain a manifold with $\mathcal{SW} = \Delta_L(t_1, \ldots, t_n)$, the multivariable Alexander polynomial of the link. Unfortunately as we will show in the next section, these invariants are not complete.

9.5 The Failure of the Complete Smooth Classification

We now look at an example of a pair (M_1, M_2) of (symplectic) 4-manifolds with M_1 homeomorphic to M_2, $\mathcal{SW}_{M_1} = \mathcal{SW}_{M_2}$, but M_1 is not diffeomorphic to M_2 establishing the fact:

Seiberg-Witten invariants do not completely distinguish differential structures on homeomorphic 4-manifolds.

Following Fintushel and Stern [Fintushel and Stern (1998a)] choose a pair of fibered 2-bridge knots $K(\alpha, \beta_1)$ and $K(\alpha, \beta_2)$ with the same Alexander polynomials. For example, $K_1 = K(105, 64)$ and $K_2 = K(105, 76)$ with Alexander polynomial
$$\Delta_K(t) = t^{-4} - 5t^{-3} + 13t^{-2} - 21t^{-1} + 25 - 21t + 13t^2 - 5t^3 + t^4.$$
Although these knots have the same Alexander polynomial, they can be distinguished by the fact that their branch covers are the lens spaces $L(\alpha, \beta_1)$ and $L(\alpha, \beta_2)$ which are distinct. In this specific case $L(105, 64)$ is not diffeomorphic to $L(105, 76)$.

Perform the knot surgery construction of section 9.4 on the $K3$ surface, replacing $T^2 \times D^2$ with $S^1 \times (S_{K_j} \setminus \tilde{K}_j)$ with resulting 4-manifolds $M_i, i = 1, 2$. The M_i are not simply connected (but are homeomorphic). In particular, $\pi_1(M_1) = \pi_1(M_2) = \mathbb{Z}_\alpha$, and the α-fold covers $\tilde{M}_1$ and $\tilde{M}_2$ of M_1 and M_2 are not diffeomorphic. It follows that
$$\mathrm{SW}_{\tilde{M}_i} = \Delta_{L_i}(t_1, \ldots, t_\alpha) \cdot \prod_{j=1}^{\alpha} (t_j^{1/2} - t_j^{-1/2})$$

Since the linking numbers of the links L_1 and L_2 are different, it can be shown that the Alexander polynomials are different too and $\tilde{M}_1$ is not diffeomorphic to $\tilde{M}_2$ completing the counterexample.

9.6 Beyond Seiberg-Witten: The Cohomotopy Approach

We close this chapter with a brief mention of approaches initiated by Furata [Furata (2001)] on the $\frac{11}{8}$-conjecture by using the monopole mapping in the Seiberg-Witten formalism. Independently Bauer and Furata [Furata (1998)] defined a cohomotopy invariant using the Pontrjagin-Thom construction for infinite-dimensional Hilbert spaces. Later both authors unified the approaches in a common paper [Bauer and Furuta (2002)]. In another paper [Bauer (2002)], Bauer showed a simple relation between the connected sum of two 4-manifolds and the corresponding cohomotopy invariants.

First, look at the construction of the monopole map. Let $S^\pm$ denote the Spin$_C$ bundles over the 4-manifold M and let L denote their determinant line bundle. Furthermore, let $\mathcal{A}$ be the space of Spin$_C$-connections. The map $\tilde{\mu}$ is given by:

$$\tilde{\mu}: \mathcal{A} \times (\Gamma(S^+) \oplus \Omega^1(M) \oplus H^0(M)) \to \mathcal{A} \times (\Gamma(S^-) \oplus \Omega^+(M) \oplus \Omega^0(M) \oplus H^1(M))$$
$$(A, \phi, a, f) \mapsto (A, \partial\!\!\!/_{A+a}\phi, F^+_{A+a} - \sigma(\phi), \delta a + f, [a])$$

using the notation of section 9.1. Here $[a]$ denotes the cohomology class of the 1-form a, δ is the co-differential introduced in section 3.3 and $\partial\!\!\!/_A$ is the Dirac operator with respect to the connection A. This map can be simply interpreted as the perturbation of the Seiberg-Witten equations (as introduced in section 9.1). The gauge group $\mathcal{G}$ is the set of maps $M \to S^1$. Fix a base point $* \in M$ to define the based gauge group $\mathcal{G}_0$. Let A be a fixed connection, then the subspace $A + \ker(d) \subset \mathcal{A}$ is invariant under the action of the based gauge group with quotient space isomorphic to

$$Pic^s(M) = \frac{H^1(M, \mathbb{R})}{H^1(M, \mathbb{Z})}$$

the so-called Picard group. Furthermore define

$$\mathcal{E} = (A + \ker d) \times (\Gamma(S^+) \oplus \Omega^1(M) \oplus H^0(M))/\mathcal{G}_0$$
$$\mathcal{F} = (A + \ker d) \times (\Gamma(S^-) \oplus \Omega^+(M) \oplus \Omega^0(M) \oplus H^1(M))/\mathcal{G}_0$$

which are bundles over $Pic^s(M)$. Then the quotient $\mu = \tilde{\mu}/\mathcal{G}_0 : \mathcal{E} \to \mathcal{F}$ is the *monopole map*, a S^1-equivariant map over $Pic^s(M)$.

Forgoing the details of a very complicated construction, we remark only that it results in a structure $\pi_{S^1,H}^{b^+}(Pic^s(M); ind(\partial\!\!\!/))$ which is a cohomotopy.[1]

All of this lead up to the following theorem on the existence of the equivariant cohomotopy invariant induced by the monopole map.

Theorem 9.5. *The monopole map μ defines an element in an equivariant stable cohomotopy group*[2]

$$[\mu] \in \pi_{S^1,H}^{b^+}(Pic^s(M); ind(\partial\!\!\!/)),$$

which is independent of the chosen Riemannian metric.

For $b^+ > \dim(Pic^s(M)) + 1$, a homology orientation determines a homomorphism of this stable cohomotopy group to $\mathbb{Z}$, which maps $[\mu]$ to the integer-valued Seiberg-Witten invariant.

In our applications, the most important property of this invariant is the behavior with respect to the connected sum of two 4-manifolds.

Theorem 9.6. *For a connected sum $M = M_0 \# M_1$ of 4-manifolds, the stable equivariant cohomotopy invariant is the smash product of the invariants of its summands*

$$[\mu_M] = [\mu_{M_0}] \wedge [\mu_{M_1}]$$

Remark 9.4.
The so-called smash product $M \wedge Y$ between two spaces M, Y defined by:

$$M \wedge Y = \frac{M \times Y}{(M \times \{y_0\}) \cup (\{x_0\} \times Y)},$$

is the natural product in the cohomotopy group.

Informally, the monopole map μ_M has the same stable cohomotopy invariant as the product $\mu_{M_0} \times \mu_{M_1}$. This is in strong contrast to the Seiberg-Witten invariants where gluing theorems are needed to get the same result, and then only for special manifold splittings. Other splitting theorems can be obtained. For example,

Theorem 9.7. *Let M be a connected sum of two symplectic 4-manifolds, each with $b^+ \equiv 3 \bmod 4$ and vanishing first Betti number. Then no connected sum decomposition of M contains a manifold with $b^+ \equiv 1 \bmod 4$ as a summand.*

---
[1] Recall a homotopy group $\pi_n(M)$ of a space M is a class of mappings $S^n \to M$ up to homotopy. Then a cohomotopy group $\pi^n(M)$ is a class of mappings $M \to S^n$ up to homotopy.

[2] It is possible to introduce a cogroup structure on the cohomotopy set. But we omit this complicated fact in the construction above.

The interested reader can find many of the technical details in [Bauer (2002)]. Further gluing results can be found in the work of Manolescu [Manolescu (2003b,a)] extending the work of Bauer and Furata.

Chapter 10

Physical Implications

In this chapter we will explore some possible physical implications of the existence of exotic differential structures. We begin by surveying the Principle of Relativity in the light of these new structures, recalling the identification of the mathematical notion of diffeomorphism with the generalized notion of a change of reference frame. In particular, we review possible applications of varieties of differential structures to quantum gravity in five dimensions and its 1+4 spacetime foliation. The next section considers assumptions involved in the extrapolation of metric solutions to the Einstein equations from one coordinate patch to another when the smoothness may not be trivial. Perhaps localized exoticness can act as another type of source (rather than standard matter) for an external metric. In fact in the next section we mention the work of Sładkowski who showed that the empty space given by exotic $\mathbb{R}^4$ can lead to non-trivial solutions of the Einstein vacuum equations. Finally, we close with a review of the notions of global anomalies, gravitational instantons and tunneling and the possible implications of exotic manifolds in their study.

10.1 The Principle of Relativity

In Special Relativity, Einstein taught us to think in terms of a unified spacetime model, with no preferred a priori splitting of space from time, apart from the qualitative space-like, time-like, light-like ones. The transformation group preserving these spacetime properties, the gauge group of special relativity is of course the Poincaré group, that is, the homogeneous Lorentz group plus translations.

With the many successes of special relativity, it seems that the ether has finally been put to rest. Indeed it has in this classical sense. "If you can't observe it, it doesn't exist" is the standard motto of scientific operationalism. Or to paraphrase an old axiom: "No stuff has existence until it is observed to have existence." But, should we not apply this to "stuff"=manifold properties? So, is spacetime the new ether? Clearly, it does not play the same mechanical role of "transmitter of forces," as the old mechanical one. Also, it clearly does not provide an "absolute rest reference frame." But, it does have other properties which have generally been assumed without thorough analysis of alternatives. It provides, in operationally unobservable ways, the substratum to carry the many structures used by modern theories. Perhaps these assumed spacetime structures in modern theories have built it into a "new ether."

Einstein's general relativistic gravitational theory logically begins with weakening one of the basic assumptions of special relativity, namely the existence of a flat Minkowski metric, and associated preferred inertial frames. By dropping the assumption of a pre-given flat metric, the door is open to a geometric model of gravity. Many insights led Einstein to his theory of gravity, including the Principle of General Relativity, the Equivalence Principle, and Mach's Principle. Of course, the actual history is more complicated and interesting, and the reader can consult the volume one of the Einstein Studies[Howard and Stachel (1986)], for more complete and accurate accounts of the story. For our purposes, it is sufficient to point out that Einstein was aware of the rigid structure still remaining on the spacetime of special relativity imposed by the Lorentz metric and the associated preferred set of inertial reference frames. Mach's Principle addresses the issue of why the fixed stars have constant velocity in the inertial frames, while the Principle of General Relativity proposes extending the physically acceptable frames beyond this restricted set. In other words, while special relativity had weakened the assumption of a preferred (zero) absolute-velocity-defining ether, it replaced it by a preferred (zero) absolute-acceleration-defining one. So, the next step toward generally covariant theories was a result of re-examining and loosening previous rigid structures[1]. This problem is one of finding the "correct" spacetime model.

One way to sketch the logical development of the current spacetime models used in physics is as a diagram:

$$point\ set \to topological\ space \overset{??}{\to} smooth\ manifold \to bundles, etc.$$

The middle transition has turned out to be rather mysterious and nontrivial as we have seen in the previous chapters. Einstein's theory starts

---

[1] John Norton [Norton (1984)] has given us a thorough and highly interesting analysis of how Einstein arrived at his equations of General Relativity.

with the bundle calculus and assumes that the other levels are given (or can be defined). Thus, gravitation in the "classical" sense assumes a differential structure and concentrates on the possible geometries. The moduli space can be seen as the factor space $\mathcal{M} = Riem(M)/Diff(M)$ where $Riem(M)$ is the space of Riemannian (or Lorentzian) metrics and $Diff(M)$ is the diffeomorphism group of a 4-manifold M. The physically significant aspect of a solution of the Einstein equation, metric modulo diffeomorphism, is defined by an element of this space. Now one may ask: what changes if we try to unify GRT and quantum theory to quantum gravity? Then every element in the moduli space $\mathcal{M}$ is a state of the quantum gravity, i.e. one has to integrate over all possible elements in the functional integral. Unfortunately this space is rather intractable. In the covariant approach to quantum gravity Wheeler and deWitt foliate the 4-manifold into 3-dimensional slices along the "time" axis. Later, Ashtekar [Ashtekar (1986)] introduced new variables to describe this situation and identified the space $\mathcal{M}$ with the space of loops. In that approach, $\mathcal{M}$ is described by graph-theoretical methods [Ashtekar and Lewandowski (1995)]. Then it is necessary to study the diffeomorphisms of the 3-dimensional slices where the diffeomorphism group of the 3-manifold naturally has a different topology from that of the 4-dimensional case. But, as we have been studying, 4 dimensions gives rise to the phenomenon of exotic differential structures even on topologically trivial spaces such as $\mathbb{R}^4$. This does not occur in dimension 3. The four dimensional exotica then greatly complicate the structure of the 3-dimensional slices (after a suitable foliation). Thus a naive approach to 4-dimensional quantum gravity by a path integral over all possible metrics in $\mathcal{M}$ should also include the exotic differential structures. Let g and g_{exotic} be two different elements of $\mathcal{M}$ not connected by a diffeomorphism $Diff(M)$, then these two metrics lie in different components of $\mathcal{M}$ leading to different elements in $\pi_0(\mathcal{M})$. From the fibration

$$Diff(M) \to Riem(M) \to \mathcal{M}$$

the contractibility of $Riem(M)$ leads to an isomorphism $\pi_n(\mathcal{M}) = \pi_n(Diff(M))$ for all $n \geq 0$. For the special case of $M = S^k$ with $k > 5$ we obtain the well-known relationship between isotopy classes lying in $\pi_0(Diff(M))$ and exotic spheres (see section 7.8). Finally, any quantum gravity theory must have a classical limit which would include exotic four manifolds. Thus, quantum gravity should incorporate exotic structures.

10.2 Extension of Metrics

Even though an explicit, effective coordinate patch presentation of $\mathbb{R}_\Theta^4$[2] is not available, certain additional facts about such a space, including some of a geometric and thus physical sort can be discovered. For a more complete exposition and discussion of these results see [Brans (1994b)] and [Brans (1994a)]. Here we merely review some of the results.

First, the question naturally arises concerning the given global topological coordinates, $\{p^\alpha\}$, which define the topological manifold $\mathbb{R}^4$, and their relationship to the local smooth coordinates given by the coordinate patch functions, ϕ_U^α. Both provide maps from an abstract $\mathbf{p} \in \mathbb{R}^4$, into $\mathbb{R}^4$ itself. Clearly the global topological coordinates cannot themselves be smooth everywhere since otherwise they would provide a diffeomorphism of $\mathbb{R}_\Theta^4$ onto standard $\mathbb{R}^4$. But can they be locally smooth? This is answered in the affirmative by

Theorem 10.1. *There exists a smooth copy of each $\mathbb{R}_\Theta^4$ for which the global C^0 coordinates are smooth in some neighborhood. That is, there exists a smooth copy, $\mathbb{R}_\Theta^4 = \{(p^\alpha)\}$, for which $p^\alpha \in C^\infty$ for $|\mathbf{p}| < \epsilon$.*

The implied obstruction to continuing the $\{p^\alpha\}$ as smooth beyond the ϵ limit presents a challenging issue for further investigation. Related to this is a defining feature of the early discovery work of $\mathbb{R}_\Theta^4$'s, namely the non-existence of arbitrarily large smoothly embedded three-spheres.

There are also certain natural "topological but not smooth" decompositions. For example,

Theorem 10.2. $\mathbb{R}_\Theta^4$ *is the topological, but not smooth, product, $\mathbb{R}^1 \times \mathbb{R}^3$.*

Many interesting examples can be constructed using Gompf's "end-sum" techniques [Gompf (1985)] discussed in 8.5 above. In this construction topological "ends" of non-compact smooth manifolds are glued together smoothly, $X \cup_{end} Y$. If one of the manifolds, say X, is also topological $\mathbb{R}^4$, the topology of the resultant space is unchanged, that is $\mathbb{R}^4 \cup_{end} Y$ is homeomorphic to Y. However, if X is an $\mathbb{R}_\Theta^4$ which cannot be smoothly embedded in standard $\mathbb{R}^4$, then neither can the end sum. Thus,

Theorem 10.3. Gompf's end sum result: *If $X = \mathbb{R}_\Theta^4$ cannot be smoothly embedded in standard $\mathbb{R}^4$, but Y can be, then $\mathbb{R}_\Theta^4 \cup_{end} Y$ is homeomorphic, but not diffeomorphic to Y.*

---
[2]Recall this is our notation for an exotic $\mathbb{R}^4$.

This technique will be used further below.

To do geometry we need a metric of the appropriate signature. It is a well known fact that any smooth manifold can be endowed with a smooth Riemannian metric, g_0. This follows from basic bundle theory [Steenrod (1999)]. Similarly, if the Euler number of X vanishes a globally non-zero smooth tangent vector, u exists. g_0 and u can be combined then to construct a global smooth metric of Lorentz signature, $(-,+,+,+)$, in dimension four. A generalization of this result follows also from standard bundle theory, [Steenrod (1999)].

Theorem 10.4. *If M is any smooth connected 4-manifold and A is a closed sub-manifold for which $H^4(M, A; \mathbf{Z}) = 0$, then any smooth time-orientable Lorentz signature metric defined over A can be smoothly continued to all of M.*

One immediate conclusion about certain geometries on $\mathbb{R}^4_\Theta$ can be drawn from an investigation of the exponential map of the tangent space at some point, which is standard $\mathbb{R}^4$, onto the range of the resulting geodesics. The Hadamard-Cartan theorem guarantees that this map will be a diffeomorphism onto the full manifold if it is simply connected, the geometry has non-positive curvature and is geodesically complete. Thus,

Theorem 10.5. *There can be no geodesically complete Riemannian metric with non-positive sectional curvature on a $\mathbb{R}^4_\Theta$.*

The lack of localization of the "exoticness" means that it must extend to infinity in some sense as illustrated by the lack of arbitrarily large smooth three-spheres. However, it turns out to be possible that the exoticness can be localized in a *spatial* sense as follows:

Theorem 10.6. *There exist smooth manifolds which are homeomorphic but not diffeomorphic to $\mathbb{R}^4$ and for which the global topological coordinates (t, x, y, z) are smooth for $x^2 + y^2 + z^2 \geq \epsilon^2 > 0$, but not globally. Smooth metrics exists for which the boundary of this region is time-like, so that the exoticness is spatially confined.*

The details of the construction of such manifolds are given in [Brans (1994b)]. First, Gompf's end-sum technique is used to produce a $\mathbb{R}^4_\Theta$ for which the global topological coordinates are smooth outside of the cylinder, that is, in the closed set $c_0 = \{(t, x, y, z) | x^2 + y^2 + z^2 \geq \epsilon\}$ described in the first part of the theorem. Next, a Lorentz signature metric is constructed on c_0. This metric can even be a vacuum Einstein metric. The only condition is that the $\partial/\partial t$ be time like on c_0. The cross section continuation result with $A = c_0$ then guarantees the extension of the metric over the full space consistent with the conditions of the theorem. What makes the complement of c_0 exotic is the fact that the (x, y, z, t) cannot be continued as smooth functions over all of it. This result leads to

Conjecture: *This localized exoticness can act as a source for some externally regular field, just as matter or a wormhole can.*

This conjecture was affirmed by Asselmeyer [Asselmeyer (1996)] for the compact case and by Sładkowski [Sładkowski (1999)] for the non-compact case.

10.3 Exotic Cosmology

In this section we will review some of the consequences of the existence of exotic differential structures on cosmology. Consider the exotic product, $X = \mathbb{R}_\Theta - \{0, 0, 0, 0\} = \mathbb{R} \times_\Theta S^3$, which arises from a puncturing of $\mathbb{R}^4_\Theta$. This topology, but not smoothness, is a standard model used in cosmology. It is not hard to apply the same techniques used in the previous section to show that this product can be the standard smooth one for a finite, or semi-infinite range of the first variable, say t. The resulting manifold could then be endowed with a standard cosmological metric. However, this metric, and even the variable t itself, cannot be continued as globally smooth indefinitely, because of the exotic smoothness obstruction. Nevertheless, X is still a globally smooth manifold, with some globally smooth Lorentz-signature metric on it. Other interesting topological but not smooth products can be constructed by use of the end-sum construction. One interesting example is exotic Kruskal, $X_K = \mathbb{R}^2 \times_\Theta S^2$. Using the cross section continuation theorem above, the standard vacuum Kruskal metric can be imposed on some closed set, $A \subset X_K$, and then continued to some smooth metric over the entire space. However, it cannot be continued as Kruskal, since otherwise X_K would then be standard $\mathbb{R}^2 \times S^2$. In sum,

Theorem 10.7. *On some smooth manifolds which are topologically $\mathbb{R}^2 \times S^2$, the standard Kruskal metric cannot be smoothly continued over the full range, $u^2 - v^2 < 1$.*

We will now look at some qualitative issues associated with Einstein's equation in the exotic settings. Sładkowski [Sładkowski (1999)] has provided some valuable results based on the work of Taylor [Taylor (1997, 1998)] about isometry groups of exotic $\mathbb{R}^4$. We follow the paper of Sładkowski closely. Recall several definitions. A diffeomorphism $\phi : M \mapsto M$, where M is a (pseudo-)Riemannian manifold with metric tensor g, is an isometry if and only if it preserves g, $\phi^*g = g$. Such mappings form a group called the isometry group. A smooth manifolds has *few symmetries* provided that for every choice of differentiable metric tensor, the isometry group is finite. L. R. Taylor [Taylor (1998)] presented examples of exotic $\mathbb{R}^4$'s with few symmetries. Among these are examples with nontrivial but still finite isometry groups. Taylor's results apply to Riemannian structures, but a natural extension gives further insights into the possible role of differential structures in Lorentz manifolds physics. Define proper actions of a group on manifolds as follows. Let G be a locally compact topological group acting on a metric space X. G acts properly on X if and only if for all compact subsets $Y \subset X$, the set $\{g \in G : gY \cap Y \neq \emptyset\}$ is also compact. Thus G acts non-properly on X if and only if there exist sequences $x_n \to x$ in X and $g_n \to \infty$ in G, such that $g_n x_n$ converges in X. Here $g_n \to \infty$ means that the sequence g_n has no convergent subsequence in G. Note that for many manifolds a proper G action, for G an isometry, is topologically impossible. On the other hand, non-proper G action on a Lorentz (or pseudo-Riemannian) manifolds is impossible for all but a few groups. Sładkowski used a result from the theory of transformation groups proved by Kowalsky (see [Sładkowski (1999)] for the reference).

Theorem 10.8. *Let G be Lie transformation group of a differentiable manifold X. If G acts properly on X, then G preserves a Riemannian metric on X. The converse is true if G is closed in Diff(X).*

As a special case we have:

Theorem 10.9. *Let G and X be as above, with G connected. If G acts properly on X preserving a time-orientable Lorentz metric, then G preserves a Riemannian metric and an everywhere nonzero vector field on X.*

If we combine these theorems with Taylor's results we immediately get:

Theorem 10.10. *Let G be a Lie transformation group acting properly on an exotic $\mathbb{R}^4$ with few symmetries and preserving a time-orientable Lorentz metric. Then G is finite.*

Further, due to Kowalsky, we also have:

Theorem 10.11. *Let G be a connected non-compact simple Lie group with finite center. Assume that G is not locally isomorphic to SO(n, 1) or SO(n,2). If G acts non-trivially on a manifold X preserving a Lorentz metric, then G actually acts properly on X.*

and

Theorem 10.12. *If G acts non-properly and non-trivially on X, then G must be locally isomorphic to SO(n,1) or SO(n,2) for some n.*

In many cases it is possible to describe the cover $\tilde{X}$ up to Lorentz isometry.

Now, suppose we are given an exotic $\mathbb{R}^4_\Theta$ with few symmetries. Suppose we have found some solution to the Einstein equations on $\mathbb{R}^4_\Theta$. Whatever the boundary conditions are we would face one of the two following situations.

- The isometry group G of the solution acts properly on $\mathbb{R}^4_\Theta$. Then according to Theorem 3 G is finite. There is no nontrivial Killing vector field and the solution cannot be stationary. This gravitational field must be very "complicated." Note that this conclusion is valid for any open subspace of $\mathbb{R}^4_\Theta$. This means that this phenomenon cannot be localized on such spacetimes.
- The isometry group G of the solution acts non-properly on $\mathbb{R}^4_\Theta$. Then G is locally isomorphic to SO(n,1) or SO(n,2). But the non-proper action of G on $\mathbb{R}^4_\Theta$ means that there are points infinitely close together in $\mathbb{R}^4_\Theta$ ($x_n \to x$) such that a non-convergent sequence of isometries ($g_n \to \infty$) in G maps them into infinitely close points in $\mathbb{R}^4_\Theta$ ($g_n x_n \to y \in \mathbb{R}^4_\Theta$). There must exist strong gravity centers to force such convergence (even in empty spacetimes). Such spacetimes are unlikely to be stationary.

We see that in both cases Einstein gravity is quite nontrivial even in the absence of matter. Recall that if a spacetime has a Killing vector field ζ^a, then every covering manifold admits a Killing vector field ζ'^a projected onto ζ^a by the differential of the covering map. This means that the properties discussed above are "projected" onto any space that has exotic $\mathbb{R}^4$ with few symmetries as a covering manifold, e.g., quotient manifolds obtained by a smooth action of some finite group.

This result has further consequences as discussed in [Asselmeyer-Maluga and Brans (2002)]. If there were an exotic differential structure in the past then we cannot be sure about our interpretation of current observed data arriving now, at earth. That is, we cannot distinguish gravitational lensing by stars or by the change of a differentiable structure. Thus, null geodesics arriving from distant sources may not be extrapolated back as good radial coordinate lines because of intervening coordinate patch transformations caused by global exotic smoothness. In summary, what we want to emphasize is that without changing the Einstein equations or introducing yet undiscovered forms of matter, or even without changing topology, there is a vast resource of possible explanations for recently observed surprising astrophysical data at the cosmological scale provided by differential topology.

10.4 Global Anomaly Cancellation of Witten

Anomalies present both problems and opportunities in the study of quantum field theories. Consider symmetries in a "classical" field theory generated by Lie groups. Do all such symmetries survive the quantization procedure? If not, then the theory is said to have an **anomaly**. In fact, in most gauge theories (as well as string theory), some "classical" symmetries are no longer symmetries after the quantization process. Specifically, after quantization, certain observables may change under a group action, while these same observables were not changed at the classical level of group action. This loss of symmetry is called an anomaly to express this unwanted behavior of quantum field theories. On the other hand, quantum field theorists use this as an opportunity to select possible theories. If a quantum field theory either has no anomaly, or one that can be removed satisfactorily, it is a possible candidate for physics. Otherwise, the theory is unacceptable. The cause of such anomalies can be very complicated. From the physics point of view, the path integral measure is not an invariant with respect to an anomalous symmetry. This physical observation can be expressed in terms of the non-triviality of certain bundles related to non-trivial properties of the moduli space $\mathcal{A}/\mathcal{G}$ of the Yang-Mills theory, or $Diff(M)$ for gravitational anomalies, that is, anomalies associated with diffeomorphism induced metric changes. From relativity principles, metric dependent observables must be diffeomorphism invariant. For more on the behavior of observables under gauge transformations see our discussion in Chapter 4 and Chapter 5, especially §5.8. Also, Nash [Nash (1991)], chapter X, provides an excellent summary of these issues with many more details than we present here.

It is standard to distinguish two kinds of anomalies: local and global. Refer to the discussion in Chapter 4. Global anomalies occur when the gauge group of the gauge theory (Yang-Mills or GRT) has more than one connected component. In the following we will concentrate on the global gravitational anomalies which occur for non-trivial $\pi_0(Diff^+(M))$ for a manifold M ($Diff^+$ is the group of orientation-preserving diffeomorphisms). Recall that global gauge transformations are originally defined as a group action which is constant on the base space. However, if the gauge group is disconnected, we can have different, but constant, gauges taking values in different components. Furthermore, specialize to Riemannian metrics and spheres. The argument of Cerf in section 7.8 shows that the isotopy classes of n-spheres ($= \pi_0(Diff^+(S^n))$) give the number of differential structures on S^{n+1}. Witten [Witten (1985)] fixed the manifold to be $M = S^{10}$ and restricted to the structure groups $E_8 \times E_8$ or $O(32)$ of string theory. He was able to show that for these gauge groups there are no global anomalies.

Remark 10.1.
The gauge group $\mathcal{G}$ is homotopy equivalent to the loop space
$$\Omega^{10} G = \{\underbrace{[0,1] \times \ldots \times [0,1]}_{10} \to G\}$$
by an argument in [Nash (1991)], p.219. Then:
$$\pi_0(\mathcal{G}) = \pi_0(\Omega^{10} G) = \pi_{10}(G)$$
which is zero for $G = E_8 \times E_8$ or $O(32)$.

Because of the fact
$$\pi_0(Diff^+(S^{10})) = \mathbb{Z}_{992}$$
global gravitational anomalies might possibly occur. Thus, for our purposes, it is significant that these possible global gravitational anomalies on the 10-sphere S^{10} are connected with the exotic differential structures on S^{11}.

We now briefly sketch the outline of this anomaly problem and Witten's resolution of it. Recall that the key element will be the behavior of the functional integral involving the action, as in §5.8, especially (5.33). The functional integral in this equation is over all values of all fields. In particular focus on an action such as given in (4.92) with a Dirac field, ψ, with $m = 0$ and a gauge fields A, with curvature denoted in that equation by Ω. The partition function from which the expectation value of all observables can be obtained involves a path integration over all fields of $\exp(-S)$. For the Dirac part, this involves not only ψ but the connection, A, and a metric, ρ, on the base manifold, S^{10} here. It turns out that the Dirac part of the integration gives

$$Z_{dirac} = \int D[\psi] \exp(-S(\psi, \rho, A)) = \sqrt{det(\slashed{\partial}_\rho \slashed{\partial}_\rho)}. \quad (10.1)$$

The changes of the determinant $\det(\bar{\partial}^*_\rho \bar{\partial}_\rho)$ under quantization expresses the anomaly.[3] Thus we study the variation of the quantized version of the Dirac determinant $\det(\bar{\partial}^*_\rho \bar{\partial}_\rho)$ as the metric ρ changes to $f \cdot \rho$ where f is a global diffeomorphism belonging to some component of $Diff^+(S^{10})$. To express the variation we consider the 1-parameter family of metrics $\rho^t = (1-t)\rho + t f \cdot \rho$. Recall that a gravitational anomaly is one that appears in the partition function when the metric is changed by a diffeomorphism. Then we consider the 11-dimensional manifold N with the line element $ds^2 = dt^2 + \rho^t dx^2$. Witten [Witten (1985)] constructed N as the total space of a fiber bundle over S^1 with fiber S^{10} and structure group $Diff^+(S^{10})$. Remember, the classification of such bundles is given by the elements of the group $\pi_0(Diff^+(S^{10}))$.

Remark 10.2.
Usually, in the classification such bundles is given by the set of homotopy classes from the base manifold to some classifying space. The above bundle N is a $Diff^+(S^{10})$-principal bundle over S^1. Thus $[S^1, BDiff^+(S^{10})]$ is the set of equivalence classes of bundles N. Because of the isomorphism $\pi_n(G) = \pi_{n+1}(BG)$ for all Lie groups G we obtain

$$[S^1, BDiff^+(S^{10})] = \pi_1(BDiff^+(S^{10})) = \pi_0(Diff^+(S^{10}))$$

and the set $\pi_0(Diff^+(S^{10}))$ classifies the bundles N. The group structure comes from the fact that $BDiff^+(S^{10})$ is an H-space.

That means, there are precisely 992 bundles or manifolds N parameterizing the exotic S^{10}'s. To calculate the anomaly we must translate the data on S^{10} into the data on N. Extend the Dirac operator $\bar{\partial}$ to the corresponding operator $\slashed{D}$ on N with respect to the extended metric $\bar{\rho}$ which is nothing else then the family ρ^t. Now the main result of Witten [Witten (1985)] was the calculation of the map

$$\det \bar{\partial}^*_\rho \bar{\partial}_\rho \mapsto \det \bar{\partial}^*_{f \cdot \rho} \bar{\partial}_{f \cdot \rho} = \det \bar{\partial}^*_\rho \bar{\partial}_\rho \exp(i\pi \eta_{\slashed{D}}(0))$$

where $\eta_{\slashed{D}}(0)$ is the eta invariant of the operator $\slashed{D}$. The anomaly is absent if the square root form $\sqrt{\det \bar{\partial}^*_\rho \bar{\partial}_\rho}$ is independent of f which means that $\eta_{\slashed{D}}(0) = 4k$ for any $k \in \mathbb{Z}$. With the input data of the string theory, Witten calculated the eta invariant to show that indeed for the gauge groups $E_8 \times E_8$ or $O(32)$ the theory is free from global anomalies.

The general analysis of global anomalies is rather long and technical. Under certain conditions, however, global anomalies have a rather simple manifestation: they show up in the existence of an instanton field in

---
[3]Nash [Nash (1991)], page 271, points out that while $det(AB) = det(A)det(B)$ in finite dimensions, this is not necessarily true in the infinite dimensions of quantum space. This can be regarded as the source of this anomaly.

which there are an odd number of fermion modes. In discussions of gravitational instantons, a basic fact about Yang-Mills instantons is sometimes overlooked. Let I be a Yang-Mills instanton in an Euclidean space M of dimension d. Then there is always an anti-instanton $\bar{I}$ with the property that $I+\bar{I}$ can be reached continuously from $A_\mu^a = 0$. Here $I+\bar{I}$ is a pair consisting of an instanton and a widely separated anti-instanton. Because of the separation, the effect coming from $I + \bar{I}$ is the same as the effect from an isolated instanton. This is why instantons play a role in Yang-Mills theory.

Consider an asymptotically Euclidean n-dimensional space with a localized topological defect which we wish to regard as a gravitational instanton. Except for very special cases, there does not exist an anti-instanton $\bar{J}$ such that $J + \bar{J}$ is diffeomorphic to $\mathbb{R}^n$. If $\bar{J}$ does not exist there is no need to include J in path integrals.

Now, Witten was able to determine when $\bar{J}$ exists. If one of the Betti numbers b_i of J (seen as topological effect) is non-zero for $1 \leq i \leq n-1$, then $\bar{J}$ cannot exist. Because of the non-negativeness of the Betti numbers, we obtain for $b_i(J + \bar{J}) = b_i(J) + b_i(\bar{J}) \geq b_i(J)$. For $J + \bar{J} \simeq \mathbb{R}^n$ we have $b_i(J + \bar{J}) = 0$ and thus $b_i(J) = 0$ for all $1 \leq i \leq n-1$. Now, perform a one-point compactification X of J and thus $J + \bar{J} \simeq S^n$. If n is 2 or 3, $\bar{J}$ only exists if $J = \mathbb{R}^n$ which follows from the classification of 2-manifolds and from the prime decomposition theorem for 3-manifolds. For $n \geq 4$ we obtain from the generalized Poincaré conjecture that X must be homeomorphic (but not diffeomorphic) to S^n if $b_i(J) = 0$ for $1 \leq i \leq n-1$. The existence of $\bar{J}$ is thus connected to the differential structure of S^n. For $n = 4$ it is not known whether there is an exotic differential structure on S^4, so focus on $n \geq 5$. Let π be a fixed diffeomorphism of S^{n-1} and S_π^n the topological n-sphere with the induced differential structure from π. We know that the order of the group $\pi_0(Diff(S^{n-1}))$ is the number of non-diffeomorphic differential structures on S^n. Let π^{-1} be the inverse diffeomorphism of π and let $\overline{X} = S_{\pi^{-1}}^n$. Then $X + \overline{X}$ is diffeomorphic to S^n. If $\bar{J}$ is related to $\overline{X}$ the way J is related to X (by removing a point and making the standard conformal change of metric to an asymptotically Euclidean space), then $\bar{J}$ is the desired. Thus it follows that exotic spheres are the only gravitational instantons for which there is a sound basis within the presently understood framework of anomaly-free Yang-Mills fermion quantum field theory.

At the end of this section, we briefly mention an interesting interpretation of exotic spheres as tunneling events, just like Yang-Mills instantons. Let $g_{\mu\nu}$ be the Euclidean metric of $\mathbb{R}^{n-1}$ in some coordinate system, and $g_{\mu\nu}^\pi$ its conjugate under π (which, of course, is the Euclidean metric in the transformed coordinates). Now, consider the metric

$$ds^2 = dt^2 + [(1 - \lambda(t))g_{\mu\nu} + \lambda(t)g^\pi_{\mu\nu}]dx^\mu dx^\nu ,$$

where t is "time" and λ is a strictly monotonic function with $\lambda(-\infty) = 0$, $\lambda(+\infty) = 0$. The one point compactification of this space is the exotic sphere $S^n_{\pi-1}$. Thus, the instanton connected with $S^n_{\pi-1}$ is a tunneling event from g to g^π.

Chapter 11

From Differential Structures to Operator Algebras and Geometric Structures

This chapter surveys some of the interesting interplay of exotic smoothness with other areas of mathematics and physics. In the first section we consider the "change" of a differential structure on a given TOP manifold to a differential structure on a second manifold homeomorphic but not diffeomorphic to the first one. Harvey and Lawson introduced the notion of singular bundle maps and connections to study this problem. This leads to speculations that such a process could give rise to singular string-like sources to the Einstein equations of General Relativity, including torsion. The next section deals with formal properties of a connection change and its relation to cyclic cohomology, providing a relationship between Casson handles and Ocneanus string algebra. This approach motivates introduction of the hyperfinite $I\!I_1$ factor C^* algebra $\mathcal{T}$ leading to the conjecture that the differential structures are classified by the homotopy classes $[M, BGl(\mathcal{T})^+]$. This conjecture may have some significance for the the 4-dimensional, smooth Poincaré conjecture. The last section introduces a conjecture relating differential structures on 4-manifolds and geometric structures of homology 3-spheres naturally embedded in them.

11.1 Exotic Smooth Structures and General Relativity

As discussed in chapter 10, exotic differential structures should certainly be included in any theory of quantum gravity. But to study the effect of exotic differential structures on quantum gravity we must first study the "change of a differential structure" on classical general relativity. We begin with a discussion of the notion of "changing" a smoothness structure and present conjectures on the effect of such a change on the tangent bundle as well on Einstein field equations.

Consider two homeomorphic but non-diffeomorphic smooth 4-manifolds M, M' with homeomorphism $h : M \to M'$. Thus, h allows us to identify M and M' in the TOP category. Since h also provides an isomorphism of point sets, we can say that the smoothness structure on $M' = h(M)$ is a **change** from that on M under the identification of points, $m \sim m' = h(m)$. Consider a smooth map

$$f : M \to M'$$

which is near a homeomorphism[1]. Since f cannot be a diffeomorphism, it must have singularities, that is, there is a subset $\Sigma \subset M$ on which the differential $df : TM \to TM'$ is not of maximal rank 4, $\Sigma = \{x \in M \mid rank(df_x) < 4\}$ so df has no inverse on this set. The following remark looks into this in more detail.

Remark 11.1.
See [Golubitsky and Guillemin (1973)] for an introduction to the theory of singular maps. Begin with an equivalence relation on the set $C^\infty(M, M')$ of smooth maps $M \to M'$. Define two smooth maps, $f, g \in C^\infty(M, M')$, to be equivalent $f \sim g$ if and only if there are diffeomorphisms $h_M : M \to M$ and $h_{M'} : M' \to M'$ such that the diagram

$$\begin{array}{ccc} M & \xrightarrow{f} & M' \\ h_M \downarrow & & \downarrow h_{M'} \\ M & \xrightarrow{g} & M' \end{array}$$

commutes. Consider the set $C^0(M, M')$ of homeomorphisms with the compact-open topology. By a fundamental theorem (see [Hirsch (1976)] Theorem 2.6) the smooth maps $C^\infty(M, M')$ are a dense subset of $C^0(M, M')$, i.e., in every neighborhood of a homeomorphism (in the sense of the compact-open topology) there exists a smooth map which approximates the homeomorphism. We say that such a smooth, surjective map $f : M \to M'$ is *near a homeomorphism*. Now we will give a criteria for these maps. Let p_M and $p_{M'}$ be the Pontrjagin class of the 4-manifolds M and M', respectively. Stingley proved in his PhD thesis [Stingley (1995)] a 4-dimensional version of the Riemann-Hurwitz theorem which states that homeomorphic 4-manifolds with a smooth map $f : M \to M'$ near a homeomorphism fulfills the relation

$$f^* p_{M'} = p_M \quad . \tag{11.1}$$

From the Hirzebruch signature theorem this agrees with the classification result of Freedman [Freedman (1982)], that is, the two signatures of the intersection forms corresponding to M and M' agree.

The fundamental question of singularity theory is then how to deform the manifolds M, M' so as to remove the singularities of f, replacing it by $\tilde{f}$. Such a procedure is called *unfolding of f* and Hironaka [Hironaka (1964)] proved the general theorem that for every singular map f between homeomorphic manifolds there is a sequence of operations which unfolds f. These operations are usually called *blow-up* and *blow-down*. In our case $f : M \to M'$ a blow-up leads to a map $\tilde{f} : M \# \mathbb{C}P^2 \to M' \# \mathbb{C}P^2$ and a blow-down to $\tilde{f} : M \# \overline{\mathbb{C}P}^2 \to M' \# \overline{\mathbb{C}P}^2$. Let $C(n)$ and $\bar{C}(m)$ be the connected sums

$$C(n) = \underbrace{\mathbb{C}P^2 \# \ldots \# \mathbb{C}P^2}_{n} \qquad \bar{C}(m) = \underbrace{\overline{\mathbb{C}P}^2 \# \ldots \# \overline{\mathbb{C}P}^2}_{m},$$

then Hironaka's theorem stated that the unfolding of f leads to a diffeomorphism

$$\tilde{f} : M \# C(n) \# \bar{C}(m) \to M' \# C(n) \# \bar{C}(m).$$

---

[1] It is known that all smooth maps are dense in the set of continuous maps.

Using the equivalence (see [Kirby (1989)] p.11)
$$(S^2 \times S^2) \# \mathbb{C}P^2 = \mathbb{C}P^2 \# \overline{\mathbb{C}P}^2 \# \mathbb{C}P^2$$
we obtain a diffeomorphism
$$\tilde{f} : M \underbrace{\# S^2 \times S^2 \cdots \# S^2 \times S^2}_{m} \# C(n-m) \to M' \# \underbrace{S^2 \times S^2 \# \cdots \# S^2 \times S^2}_{m} \# C(n-m)$$
where we assume without loss of generality $m < n$. This is a weaker version of the famous theorem of Wall about diffeomorphisms between 4-manifolds (see [Wall (1964a,b); Kirby (1989)]).

Another important concept of use in determining the structure of the singular set Σ is that of a stable mapping. Let $f \in C^\infty(M, M')$ be a smooth mapping $f : M \to M'$. Then f is stable if there is a neighborhood W_f of f in $C^\infty(M, M')$ (we use the compact-open topology for that space) such that each f' in W_f is equivalent (in the above sense) to f. According to Mather [Mather (1971)] stable smooth mappings between 4-manifolds are dense in the set of smooth mappings. Thus according to Stingley [Stingley (1995)] one can focus on stable maps between homeomorphic but non-diffeomorphic 4-manifolds. Locally such maps are given by stable maps between $\mathbb{R}^4 \to \mathbb{R}^4$. For such a case there are two maps (rank 2 singularities) with a 2-dimensional singular subset and five maps (Morin singularities or rank 3 singularities) with a 3-dimensional singular subset. Here the "number" of maps with rank n singularities, means the number of equivalence classes of maps, called map germs. Stingley [Stingley (1995)] extended this result beyond $\mathbb{R}^4$ to all smooth 4-manifolds and showed that the rank 2 singularities can be killed by an isotopy for maps $f : M \to M'$ between two homeomorphic but non-diffeomorphic 4-manifolds.

From this last result in the remark, we are only left with the rank 3 singularities. If the map has a local representation
$$f : (x, y, z, t) \mapsto (x, y, z, g(x, y, z, t))$$
then the rank 3 singularities singularities look locally like the set of points, (x, y, z, t), for which all first derivatives of $g(x, y, z, t)$ vanish. That is, the singular set itself, which is in the domain of f, is defined as the locus of points where $\nabla g = 0$. For example, if $g = 0$, the rank 3 singular set is 4 dimensional. If $g = t^2$, the singular set is the 3-space, $t = 0$, etc. So, in general, the dimension of the rank 3 singularity set can be any number from 0 to $\dim(M)$. For our purposes, we will restrict our considerations to rank 3 singularities for which the singular subset $\Sigma = \{x \in M \mid rank(df)_x = 3\}$ is 3-dimensional. These are exactly the Morin singularities defined by a map with local representation
$$f : (x, y, z, t) \mapsto (x, y, z, g(t))$$
so that $g'(t) = 0$ defines the 3-dimensional singular subset if there is only one isolated root, as we assume. The theorem 6.16 in chapter 6 addresses

the question of the singular subset of the smooth map $f : M \to M'$ encoded in the non-triviality of the h-cobordism, W, with boundary the disjoint union of M and M', each a deformation retract of W. In the theorem 6.16 we have contractible submanifolds $A_1 \subset M, A_2 \subset M'$ and a smoothly non-trivial h-cobordism V, called the Akbulut cork. In the following we will also denote the contractible pieces A_1, A_2 as Akbulut corks but the meaning of the word will be clear from the context. Then the theorem implies[2] that the singular set Σ of f has to be 3-dimensional and the boundary of the Akbulut cork A_1 and A_2.

But the Akbulut cork A_1, A_2 is contractible and thus the boundary is homology 3-sphere, which is a compact, closed 3-manifold having the same homology as the 3-sphere S^3. Such a 3-manifold is built from 1- and 2-handles and we obtain suitable surfaces as attaching regions. The complexity of the 3-manifold contains information about the "difference" between the differential structures.

In section 9.4 we describe the change of the smooth structure by using a surgery along knots and links which is a modification of an embedded torus in a 4-manifold. But that approach leaves one important question open: What is the relation between the map $f : M \to M'$ discussed above and the surgery described in section 9.4? Thus we are looking for another description of the 3-manifold Σ by using a kind of fibration, also called branched covering. A **branched covering** of 3-manifolds is defined as a continuous map $p : M^3 \to N^3$ such that there exists a one-dimensional subcomplex L^1 in N^3 whose inverse image $p^{-1}(L^1)$ is a one-dimensional subcomplex on the complement to which, $M^3 \setminus p^{-1}(L^1)$, the restriction of p is a covering. A central theorem in that approach states that for any closed, compact, 3-manifold M^3 there exists a 3-fold covering $p : M^3 \to S^3$ of the 3-sphere by this manifold branched along a knot (as the 1-dimensional subcomplex L^1). The proof of this theorem starts with a 3-fold covering $S^3 \to S^3$. Then a solid 2-torus $D^2 \times S^1$ is cut out, and glued back in using a non-trivial map. A simple example of a 2-dimensional branched covering is given by a map $z \in D^2 \to z^2 \in D^2$, $z \in \mathbb{C}$. Except for the point $z = 0$, called the branch point, this is a 2-fold covering. On 3-manifolds the map branches not on points but rather on closed knotted curves. The link between the two approaches is part of the following conjecture:

Conjecture: For every surgery along a link L there is a map $f : M \to M'$

---

[2]This cannot be seen easily but we remark that the differential structure changes if we use an involution of the boundary of the Akbulut cork. Thus the boundary of the Akbulut cork contains the main information about the differential structure.

which is singular along a compact, closed 3-manifold Σ so that the branched covering of this 3-manifold branches along the link L.
This conjecture relates the two approaches. In general the map $f : M \to M'$ may not be explicitly available. However, the conjecture would provide a substitute approach which could be used to calculate the effect of the change of the smooth structure on the Levi-Civita connection. To preview this approach, consider a simple example of an n-fold branched cover on D^2 given by a map $z \mapsto p(z) = z^n$. The tangent bundle of the disk is a complex line bundle. Thus the map p induces a bundle map which has a singularity at $z = 0$. Harvey and Lawson [Harvey and Lawson (1993)] developed techniques to study this problem. We then extend this theory to the 3-dimensional case in which the singularity is located along a closed curve, a knot for short. This leads to a description of the non-trivial 3-manifold Σ by using a branched covering $p : \Sigma \to S^3$ which is determined by the map f via the h-cobordism.

We proceed with the theory of singular bundle maps which we believe will be useful in applications to physics.

Singular connections of complex line bundles

Harvey and Lawson [Harvey and Lawson (1993)] described a theory of singular bundle maps used in [Stingley (1995)] to obtain a Riemann-Hurwitz type formula for stable maps between 4-manifolds. Here we use this theory to obtain the change of the connection apart from a gauge transformation after performing a logarithmic transform of multiplicity p.

To illustrate of the method of Harvey and Lawson, consider the special example of a complex line bundle. Let L_1 and L_2 be two non-isomorphic complex line bundles over the same oriented manifold M of dimension 4. Locally fix two frames e and f for L_1 and L_2, respectively. Consider a smooth bundle map $\alpha : L_1 \to L_2$. If α vanishes nondegenerately we define its divisor to be the current $Div(\alpha) = \delta_\Gamma$, i.e. the delta function with support Γ, associated to the oriented codimension-2 submanifold $\Gamma = \{x \in M : \alpha_x = 0\}$. Thus α vanishes on a 2-dimensional subset $\Gamma \subset M$ where the rank of α is less than maximal, *so the singular set Γ is 2-dimensional.* Let D_{L_2} be a connection on L_2 with the corresponding 1-form defined by $D_{L_2} f = \omega_2 f$. Then there is a complex function a so that the bundle map can be written as $\alpha(e) = af$. The connection

$$D = \alpha^{-1} \circ D_{L_2} \circ \alpha$$

is the pull-back connection of D_{L_2}. Outside of the singular set Γ

$$De = (\alpha^{-1} \circ D_{L_2} \circ \alpha)e \qquad (11.2)$$

$$= \left(\frac{da}{a} + \omega_2\right)e \qquad (11.3)$$

determines the connection ω apart from a gauge transformation. It is a surprising fact that Harvey and Lawson [Harvey and Lawson (1993)] were able to extend this formula to the singular set Γ as well. In particular, they made sense of the expression

$$Div(\alpha) = d\left(\frac{da}{a}\right) \quad .$$

As a simple example, consider the complex function $a : M \to \mathbb{C}$ and define a 1-form $\frac{dz}{z}$ on $\mathbb{C}\setminus\{0\}$. From the theorem of Cauchy we obtain

$$\frac{1}{2i\pi}\int_\gamma \frac{dz}{z} = \int_S \delta_S(z) dz \wedge d\bar{z} = 1 \quad .$$

for a closed curve γ around $z = 0$ which is the boundary $\gamma = \partial S$ of a surface $S \subset \mathbb{C}$. Here the delta function δ_S has the support $supp(\delta_S) \subset S$. The pull-back via the function a leads to the expression $a^*(\frac{dz}{z}) = \frac{da}{a}$ which is a 1-form. Let $S_M \subset M$ be a two-dimensional submanifold of M with boundary γ_M such that $a(S_M) = S$, $a(\gamma_M) = \gamma$. Furthermore, assume S_M intersects Γ only in a finite number of isolated points. In the following we will refer to S_M as the "domain" of the singular form, so $dom(da/a) = S_M$. This curve γ_M is unique up to homotopy and satisfies,

$$\frac{1}{2i\pi}\int_{\gamma_M} \frac{da}{a} = \int_{S_M} \delta_{S_M} dvol(S_M) \quad .$$

where δ_{S_M} has support S_M. Note that the integral of the 1-form da/a can be regarded as the degree of the function a. The expression $Div(\alpha)$ denotes the curvature of the map α which is singular on the 2-dimensional manifold S_M i.e. a delta-function with support in S_M.

Connection change by a logarithmic transform

Now apply the previous techniques to calculate the connection changes caused by logarithmic transformations of embedded tori. Specifically, a logarithmic transformation of a smooth, simply connected, closed, oriented 4-manifold M is defined by the following procedure: Let M contain a smoothly embedded 2-torus $i : T^2 \hookrightarrow M$ with trivial self-intersection number. The image of the embedding i is denoted by $i(T^2) = F \subset M$. Using methods similar to those of Dehn surgery on knots in 3-manifolds, a *generalized logarithmic transformation* of M along T^2 is defined by deleting a tubular neighborhood νF of F from M and gluing it back using a diffeomorphism $\phi : \partial(\nu F) \to \partial(M \setminus \nu F)$. The map ϕ is defined by a composition $\phi = \beta \circ \alpha$ where $\beta : \partial(\nu F) \to \partial(M \setminus \nu F)$ is a degree-one map and $\alpha : \partial(\nu F) \to \partial(\nu F)$. If the map α has degree p, then ϕ has degree p, and the manifold is M_p. The manifolds M and M_p are homeomorphic but in general not diffeomorphic. Now calculate the connection change induced by the degree map of the boundary of νF. First, define a smooth, surjective map $f : M \to M_p$ by

$$f = \begin{cases} h & \forall x \in M \setminus \nu F \\ \overline{\alpha} & \forall x \in \nu F \end{cases}$$

where the diffeomorphism h is defined in [Griffiths and Harris (1994)] p.566 for the case of elliptic surfaces and $\overline{\alpha} : \nu F \to \nu F$ is the extension of α to the whole tubular neighborhood νF as explained in the next remark. Outside of the tubular neighborhood νF, this map is a diffeomorphism h. There the connection change is trivial, given by a pure gauge transformation $g^{-1}dg$ where g is the group element associated to h. So we need only consider the map $f|_{\nu F}$. Next, consider the tangent bundles $T(\nu F) = T(D^2) \times T(T^2) = L_0 \times T(T^2)$ and $T(\overline{\alpha}(\nu F)) = L_p \times T(T^2)$, where L_p is the tangent bundle over the p-fold cover of D^2 and $\overline{\alpha}$ is a map $\nu F \to \nu F$ defined in the remark below. This defines the two complex line bundles L_0 and L_p, respectively, both over D^2. After an explicit description of the logarithmic transform we will relate this construction to the singular set of the singular, smooth function $f : M \to M'$ in the spirit of the conjecture above, the theory of branched coverings. From this procedure, 3-manifolds can be constructed using 2-dimensional submanifolds and knots. By a general position argument we can arrange that the tubular neighborhood of the torus $D^2 \times T^2$ is embedded in the Akbulut cork $A \subset M$ so that the 3-dimensional submanifold $D^2 \times S^1$ lies on the boundary $\partial A = \Sigma$ of the Akbulut cork A. A logarithmic transform produces a branched covering of $D^2 \times S^1$ from the trivial fibration $D^2 \times S^1$. The covering branches along a closed 1-manifold lying at the center of the disk D^2. But how can we describe that branching at the level of tangent bundles? The branched covering of a 3-manifold can be derived by a branched covering of a particular 2-dimensional submanifold. Thus we have a smooth, singular mapping $f : M \to M'$ with a 3-dimensional singular set where the essential part of the singularity is located at a 2-dimensional subset. In case of a logarithmic transform the essential part is given by the disc D^2 as base of the tubular neighborhood $D^2 \times T^2$. The tangent bundle $T(D^2)$ is a trivial complex line bundle L_0. The p-fold mapping of that disc $z \mapsto z^p$ for $z \in D^2$ produces a new complex line bundle L_p. A bundle mapping $L_0 \to L_p$ can be described by using the work of Harvey and Lawson to obtain the singular connection $\frac{dz}{pz}$. Some details of that construction are set out in the following remark.

Remark 11.2.
Identify the tubular neighborhood νF with $D^2 \times T^2$ with boundary $\partial(\nu F) = S^1 \times T^2$. Consider the splitting of the map ϕ into two maps $\phi = \beta \circ \alpha$ as defined above. The complexity of ϕ is concentrated in the map α which we choose to be $\alpha(w, r) = (w^p, pr) = (z, pr) \in S^1 \times T^2 = \partial(\nu F)$ where $w \in \partial D^2 = S^1$ and $r \in T^2$. Here we are using complex notation for $D^2 \sim \mathbb{C}$. Now we extend α to the whole tubular neighborhood to obtain $\overline{\alpha}: D^2 \times T^2 \to D^2 \times T^2$ as expression for $f|_{\nu F}$, using an extension of the coordinates $w, z \in S^1$ to $w, z \in D^2$. This map is non-singular outside of $z = 0$ and singular at $z = 0$. This defines a fibration of the tubular neighborhood into the fibers F and the base D^2. All fibers are diffeomorphic to T^2 except for the point $z = 0$. The fiber over $z = 0$ is the p-fold cover of T^2 and is called the **singular fiber** S. Consider the map $\overline{\alpha}$ well defined outside of the singular fiber. The change of the frames can be written

$$d\overline{\alpha}^{-1}\left(\frac{\partial}{\partial z}, \frac{\partial}{\partial r}\right) = \left(\frac{1}{p}w^{1-p}\frac{\partial}{\partial w}, \frac{1}{p}\frac{\partial}{\partial r}\right) . \tag{11.4}$$

Let $e = \{\frac{\partial}{\partial z}, \frac{\partial}{\partial r}\}$ and $f = \{\frac{\partial}{\partial w}, \frac{\partial}{\partial r}\}$ be the frames of tangent bundles $T(0) = L_0 \times T(T^2)$ and $T(p) = L_p \times T(T^2)$, respectively. The formula (11.4) can be expressed by $d\overline{\alpha}^{-1}(e) = a(w) \cdot f$ with $a(w) = (w^{1-p}/p, 1/p)$. According to formula (11.2) and (11.3) we obtain for the pullback-connection of $D_{T(p)}$

$$D_{T(p)}f = \omega_{T(p)}f = \left(\omega_{T(0)} - \frac{da}{a}\right)f . \tag{11.5}$$

From this, we obtain

$$\omega_{T(p)}f = \left(\omega_{T(0)} + \begin{pmatrix} \frac{p-1}{p}\frac{dw}{w} & 0 \\ 0 & 0 \end{pmatrix}\right)f . \tag{11.6}$$

Note that the notation for the frames, e, f involves , $z \in D^2$, and $r \in T^2$, so the real dimension of sets (z, r) is four. Thus, in real notation ω must be a four by four matrix, and in 11.6, $\frac{dw}{w}$ is a real two by two matrix. By a slight abuse of notation, identify a with the non trivial part, $\frac{(p-1)}{p}\frac{dw}{w}$. The work of Harvey and Lawson [Harvey and Lawson (1993)] helps us to interpret this for $w = 0$. From this, we see that the curvature is changed by an expression which is singular like Dirac's delta function along the singular fiber S.

Note also that the class dw/w represents the non-triviality of the embedded torus, i.e. the class

$$\frac{i}{2\pi}d\frac{dw}{w} = \Omega$$

represents the cohomology class $\Omega \in H^2(X, \mathbb{Z})$ which is nothing else then the Poincaré dual of the homology class $PD(\Omega) = [F] \in H_2(X, \mathbb{Z})$ of the regular fiber F.

This discussion relates the effect of a logarithmic transformation on the connection in a line bundle. But what can be said about such a transformation in terms of the four dimensional metric connection used in Einstein's theory of general relativity?

Application to the theory of general relativity

Now we will follow closely the article[3] [Asselmeyer (1996)] and look at some physical questions raised by non-diffeomorphic differential structures in the theory of general relativity. Einstein's theory requires a given differential structure on a 4-manifold to express the field equations describing the gravitational field. From the beginning Einstein questioned the need to find a separate source for gravitation which ultimately turned out to be the stress-energy tensor of the system. This of course. is not determined by relativity theory itself. So we now look to the singularities associated with change of differential structures as possible "sources."

Choose two homeomorphic 4-manifolds M and M' with different differential structures and thus non-diffeomorphic tangential bundles TM and TM'. Consider a connection ∇ on the tangential bundle TM which produces the Riemannian curvature $R(X,Y)Z = \nabla_X \nabla_Y Z - \nabla_Y \nabla_X Z + \nabla_{[X,Y]} Z$. Here $X, Y, Z \in \Gamma(TM)$ are vector fields and $\Gamma(TM)$ denotes the set of all vector fields of M. Let $Ric(X,Y)$ be the Ricci tensor, R the curvature scalar and $g(X,Y)$ the metric defined for all $X, Y \in \Gamma(TM)$. Then Einstein's vacuum field equations are

$$Ric(X,Y) - \frac{1}{2} g(X,Y) R = 0 \quad \text{in} \quad M \tag{11.7}$$

or simply

$$Ric(X,Y) = 0. \tag{11.8}$$

For (11.7) above we need a Levi-Civita connection, i.e. a connection 1-form $\Gamma_j^k = \Gamma_{ij}^k dx^i$ with $i, j, k = 1, 2, 3, 4$ determined by a metric, $g(X,Y)$. From the discussion above we know that the change of the differential structure can be interpreted in terms of a change (11.6) of the connection leading to a change in the curvature, at least for line bundles. But more is known. Consider the smooth, singular map $f : M \to M'$ with a singular 3-manifold Σ. In a neighborhood $\Sigma \times [0,1]$ of the singular 3-manifold we obtain a splitting of the tangent bundle $T(\Sigma \times [0,1]) = T\Sigma \oplus T[0,1]$ into a 1-dimensional and 3-dimensional subbundle. Such a splitting is equivalent to the existence of a Lorentzian structure on $\Sigma \times [0,1]$. We call the coordinates of Σ space-like coordinates and the coordinate of $[0,1]$ time-like coordinates. In the following we assume that this splitting induces a foliation of the whole 4-manifold.

According to the discussion above (see theorem 6.16 in chapter 6), the change of the differential structure of M (leading to M') is given by a local

---
[3]The article contains some errors in the argumentation which will be corrected now.

modification of the 3-manifold Σ, i.e. the singular 3-manifold of the map f. Furthermore we know that this local modification is essentially defined on a 2-manifold. In the case of the logarithmic transform it is the disc D^2 which is modified using a p-fold cover, i.e., $z \mapsto z^p$ for $z \in \mathbb{C}, |z| \leq 1$. Let D^2 be the usual disc with metric $(g_{\mu\nu}) = diag(1,1)$ and $\mu, \nu = 1,2$ in the coordinate system (u, v) with $z = u + iv$. Then, let D_p^2 be the p-fold cover of D^2 with metric $g'_{\mu\nu}$ and coordinates (x, y) and $x = \Re(z^p), y = \Im(z^p)$. By solving the equations $x = \Re((u+iv)^p)$, $y = \Im((u+iv)^p)$ we obtain functions $h_1(x,y) = u, h_2(x,y) = v$. Then for the disc D_p we obtain the new metric

$$(g'_{\mu\nu}) = \begin{pmatrix} b_1(x,y) & 0 \\ 0 & b_2(x,y) \end{pmatrix} \tag{11.9}$$

with functions $b_1(x,y) = \left(\frac{\partial h_1}{\partial x}\right)^2$, $b_2(x,y) = \left(\frac{\partial h_2}{\partial y}\right)^2$ having a singularity at $x = y = 0$. Extend this to the whole tubular neighborhood $D^2 \times T^2$ to obtain the new metric

$$(g'_{ik}) = \begin{pmatrix} b_1(x,y) & 0 & 0 & 0 \\ 0 & b_2(x,y) & 0 & 0 \\ 0 & 0 & 1 & 0 \\ 0 & 0 & 0 & -1 \end{pmatrix} \tag{11.10}$$

for the modified manifold $D_p^2 \times T^2$. From the metric (11.10) we obtain for the Ricci tensor

$$-\Delta Ric = \begin{pmatrix} \frac{x}{4}R(x,y) & 0 & 0 & 0 \\ 0 & \frac{y}{4}R(x,y) & 0 & 0 \\ 0 & 0 & 0 & 0 \\ 0 & 0 & 0 & 0 \end{pmatrix}$$

with the function

$$R(x,y) = -2\frac{\partial^2 b_1}{\partial y^2} - 2\frac{\partial^2 b_2}{\partial x^2} + y\left(\frac{\partial b_1}{\partial y}\right)^2 + y\left(\frac{\partial b_1}{\partial x}\frac{\partial b_2}{\partial x}\right) + x\left(\frac{\partial b_2}{\partial x}\right)^2 + x\left(\frac{\partial b_1}{\partial y}\frac{\partial b_2}{\partial y}\right).$$

By using the metric (11.10), we can calculate the scalar curvature ΔR from ΔRic. So, we obtain for $p > 1$

$$Ric(X,Y) - \frac{1}{2}g(X,Y)R = \Delta Ric(X,Y) - \frac{1}{2}g(X,Y)\Delta R \neq 0 \quad \text{in} \quad M'. \tag{11.11}$$

where the right hand side of this equation represents the source of the gravitational field a singular energy-momentum tensor. The conservation law (with singularities) for this tensor follows from its construction as an Einstein tensor.

11.2 Differential Structures: From Operator Algebras to Geometric Structures on 3-manifolds

In this section we present some ideas about the possible classification of smooth structures on 4-manifolds inspired by physical considerations. At this time we do not have a rigorous formulation, but the idea is to understand the relationship between smooth structures and purely algebraic topics like algebraic K-theory. The main instrument is the structure theorem describing the h-cobordism between two homeomorphic, but non-diffeomorphic 4-manifolds (see theorem 6.16). This theorem shows that the information about the differential structure can be localized into a contractible 4-dimensional submanifold, called an Akbulut cork. The difference between two non-diffeomorphic differential structures is encoded in the non-trivial h-cobordism between these contractible submanifolds. By the work of Freedman we know that the non-triviality of the h-cobordism is related to the existence of a Casson handle defining the h-cobordism. The set of all Casson handles can be described by a binary tree, i.e., every path in that tree defines a specific Casson handle. For the smoothness change we need a pair of Casson handles or a pair of paths in the tree. But a pair of paths has the structure of an algebra $\mathcal{T}$ as was pointed out by Ocneanu. Thus we obtain the description of a differential structure specified by an Akbulut cork which is determined by the Casson handle construction. Algebraic methods can be used to study $\mathcal{T}$. First, this algebra can be related to the Clifford algebra of the infinite Euclidean space. Alternatively, this algebra is a C^* algebra with finite valued trace (or a factor II_1 for short). Then the differential structure is conjectured to be determined by a projective $\mathcal{T}$-module bundle over the 4-manifold M classified by the homotopy classes $[M, BGL(\mathcal{T})^+] = K_{\mathcal{T}}(M)$ and by fixing a class in $[S^3, BGL(\mathcal{T})^+] = K_3(\mathcal{T})$ which determines the boundary of the Akbulut cork. This last structure is well-known in abstract algebra as algebraic K-theory. The first structure $K_{\mathcal{T}}(M)$ determines the Casson handle but how can we specify an element in $K_3(\mathcal{T})$? By using homotopy theory one can show that an element of $K_3(\mathcal{T})$ is specified by a map $\pi_1(\Sigma) \to GL(\mathcal{T})$ from the fundamental group of a homology 3-sphere Σ (or the boundary of the Akbulut cork) to the linear group over the algebra $\mathcal{T}$. We know that the algebra $\mathcal{T}$ is a complex algebra and thus it is enough to have a representation $\pi_1(\Sigma) \to SL(2,\mathbb{C})$. Such a map determines a geometric structure (see subsection 6.2.2 for the definition) and/or a codimension-1 foliation of the homology 3-sphere Σ.

In the first subsection we construct the algebra of the change of the differential structure by using formal properties of the connection in the tangent bundle. Then we describe a Casson handle determining the Akbulut cork by using a pair of paths in the tree. This approach leads to an unexpected relationship to the algebraic K-theory of a certain factor II_1 operator algebra $\mathcal{T}$. The main conjecture is then that the pair of homotopy classes $[M, BGL(\mathcal{T})^+]$ and $[S^3, BGL(\mathcal{T})^+] = K_3(\mathcal{T})$ for a compact manifold determines the differential structures on M. In the last subsection we study the classes in $K_3(\mathcal{T})$ and relate them to geometric structures and foliations on the boundary of the Akbulut cork.

11.2.1 Differential structures and operator algebras

In this subsection we will construct the operator algebra which is associated with the change of the differential structure. More details can be found in the publication [Asselmeyer-Maluga and Rosé (2005)].

Start with a purely *formal discussion* of the change of the smooth structure by using singular maps. A singular smooth map $f \colon M \to M'$ can be interpreted as a change of the differential structure as discussed above. Let D_M and $D_{M'}$ be the covariant derivatives of the tangent bundles TM and TM'. Any covariant derivative can be decomposed by $D = d + \omega$ with the connection 1-form ω having values in some group such as $SO(p,q)$ with $p + q = 4$. We get the transformation

$$D_M = f_*^{-1} D_{M'} f_*$$

of D_M to $D_{M'}$ with the element f_* of the gauge group $M \to SO(p,q)$ induced by the differential $df \colon TM \to TM'$. The connection 1-form transforms formally by

$$\omega_M = f_*^{-1} \omega_{M'} f_* + f_*^{-1} df_*.$$

The inhomogeneous contribution of the change of the connection is given by the *singular 1-form*

$$\varphi = f_*^{-1} df_*. \tag{11.12}$$

This 1-form is matrix-valued, taking values in $\mathfrak{g}$, the Lie algebra of the structure group G for the tangent bundle TM. *But this singular 1-form φ is ill-defined yet.* Assuming the conjecture above, we can construct a 1-form with domain a 2-dimensional manifold having point singularities in the domain. This form is given by a similar calculation as in the previous section using surgery along a link. Here we are not interested in the explicit expression of that 1-form. According to our conjecture above, *to every* *singular 1-form constructed above with 2-dimensional domain there is a singular map with singular set a 3-dimensional manifold.* This means we associate to every singular 1-form with 2-dimensional domain (in the sense of Harvey and Lawson) a change of the smooth structure given by a smooth map $f : M \to M'$ with 3-dimensional singular set. Thus, we define the *singular support singsupp(φ) of φ* as

$$singsupp(\varphi) = \{x \in M \mid rank(df_x) < 4\} = \Sigma_\varphi$$

Before proceeding with the definition of a singular 1-form, recall the action of a diffeomorphism on a connection. Let $g: M \to M$ be a diffeomorphism that induces a map $dg: TM \to TM$, also a diffeomorphism. The application of g changes the derivative D to $g_*^{-1} D g_*$ and keeps the trace $Tr_M(D)$ (defined in the remark below) of operator D over M invariant: $Tr_M(g_*^{-1} D g_*) = Tr_M(D)$ for all diffeomorphisms g. The last relation is only true for diffeomorphisms and not for singular maps like f. Thus instead of using the operator D, which is not diffeomorphism-invariant, we consider the diffeomorphism-invariant trace $Tr_M(D)$.

Remark 11.3.
The definition of the trace $Tr_M(D)$ is not simple. We start with the remark that the covariant derivative D_{e_i} along the basis element e_i can be made into a Dirac operator $\gamma^i D_{e_i}$. Then we can form the square of the Dirac operator and its L^2 norm with respect to spinor fields. Let S be the spinor (or $Spin_C$) bundle over M with the set of sections $\Gamma(S, M)$. Furthermore, let $\langle \psi_1, \psi_2 \rangle$ be the natural scalar product between two spinor fields $\psi_1, \psi_2 \in \Gamma(S, M)$ (induced by the Clifford product). Then we obtain for the trace

$$Tr_M(D) = \sqrt{\inf_{\psi \in \Gamma(S,M)} \int_M \langle \gamma^i D_{e_i} \psi, \gamma^i D_{e_i} \psi \rangle dvol(M)} \qquad (11.13)$$

The square of the Dirac operator guarantees the convergence of the integral. The heat kernel expansion formalism establishes the diffeomorphism invariance of the expression.

Now define an algebraic structure on the set $\mathcal{S}$ of singular 1-forms φ. The operations, sum and product, are related to the union and intersection of the singular supports. Start with two smooth maps with singularities, $f : M \to M'$, $g : M \to M''$ with non-homeomorphic singular supports and singular 1-forms by φ and ψ, respectively, with $\Sigma_\varphi \neq \Sigma_\psi$. Now we ask for a new singular 1-form χ which defines a smooth map with singularities, $h : M \to M'$, uniquely up to a diffeomorphism. Define the *sum* by

$$\varphi + \psi = \chi : \varphi, \psi, \chi \in \mathcal{S}, \ singsupp(\chi) = singsupp(\varphi) \cup singsupp(\psi),$$

using the algebraic sum of one-forms. The construction of the product $\varphi \cdot \psi$ is a bit more complicated. Consider two forms φ, ψ with their singular supports $\Sigma_\varphi, \Sigma_\psi$. Any 1-form is naturally associated with a closed curve in the domain of the singular 1-form: This *associated curve* C_φ is defined by the dual of the 1-form φ. Recall that a singular 1-form φ is a a 1-form with point singularities in its domain and taking values in the Lie algebra of the structure group of the tangent bundle. Given a closed curve γ in M which does not intersect the singular points in the domain of φ then we can integrate the 1-form in the exponential

$$\exp\left(\int_\gamma \varphi\right),$$

defining the holonomy of φ (seen as connection 1-form) along γ. From the holonomy we get a representation of the fundamental group $\gamma \in \pi_1(\Sigma_\varphi)$ into the structure group $G = SO(p,q)$ of the tangent bundle. The technical construction is given by

$$\frac{1}{2i\pi} ln\, Tr\left(\exp\int_{C_\varphi} \varphi\right) = 1$$

where we take the mean value for the logarithm ln and Tr is the trace with respect to a suitable representation of the structure group G. The corresponding curve C_φ of φ is determined only up to homotopy, but that is all that is needed in the following. Thus C_φ and C_ψ are the associated curves to the singular forms φ and ψ, respectively. Furthermore, let S_φ be the surface (unique up to homotopy) with boundary $\partial S_\varphi = C_\varphi$ also known as Seifert surface.

Because of the possible knotting of the curve C_φ, the Seifert surface can be very complicated (see [Rolfson (1976)]). The two curves C_φ, C_ψ are *linked* if their corresponding surfaces intersect transversely. We denote linked curves by $C_\varphi \between C_\psi$ and the transversely intersection by $S_\varphi \pitchfork S_\psi$. By this, we are able to define the *product*: The product between the singular 1-forms φ and ψ is a new singular 1-form $\chi = \varphi \cdot \psi$ with singular support $singsupp(\chi) = singsupp(\varphi) \cap singsupp(\psi)$ where the associated curve C_χ of the new singular form χ is given by

$$C_\chi = \begin{cases} C_\varphi \between C_\psi : S_\varphi \pitchfork S_\psi \\ C_\varphi \sqcup C_\psi : S_\varphi \cap S_\psi = \emptyset \end{cases}$$

The first case $C_{\varphi\cdot\psi} = C_\varphi \between C_\psi$ represents a *non-commutative* product $\varphi \cdot \psi \neq \psi \cdot \varphi$ because the link $C_{\psi\cdot\varphi} = C_\psi \between C_\varphi$ is different from $C_{\varphi\cdot\psi}$. In knot theory one calls $C_{\psi\cdot\varphi}$ the mirror link of $C_{\varphi\cdot\psi}$. In the second case one gets from $C_{\varphi\cdot\psi} = C_\varphi \sqcup C_\psi$ by using $C_\varphi \sqcup C_\psi = C_\psi \sqcup C_\varphi$ the relation

$C_{\varphi \cdot \psi} = C_{\psi \cdot \varphi}$. Thus, in this case the product is *commutative* $\varphi \cdot \psi = \psi \cdot \varphi$. Finally the product with a number field K is induced from the corresponding operation of differential forms. That completes the construction. The set of singular 1-forms $\mathcal{S}$ endowed with these two operations forms $\mathcal{T} = (\mathcal{S}, +, \cdot)$ which is conjectured to provide an *algebra of transitions of the differential structure of space-time*, i.e., of transitions between non-diffeomorphic reference frames. For each singular 1-form $\varphi \in \mathcal{T}$ there is a natural operator $D_\varphi \colon \mathcal{T} \to \Omega^1(\mathcal{T})$ – called the *covariant derivative* with respect to φ – mapping an element ψ of $\mathcal{T}$ to an 1-form $D_\varphi \psi \in \Omega^1(\mathcal{T})$ over $\mathcal{T}$ satisfying the Leibniz rule $D_\varphi(\psi \cdot \chi) = (D_\varphi \psi) \cdot \chi + \psi \cdot (D_\varphi \chi)$. The expression

$$\Phi = D_\varphi \varphi, \ \Phi \in \Omega^1(\mathcal{T})$$

is called the *curvature* of φ. According to [Bourbaki (1989)], every algebra admits an universal derivative L with $L^2 = 0$ which also makes $\mathcal{T}$ to a differential algebra. Furthermore, we can define formal differential forms $\Omega^p(\mathcal{T})$ on $\mathcal{T}$ by using L. Then, a p-form is generated by $\varphi_0 L \varphi_1 \cdots L \varphi_p$ or $1 L \varphi_1 \cdots L \varphi_p$ (see [Connes (1995)]). Every two elements $\varphi, \psi \in \mathcal{T}$ define an element of $\Omega^1(\mathcal{T})$ by $D_\varphi \psi$. By the universal property of the derivative L, $\Omega^1(\mathcal{T})$ is generated by forms like $\varphi L \psi$ or $1 L \varphi$. Thus, also $D_\varphi \psi$ must be given by a form like this and we may choose

$$D_\varphi \psi = \varphi L \psi. \qquad (11.14)$$

In contrast to the usual representation we do not have a wedge product $\wedge$ and thus we cannot define the covariant derivative as $D_\varphi = L + \varphi \wedge$. But we can choose the derivative L in such a way (using the universality property of L) that the relation $D_\varphi \psi = \varphi L \psi$ is fulfilled. Then the curvature can be written as $\Phi = D_\varphi \varphi = \varphi L \varphi$.

Furthermore, we can introduce a *trace* of a singular form $\varphi = f_*^{-1} df_*$ by the integral

$$Tr(\varphi, C) := \int_C \varphi < \infty$$

with respect to a suitable curve C. The universal derivative L extends the trace over $\mathcal{T}$ to all forms $\Omega^p(\mathcal{T})$ by relation (11.14). Using general properties of L, one obtains the relation

$$Tr(\Phi\Psi, C) = (-1)^{pq} Tr(\Psi\Phi, C), \ \Phi \in \Omega^p(\mathcal{T}), \Psi \in \Omega^q(\mathcal{T}) \qquad (11.15)$$

for the trace on $\mathcal{T}$. The construction of one operation is unsettled: the star operation $*$ which makes $\mathcal{T}$ to a $*$-algebra. A $*$-operation has to fulfill the two relations: $(\varphi^*)^* = \varphi$ and $(\varphi \cdot \psi)^* = \psi^* \cdot \varphi^*$. Let φ be a singular 1-form and C_φ the associated curve. We define φ^* to be a 1-form with curve $C(\varphi^*) = \overline{C_\varphi}$, i.e. the curve C_φ with the opposite orientation. Then the first relation $(\varphi^*)^* = \varphi$ is obvious. The second relation $(\varphi \cdot \psi)^* = \psi^* \cdot \varphi^*$ is a standard fact from knot theory: the change of the orientation of a link transforms the link to the mirror link. Let $\varphi \cdot \psi$ be a link then $(\varphi \cdot \psi)^*$ is the mirror link. By definition, $\psi^* \cdot \varphi^*$ is also the mirror link. That completes the construction of the algebra $\mathcal{T}$.

Finally we obtain:

(1) $\mathcal{T}$ is a differential $*$-algebra $\mathcal{T} = (\mathcal{S}, +, \cdot, L)$,
(2) the covariant derivative D_φ is related to L by $D_\varphi \psi = \varphi L \psi$,
(3) the finite trace $Tr(\varphi, C) = \int_C \varphi$ fulfills relation (11.15).

In the sequel we will show that the algebra $\mathcal{T}$ is a Temperley-Lieb algebra over the complex numbers.

A *Temperley-Lieb algebra* is an algebra with unit element **1** over a number field K generated by a countable set of generators $\{e_1, e_2, \ldots\}$ with the defining relations

$$e_i^2 = \tau \cdot e_i, \quad e_i e_j = e_j e_i : |i - j| > 1,$$
$$e_i e_{i+1} e_i = \tau e_i, \; e_{i+1} e_i e_{i+1} = \tau e_{i+1}, \; e_i^* = e_i \qquad (11.16)$$

where τ is a real number in $(0, 1]$. By [Jones (1983)], the Temperley-Lieb algebra has a uniquely defined trace Tr which is normalized to lie in the interval $[0, 1]$.

This leads to:

Let $\mathcal{T}$ be the differential $*$-algebra of singular 1-forms forming the set of transitions of the differential structure. Furthermore, let D_φ be the covariant derivative associated to the singular form φ and the universal derivative L of $\mathcal{T}$ defined by $\varphi L \psi := D_\varphi \psi$. Then the algebra $\mathcal{T}$ has the structure of a Temperley-Lieb algebra. The algebra of transitions $\mathcal{T}$ is extensible to a C^*-algebra comprising the algebra of field operators of particles. The complex Hilbert space is induced by $\mathcal{T}$ via the GNS construction.

The details and motivation of the proof can be found in [Asselmeyer-Maluga and Rosé (2005)]. In brief, the proof is carried out in the following steps:

- The support of the singular 1-form is a 3-manifold which splits into irreducible pieces. The generators of the algebra are projection operators which correspond to these pieces.

From Differential Structures to Operator Algebras and Geometric Structures 297

- It is possible to introduce an order structure on the set of generators by rearranging the pieces of the 3-manifold.
- The curvature (corresponding to a singular 1-form) for the sum and the product of two elements induces further relations in the algebra.
- The Bianchi-identity for the curvature supports the fact that the basis elements are projection operators.
- Finally, it is possible to introduce a star operation and to show that the number field is the set of complex numbers.

Now we make a few remarks about the relationship between the algebra $\mathcal{T}$ and the differential structure of the 4-manifold M. Every smooth map $f: M \to M'$ with singularities produces a singular 1-form φ defined by an element of the algebra $\mathcal{T}$ given as a linear combination of the generators. The coefficients of a singular 1-form φ are a countable subset of the complex numbers. For example, consider a 4-manifold M and make a logarithmic transform to get M_{p_1}. This procedure can be repeated to get $M_{p_1 p_2 \ldots p_n}$. To every singular fiber of $M_{p_1 p_2 \ldots p_n}$ corresponds a basis element e_i and every number p_i determines a coefficient a_{p_i}. A more detailed analysis shows that the basis elements for $M_{p_1 p_2 \ldots p_n}$ commute with each other, i.e., we choose e_{2i-1} as generators. Then the singular 1-form φ associated to the map $M \to M_{p_1 p_2 \ldots p_n}$ is given by

$$\varphi = \sum_{i=1}^{n} a_{p_i} e_{2i-1}$$

Other generators are needed for the knot surgery of Fintushel and Stern (see section 9.4). Thus we have established a relationship between the elements of the algebra $\mathcal{T}$ and the differential structures on a given 4-manifold M.

11.2.2 *From Akbulut corks to operator algebras*

Changes of differential structures can be understood in terms of the theorem 6.16 on the existence of non-product subcobordism. Consider a 5-dimensional TOP cobordism W between two homeomorphic but non-diffeomorphic simply-connected 4-manifolds M, M' (i.e. $\partial W = M \sqcup M'$). Then there are (contractible) submanifolds $V \subset M$ and $V' \subset M'$ where $M \setminus V$ and $M' \setminus V'$ are diffeomorphic (i.e. the cobordism between these two manifolds is a DIFF product) while the subcobordism between V and V' is not a DIFF product. From the topological point of view, everything is trivial and the cobordism W is a product. From section 6.6 we know the reason for this the failure of an extension of this from the TOP to the

DIFF category: the Whitney trick used to cancel pairs of k and $k+1$ handles valid for TOP cannot be used for the DIFF case. In our situation, it is the 2-/3-handle pair which cannot be canceled. As Freedman shows in the topological case, one can embed a Casson handle which contains an embedded topological disk in such a manner that the 2-/3-handle pair can be canceled. Unfortunately, this disk is not smoothly embedded and so the 2-/3-handle pair is left. In some sense we can say that the DIFF difference between the two manifolds M, M' is given by the non-canceling 2-/3-handle pair represented by a non-trivial Casson handle.

According to Freedman ([Freedman (1982)] p.393), a Casson handle can be represented by a labeled finitely-branching tree Q with base point $\star$, having all edge paths infinitely extendable away from $\star$. Each edge has a label $+$ or $-$ and each vertex corresponds to a kinky handle where the self-plumbing number of that kinky handle equals the number of branches leaving the vertex. The sign on each branch corresponds to the sign of the associated self plumbing (see section 6.4 for the construction of a Casson handle from that data). The work of Freedman involves the construction of a special embedded disk (Whitney disk) inside of the Casson handle which is used to cancel the 2-/3-handle pair topologically. The homeomorphism type of the Casson handle is described by re-embedding of Casson handles into a given one by using the re-embedding theorem showing that every 7-stage tower embeds into a 6-stage tower. To describe the diffeomorphism type in this construction, Freedman ([Freedman (1982)] p.398) constructs another labeled tree $S(Q)$ from the tree Q. There is a base point from which a single edge (called "decimal point") emerges. The tree is binary: one edge enters and two edges leaving a vertex. The edges are named by initial segments of infinite base 3-decimals representing numbers in the standard "middle third" Cantor set[4] $C.s. \subset [0,1]$. Each edge e of $S(Q)$ carries a label τ_e where τ_e is an ordered finite disjoint union of 6-stage towers together with an ordered collection of standard loops generating the fundamental group. There are three constraints on the labels which lead to the correspondence between the $\pm$ labeled tree Q and the (associated) τ-labeled tree $S(Q)$.

Now we will relate this formalism to the algebra of the previous subsec-

---

[4]This kind of Cantor set is given by the following construction: Start with the unit Interval $S_0 = [0,1]$ and remove from that set the middle third and set $S_1 = S_0 \setminus (1/3, 2/3)$ Continue in this fashion, where $S_{n+1} = S_n \setminus \{$middle thirds of subintervals of $S_n\}$. Then the Cantor set $C.s.$ is defined as $C.s. = \cap_n S_n$. In other words, if we use a ternary system (a number system with base 3), then we can write the Cantor set as $C.s. = \{x : x = (0.a_1 a_2 a_3 \ldots)$ where each $a_i = 0$ or $2\}$.

tion. Start with the tree $S(Q)$. Every path in this tree represents one tree Q leading to a Casson handle. Any subtree represents a Casson-handle which embeds in Q. Two different paths in the tree represent two homeomorphic, but smoothly different Casson handles. A change of the differential structure can be defined relative to a reference Casson handle. Consider two paths in the tree $S(Q)$, the reference path for the given differential structure and a path for the new differential structure. Thus, a pair of paths corresponds to one element of the algebra. This algebra is isomorphic to $\mathcal{T}$ constructed in the previous subsection. Thus we have another construction of $\mathcal{T}$.

Now consider the so-called string algebra according to Ocneanu [Ocneanu (1988)]. Define a non-negative function $\mu : Edges \to \mathbb{C}$ together with the adjacency matrix $\triangle$ acting on μ by

$$\triangle \mu(x) = \sum_{\substack{v \in Edges \\ s(v)=x \\ r(v)=y}} \mu(y)$$

where $s(v)$ and $r(v)$ denote the source and the range of an edge v. A path in the tree is a succession of edges $\xi = (v_1, v_2, \ldots, v_n)$ where $r(v_i) = s(v_{i+1})$ and write $\tilde{v}$ for the edge v with the reversed orientation. Then, a string on the tree is a pair of paths $\rho = (\rho_+, \rho_-)$, with $s(\rho_+) = s(\rho_-)$, $r(\rho_+) \sim r(\rho_-)$ which means that $r(\rho_+)$ and $r(\rho_-)$ end on the same level in the tree and ρ_+, ρ_- have equal lengths i.e., $|\rho_+| = |\rho_-|$. Define an algebra $String^{(n)}$ with the linear basis of the n-strings, i.e., strings with length n and the additional operations:

$$(\rho_+, \rho_-) \cdot (\eta_+, \eta_-) = \delta_{\rho_-, \eta_+}(\rho_+, \eta_-)$$
$$(\rho_+, \rho_-)^* = (\rho_-, \rho_+)$$

where $\cdot$ can be seen as the concatenation of paths. We normalize the function μ by $\mu(root) = 1$. Now we choose a function μ in such a manner that

$$\triangle \mu = \beta \mu \qquad (11.17)$$

for a complex number β. Then we can construct elements e_n in the algebra $String^{(n+1)}$ by

$$e_n = \sum_{\substack{|\alpha|=n-1 \\ |v|=|w|=1}} \frac{\sqrt{\mu(r(v))\mu(r(w))}}{\mu(r(\alpha))} (\alpha \cdot v \cdot \tilde{v}, \alpha \cdot w \cdot \tilde{w}) \qquad (11.18)$$

fulfilling the algebraic relations (11.16) where $\tau = \beta^{-2}$. The trace of the string algebra given by

$$tr(\rho) = \delta_{\rho_+, \rho_-} \beta^{-|\rho|} \mu(r(\rho))$$

defines on $A_\infty = (\bigcup_n String^{(n)}, tr)$ an inner product by $\langle x, y \rangle = tr(xy^*)$ which is, after completion, the Hilbert space $L^2(A_\infty, tr)$.

Now consider the parameter τ. Originally, Ocneanu introduced the string algebra to classify the splittings of modules over an operator algebra. Thus, to determine the parameter τ we look for the simplest generating structure in the tree. The simplest structure in our tree is one edge which is connected with two other edges. This graph is represented by the following adjacency matrix

$$\begin{pmatrix} 0 & 1 & 1 \\ 1 & 0 & 0 \\ 1 & 0 & 0 \end{pmatrix},$$

having eigenvalues $0, \sqrt{2}, -\sqrt{2}$. According to our definition above, β is given by the greatest eigenvalue of this adjacency matrix, i.e., $\beta = \sqrt{2}$ and thus $\tau = \beta^{-2} = \frac{1}{2}$. Then, without proof, we state that the algebra $\mathcal{T}$ is given by the Clifford algebra on $\mathbb{R}^\infty$.

Now consider the projections e_n in more detail. Every e_n represents a tree with n levels, which is the same as a collection of Casson handles. All of these Casson handles are homeomorphic to each other. Thus every projection represents an equivalence class of Casson handles. To get a further interpretation of the projection we need another important tool in Freedman's proof [Freedman (1982)] of the topological equivalence of all Casson handles: the re-embedding theorems. Bizaca [Bizaca (1994)] presents an algorithm based on these theorems which explicitly calculates lower and upper bounds for the number of self-intersections of Casson towers. One of the rules in building the tree $S(Q)$ is that the number of self-intersections of the 6th tower increases from level to level. By Bizaca's algorithm there is a relationship between the number of levels and the number of self-intersections. In [Gompf (1984)] Gompf describes the diffeomorphism type of the Casson handles by considering the self-intersections of the so-called Casson disk embedded in the Casson handle which can be used for the Whitney trick. In general there are uncountably many diffeomorphism types (see [Gompf (1989b)]) but we can embed the Casson handle into a compact manifold and so we need only countably many described by these self-intersections. Thus, we obtain a connection between the order structure of the projections e_n and the self-intersections of the Casson disk. Then the product of two projections $e_i e_j$ is a Casson handle with a disk of i self-intersections which embeds in a Casson handle with a disk of j self-intersections. The weighted sum $a \cdot e_i + b \cdot e_j$ of two projections is a disjoint union of two Casson handle having disks with i and j self-intersections, respectively. Every Casson handle is determined by its 6th Casson tower and the two factors express the self-intersections in the tower. So, if both Casson handles are identical, then not only the projections but also the two factors a, b must be identical. The only function which determines these factors is given by $\mu : Edges \to \mathbb{C}$.

So the most important thing seems to be the definition of the function $\mu : Edges \to \mathbb{C}$ where the complex structure comes into play. In the tree $S(Q)$ the edges are labeled by a 6-stage tower. We can interpret this function as the evaluation of the procedure of adding special embedded singular disks. The reason for the complex numbers comes from the fact that every complex line bundle over a 4-manifold is classified by the first Chern class given by an element in $H^2(M, \mathbb{Z})$ having the Poincaré dual lying in $H_2(M, \mathbb{Z})$ represented by a disk. By definition the number of edges in Q from a given vertex is equal to the singular points of the disk (or the kinks of the kinky handle). So we define this function μ by the value of the section of the complex line bundle in the singular point to which the edge corresponds. In the next subsection we will give another construction for this number based on global information.

11.2.3 Algebraic K-theory and exotic smooth structures

Now we start again with the same situation as in the previous subsection. Consider two homeomorphic but non-diffeomorphic simply-connected 4-manifolds M, M' and a 5-dimensional cobordism W (i.e. $\partial W = M \sqcup M'$) between them. Then there are (contractible) sub-manifolds $V \subset M$ and $V' \subset M'$ where $M \setminus V$ and $M' \setminus V'$ are diffeomorphic (i.e. the cobordism between these two manifolds is a DIFF product) and the sub-cobordism Y between V and V' is not a DIFF product. Both manifolds V, V' are contractible and have (according to Freedman [Freedman (1982)]) the boundary of a homology 3-sphere. There are many possible ways to describe this situation globally. As remarked above, the difference between M and M' is expressed by non-canceling pairs of 2-/3-handles.

We know from the work of Freedman [Freedman (1982)] that two homotopy-equivalent 4-manifolds M, N are also homeomorphic. That means, if we have a space B which is chosen to be a so-called loop space and sets $[M, B], [N, B]$ of homotopy classes of maps $M, N \to B$ then the homotopy-equivalence $M \simeq N$ means that the sets are isomorphic

$[M, B] \simeq [N, B]$ with respect to the algebraic structure induced by the loop space structure of B. Now we will construct such a space B. Consider a simply-connected 4-manifold M with a contractible sub-manifold V containing the information about the differential structure and having a homology 3-sphere ∂V as boundary. Next we deal with the cofibration $\partial V \subset M$ (∂V is closed in M) induced by the inclusion $i : \partial V \hookrightarrow M$ with quotient map $p : M \to M/\partial V$. From standard homotopy theory we obtain the exact sequence

$$[M/\partial V, B] \xrightarrow{p^*} [M, B] \xrightarrow{i^*} [\partial V, B]$$

which determines the set $[M/\partial V, B]$ in terms of $[M, B]$ and $[\partial V, B]$. At first we will concentrate on the set $[\partial V, B]$ and consider the $I\!I_1$ algebra $\mathcal{T}$ representing pairs of Casson handles. In the topologically trivial sub-cobordism Y we need an embedded Casson handle for the topological trivialization. The interior of every Casson handle CH is diffeomorphic to the $\mathbb{R}^4$ and thus non-compact. We need a substitute for the boundary, also called a frontier $Fr(CH)$, defined by $Fr(CH) = \text{closure}(\text{closure}(CH) \setminus CH)$. Embed the Casson handle CH into the sub-cobordism Y in such a manner that the interior of CH lies in the interior of Y and the frontier $Fr(CH)$ is mapped to the cobordism between the boundaries $\partial V, \partial V'$. The tree $S(Q)$ constructed above represents the frontiers of the Casson handle Q (see [Freedman (1982)] p.398). In the previous subsection we constructed the algebra $\mathcal{T}$ with every element of the algebra representing one or more Casson handles. Now consider $\mathcal{T} \subset R_0$ the algebra $\mathcal{T}$ as a sub-algebra of the hyperfinite factor $I\!I_1$ algebra R_0 with index $[R_0 : \mathcal{T}] = 2$ (see [Jones (1983)] for the definition). The hyperfinite factor R_0 can be seen as a projective $\mathcal{T}$-module. The reason for this embedding of $\mathcal{T}$ into R_0 is the fact that the tree is binary leading to the coefficient $\tau = 1/2$ (in the relations (11.16)). Thus, the set of all possible, embedded Casson handles, CH, in Y can be described by a projective $\mathcal{T}$-module bundle over ∂V. The equivalence classes of such bundles is given by the variety of group representations $\pi_1(\partial V) \to GL(\mathcal{T})$ up to conjugacy. Unfortunately, the differential structure on M depends on the structure of V. Recall, the boundary of V is a homology 3-sphere. To eliminate this ambiguity, consider the map $\partial V \to BGL(\mathcal{T})$ up to homotopy induced from $\pi_1(\partial V) \to GL(\mathcal{T})$. Then, the +-construction on both sides leads to map $(\partial V)^+ = S^3 \to BGL(\mathcal{T})^+$ or better to homotopy class $[S^3, BGL(\mathcal{T})^+] = K_3(\mathcal{T})$ also called the algebraic K-theory of the ring $\mathcal{T}$ (every algebra is trivially a ring). Thus we determine the space B above to be $BGL(\mathcal{T})^+$. This space is a loop space. Thus

From Differential Structures to Operator Algebras and Geometric Structures   303

we obtain two possible invariants of the differential structure: the group $[M, BGL(\mathcal{T})^+]$ and $K_3(\mathcal{T})$. So, if the differential structure is determined by the Casson handle, we can conjecture:
Conjecture: The differential structures on a simply-connected compact, 4-manifold M are determined by the homotopy classes $[M, BGl(\mathcal{T})^+]$ and by the algebraic K-theory $K_3(\mathcal{T})$ where $\mathcal{T}$ is the hyperfinite $I\!I_1$ factor C^*-algebra.

From a general theorem of algebraic K-theory of C^*-algebras it follows that
$$K_4(\mathcal{T}) \otimes \mathbb{Q} = \pi_4(BGl(\mathcal{T})^*) \otimes \mathbb{Q} = [S^4, BGl(\mathcal{T})^+] \otimes \mathbb{Q} = \mathbb{Q}$$
and so if this conjecture is valid there must be infinitely many distinct differential structures on S^4. This would provide a counter example to the smooth Poincaré conjecture in dimension 4. The algebraic K-theory $K_3(\mathcal{T})$ can be further simplified to uncover a relationship between 4 dimensional differential structures and geometric structures as well as codimension-1 foliations in 3 dimensions. Note the decomposition:
$$BGl(\mathcal{T})^+ = BGl(\mathcal{T}) \times_{BGl(\mathbb{C})^+} BGl(\mathbb{C})^+$$
i.e. the space $BGl(\mathcal{T})^+$ is in some sense a "$BGl(\mathbb{C})^+$-module". In the next section we will see that the hyperbolic structures on the homology 3-sphere ∂V are contained in $K_3(\mathbb{C})$.

11.2.4 Geometric structures on 3-manifolds and exotic differential structures

In this section we will partly give an answer to the question: How can we detect the exotic differential structure of a 4-manifold M by considering its boundary ∂M? Or better, which structure on the boundary can detect the exotic differential structure in the interior? Every 3-manifold has a unique differential structure. So, we have to look for another structure on the 3-manifold.

Above we introduced the two classes $[M, BGL(\mathcal{T})^+] = K_\mathcal{T}(M)$ and $K_3(\mathcal{T}) = [S^3, BGL(\mathcal{T})^+]$ to characterize the differential structure of M. Up to now we only considered the structure $K_\mathcal{T}(M)$. Now we will study the structure $K_3(\mathcal{T})$ and its main substructure $K_3(\mathbb{C})$. As we will see, an element in $K_3(\mathbb{C})$ is induced by a map from the boundary $\Sigma = \partial V$ of the Akbulut cork to the group $SL(2, \mathbb{C})$. This 3-manifold Σ is a homology 3-sphere. Now we consider a hyperbolic structure on this homology 3-sphere, i.e. a map $\alpha : \pi_1(\Sigma) \to PSL(2,\mathbb{C}) = SO(3,1)$. The group $SO(3,1)$ is familiar to physicists as the Lorentz group. At the same time, this group is also the isometry of the hyperbolic geometry of the light cone in Minkowski space. This map can be lifted to $\tilde{\alpha} : \pi_1(\Sigma) \to SL(2,\mathbb{C})$, i.e. a spin structure on Σ which gives a spin bundle S. Of course there are homology 3-spheres which do not admit a hyperbolic structure. As an example we have the Poincaré sphere which does not appear as the boundary of a smooth, contractible 4-manifold and admits only a spherical geometry.

By homotopy theoretic manipulations, this map $\tilde{\alpha}$ induces a map $\Sigma \to BGl(\mathbb{C})$. Now we use the +-construction of Quillen, to define a map $\Sigma^+ = S^3 \to BGl(\mathbb{C})^+$. The homotopy class of this map is one distinct hyperbolic structure on Σ and is related to an element of $K_3(\mathbb{C})$. In general, we have a homology 3-sphere Σ and a representation $\alpha : \pi_1(\Sigma) \to GL_N(\mathcal{T})$ for a ring $\mathcal{T}$ then the pair $[\Sigma, \alpha]$ defines an element in $K_3(\mathcal{T})$.

By the vector bundle classification in chapter 5 we know that every $SL(2,\mathbb{C})$ bundle V_α over Σ is trivial and the space of connections of such bundles is given by

$$Hom(\pi_1(\Sigma), SL(2,\mathbb{C})) = \{f : \pi_1(\Sigma) \to SL(2,\mathbb{C}) \,|\, f \text{ homomorphism}\,\} \quad .$$

Now define a twisted Dirac operator D_α

$$D_\alpha : \Gamma(S \otimes V_\alpha) \to \Gamma(S \otimes V_\alpha)$$

acting on the sections $\Gamma(S \otimes V_\alpha)$ of the twisted spin bundle $S \otimes V_\alpha$ where the bundle V_α has a flat connection (i.e. the curvature vanish) defined with respect to $\alpha \in Hom(\pi_1(\Sigma), PSL(2,\mathbb{C}))$. Of course there is also the usual Dirac operator $D : \Gamma(S) \to \Gamma(S)$ and we define the following spectral invariant $\rho(\alpha, D) = \xi(0; \alpha, D) \in \mathbb{C}/\mathbb{Z}$. Let $\{\lambda\}$ and $\{\lambda_\alpha\}$ be the eigenvalues of D and D_α, respectively. Now we define the so-called *eta* series by

$$\eta(s; D) = \sum_{\Re(\lambda)>0} \lambda^{-s} - \sum_{\Re(\lambda)<0} (-\lambda)^{-s} \qquad \eta(s; \alpha, D) = \eta(s; D_\alpha) \qquad s \in \mathbb{C}$$

and with h, h_α as the number of vanishing eigenvalues and N as the dimension of the representation α. Then the function

$$\xi(s; \alpha, D) = \frac{h_\alpha + \eta(s; \alpha, D) - Nh - N\eta(s; D)}{2}$$

is well-defined for the value $s = 0$ and defines the homotopy-invariant of D by

$$\rho(\alpha, D) = \xi(0; \alpha, D) \in \mathbb{C}/\mathbb{Z} \quad .$$

The interpretation of this invariant is rather simple: $\eta(0;D)$ measures the asymmetry in the spectrum, i.e. the difference of the number of positive and negative eigenvalues. As proved by Jones and Westbury [Jones and Westbury (1995)] $\rho(\alpha, D)$ defines a map $K_3(\mathbb{C}) \to \mathbb{C}/\mathbb{Z}$. Conversely, every homology 3-sphere defines an element in $K_3(\mathbb{C})$ by this procedure. Unfortunately, not all elements in $K_3(\mathbb{C})$ can be obtained in this way. But Jones and Westbury show that every element of finite order in $K_3(\mathbb{C})$ is of the form $[\Sigma(p,q,r), \alpha]$ with a suitable representation $\alpha : \pi_1(\Sigma(p,q,r)) \to SL_2(\mathbb{C})$ and $\Sigma(p,q,r)$ is the so-called Seifert homology sphere[5].

Now we consider a class in $K_{\mathcal{T}}(M) = [M, BGL(\mathcal{T})^+]$ for a 4-manifold and a contractible 4-submanifold V with boundary $\partial V = \Sigma$ defined by the Akbulut cork. The inclusion $\Sigma \hookrightarrow M$ induces an injective map $K_{\mathcal{T}}(M) \to K_{\mathcal{T}}(\Sigma)$ but the projection $BGL(\mathcal{T})^+ \to BGL(\mathbb{C})^+$ leads to the surjective map

$$K_{\mathcal{T}}(\Sigma) = K_{\mathcal{T}}(S^3) \to [\Sigma, BGL(\mathbb{C})^+] = [S^3, BGL(\mathbb{C})^+] = K_3(\mathbb{C})$$

where we use the functorial properties $(BGL(\mathcal{T})^+)^+ = BGL(\mathcal{T})^+$ and $\Sigma^+ = S^3$. Thus we obtain a chain of maps defining a map $K_{\mathcal{T}}(M) \to K_3(\mathbb{C})$. As we noted above, the class in $K_3(\mathbb{C})$ is induced by a representation of the fundamental group $\pi_1(\Sigma)$ of the homology 3-sphere Σ. Otherwise we know that a geometric structure on Σ is induced by a representation of the fundamental group into subgroups of $SL(2,\mathbb{C})$ like $PSL(2,\mathbb{C})$, $SO(3)$, $SL(2,\mathbb{R})$ and twisted version of them. Then, we have the following chain of arguments: a change of the differential structure on the 4-manifold M is the choice of another class in $K_{\mathcal{T}}(M)$ which changes the class in $K_3(\mathbb{C})$ coming from another possible representation $\alpha : \pi_1(\Sigma) \to SL(2,\mathbb{C})$, which is nothing other than the change of the geometric structure on Σ.

Furthermore, the class in $K_3(\mathbb{C})$ is generated by a spectral invariant $\rho(\alpha; D)$ of the Dirac operator. Thus from the map $K_{\mathcal{T}}(M) \to K_3(\mathbb{C})$ we conjecture a relation between the spectral invariant $\rho(\alpha; D)$ of the Dirac operator on Σ and a spectral invariant of the Dirac operator on the 4-manifold. Up to now, this invariant cannot determined exactly.

But more is known: Every element of the structure $K_3(\mathbb{C})$ is a flat connection (i.e. vanishing curvature) A of a $SL(2,\mathbb{C})$ bundle over a homology 3-sphere Σ. Now we can form a 3-form $\Gamma_A = Tr(A \wedge dA)$ which is an element of the group $H^3(\Sigma, \mathbb{R})$. By the theory of foliations, every 1-form θ with $d\theta \wedge \theta = 0$ defines a codimension-1-foliation of Σ induced by the

---

[5]$\Sigma(2,3,5)$ is the Poincaré 3-sphere for instance.

connection A by using the relation $d\theta = A \wedge \theta$. The invariant Γ_A is a cobordism invariant, i.e. two foliations with different invariants Γ_A define two different (=non-cobordant) foliations of Σ. But two different invariants Γ_A are induced by two different elements in $K_3(\mathbb{C})$.

To summarize, we hope to have provided support for the conjecture:
Conjecture: The differential structures on a simply-connected compact, 4-manifold M are determined by the homotopy classes $[M, BGl(T)^+]$ and by the algebraic K-theory $K_3(T)$ where T is the hyperfinite II_1 factor C^*-algebra. The classes in $K_3(T)$ are given by the geometric structure and/or a codimension-1 foliation of a homology 3-sphere in M determining the Akbulut cork of M.

From the physical point of view, this conjecture is very interesting because it connects the abstract theory of differential structures with well-known structures in physics like operator algebras or bundle theory. Perhaps such speculations may provide a geometrization of quantum mechanics or more. We close this section, and book, which these highly conjectural remarks.

Bibliography

Abresch, U., Durán, C., Püttmann, T. and Rigas, A. (2005). Wiedersehen metrics and exotic involutions of euclidean spheres, ArXiv:math.GT/0501091.
Adams, J. (1963). On the groups $J(X)$.I. *Topology* **2**, pp. 181–193.
Adams, J. (1974). *Stable homotopy and generalised homology*, Chicago Lectures in Mathematics (The University of Chicago Press, Chicago and London).
Adler, R., Baxin, M. and Schiffer, M. (1965). *Introduction to general relativity* (McGraw-Hill, New York).
Akbulut, S. (1996). Lextures on seiberg-witten invariants, *Turkish J. Math.* **20**, pp. 95–119.
Akbulut, S. (1999). A fake cusp and a fishtail, *Turk. J. of Math.* **23**, pp. 19–31.
Anderson, M. T. (2004). Geometrization of 3-manifolds via the ricci flow, *Notices Amer. Math. Soc.* **51**, pp. 184–193.
Ashtekar, A. (1986). New variables for classical and quantum gravity, *Phys. Rev. Lett.* **57**, pp. 2244–2247.
Ashtekar, A. and Lewandowski, J. (1995). Projective techniques and functional integration, *J. Math. Phys.* **36**, p. 2170.
Asselmeyer, T. (1996). Generation of source terms in general relativity by differential structures, *Class. Quant. Grav.* **14**, pp. 749 – 758.
Asselmeyer-Maluga, T. and Brans, C. (2002). Cosmological anomalies and exotic smoothness structures, *Gen. Rel. Grav.* **34**, pp. 597–607.
Asselmeyer-Maluga, T. and Rosé, H. (2005). Differential structures - the geometrization of quantum mechanics, (available as gr-qc/0511089).
Atiyah, M. (1988). New invariants of 3 and 4 dimensional manifolds, in R. Wells (ed.), *Symposium on the mathematical heritage of Hermann Weyl* (Amer. Math. Soc.), pp. 285–299.
Atiyah, M., Hitchin, N. and Singer, I. (1978). Self-duality in 4-dimensional Riemannian geometry, *Proc. Roy. Soc. London Ser. A* **362**, pp. 425–461.
Atiyah, M., Patodi, V. and Singer, I. (1973). Spectral asymmetry and Riemannian geometry, *Bull. London Math. Soc.* **5**, pp. 229–234.
Atiyah, M. F. (1979). *Geometry of Yang-Mills Fiels* (Academia nazionale det Lincei Scuola Normale Superiore, Pisa).
Barth, W., Peters, C. and van de Ven, A. (1984). *Compact complex surfaces*

(Springer).
Bauer, S. (2002). A stable cohomotopy refinement of Seiberg-Witten invariants ii, Tech. rep., math.DG/0204267.
Bauer, S. and Furuta, M. (2002). A stable cohomotopy refinement of Seiberg-Witten invariants i, Tech. rep., math.DG/0204340.
Baulieu, L. and Singer, I. (1988). Topological Yang-Mills symmetry, *Nucl. Phys. B (Proc. Suppl.)* **5B**, p. 12.
Beem, J. K., Ehrlich, P. E. and Easley, K. L. (1996). *Global Lorentzian Geometry*, 2nd edn. (Dekker, New York, Basel, Hong Kong).
Bergmann (1942). *Introduction to the theory of relativity* (Prentice-Hall, New York).
Bing, R. (1959). The Cartesian product of a certain nonmanifold and a line is e^4, *Ann. Math.(USA)* **70**, pp. 399–412.
Birmingham, D., Blau, M., Rakowski, M. and Thompson, G. (1991). Topological field theory, *Phys. Rep.* **209**, pp. 129–340.
Bizaca, Z. (1994). A reimbedding algorithm for casson handles, *Trans. Amer. Math. Soc.* **345**, pp. 435–510.
Bižaca, Ž. and Gompf, R. (1996). Elliptic surfaces and some simple exotic $\mathbb{R}^4$'s, *J. Diff. Geom.* **43**, pp. 458–504.
Borel, A. (1984). *Intersection cohomology* (Birkhäser).
Bott, R. and Tu, L. (1982). *Differential forms in algebraic topology* (Springer-Verlag, New York).
Bott, R. and Tu, L. (1995). *Differential Forms in Algebraic Topology*, Graduate Texts in Mathematics 82 (Springer-Verlag).
Bourbaki, N. (1989). *Elements of Mathematics*, commutative algebra edn. (Springer, Berlin).
Brans, C. (1980). Roles of space-time models, in A. Marlow (ed.), *Quantum theory and gravitation* (Academic Press, New York), p. 27.
Brans, C. (1994a). Exotic smoothness and physics, *J. Math. Phys.* **35**, pp. 5494–5506.
Brans, C. (1994b). Localized exotic smoothness, *Class. Quant. Grav.* **11**, pp. 1785–1792.
Brans, C. (1999). Absolulte spacetime: the twentieth century ether, *Gen. Rel. Grav.* **31**, p. 597.
Bredon, G. E. (1993). *Topology and Geometry* (Springer-Verlag, New York).
Brieskorn, E. (1970). Beispiele zur Differentialtopologie von Singularitäten, *Inv. Math.* **2**, pp. 1 – 14.
Bröcker, T. and Jänich, K. (1987). *Introduction to differential topology* (Cambridge, Cambridge).
Brooks, R., Montano, D. and Sonnenschein, J. (1988). Gauge fixing and renormalization in topological quantum field theory, *Phys. Lett. B* **214**, p. 91.
Cappell, S. and Shaneson, J. (1976). Some new four-manifolds, *Ann. Math.* **104**, pp. 61–72.
Cartan, E. (1966). *Theory of Spinors* (MIT Press, Cambridge, Massachusetts).
Casson, A. (1986). *Three lectures on new infinite constructions in 4-dimensional manifolds*, Vol. 62, progress in mathematics edn. (Birkhäuser), notes by

Lucian Guillou, first published 1973.
Cerf, J. (1968). *Sur les difféomorphismes de la sphère de dimension trois* ($\Gamma_4 = 0$), Vol. 53, springer lecture notes in math. edn. (Springer Verlag).
Cerf, J. (1970). La stratification naturelle des espaces fonctions differentiables réeles et la thèoréme de la pseudoisotopie, *Publ. Math. I.H.E.S.* **39**.
Coleman, S. and Mandula, J. (1967). All possible symmetries of the s matrix, *Phys. Rev.* **159**, pp. 1251–1256.
Connes, A. (1995). *Non-commutative geometry* (Academic Press).
Curtis, C., Freedman, M., Hsiang, W.-C. and Stong, R. (1997). A decomposition theorem for h-cobordant smooth simply connected compact 4-manifolds, *Inv. Math.* .
De Michelis, S. and Freedman, M. (1992). Uncountable many exotic r^4's in standard 4-space, *J.Diff.Geom.* **35**, pp. 219–254.
de Rham, G. (1984). *Differentiable Manifolds* (Springer-Verlag, Berlin).
DeWitt, B. (1984). *Supermanifolds* (Cambridge University Press, London).
Donaldson, S. (1983). An application of gauge theory to the topology of 4-manifolds, *J. Diff. Geom.* **18**, pp. 269–316.
Donaldson, S. (1986). Connections, cohomology, and the intersection forms of 4-manifolds, *J. Diff. Geom.* **24**, pp. 275–341.
Donaldson, S. (1987). Irrationality and the h-cobordism conjecture, *J. Diff. Geom.* **26**, pp. 141–168.
Donaldson, S. (1990). Polynomial invariants for smooth four manifolds, *Topology* **29**, pp. 257–315.
Donaldson, S. and Kronheimer, P. (1990). *The Geometry of Four-Manifolds* (Oxford Univ. Press, Oxford).
Durán, C. (2001). Pointed widersehen metrics on exotic spheres and diffeomorphisms of s^6, *Geometrica Dedicata.* **88**, pp. 199–210.
Durán, C., Mendoza, A. and Rigas, A. (2004). Blakers-massey elements and exotic diffeomorphisms of s^6 and s^{14} via geodesics, *Trans. Amer. Math. Soc.* **356**, pp. 5025–5043.
Edmonds, A. (1999). Linking pairings of 3-manifolds and intersection forms of 4-manifolds, In preparation.
Eells, J. and Kuiper, N. (1962). An invariant for certain smooth manifolds, *Ann. Mat. Pura. Appl.* **60**, 93-110.
Feehan, P. and Leness, T. (2003). On Donaldson and Seiberg-Witten invariants, in *Topology and geometry of manifolds (Athens, GA, 2001), Proceedings of Symposia in Pure Mathematics*, Vol. **71**, pp. 237–248.
Fenn, R. and Rourke, C. (1979). On Kirby's calculus of links, *Topology* **18**, pp. 1–15.
Fintushel, R. and Stern, R. (1985). Pseudofree orbifolds, *Ann. of Math.* **122**, pp. 335–364.
Fintushel, R. and Stern, R. (1995). Rational blowdowns of smooth 4-manifolds, AMS-preprint, alg-geom 9505018.
Fintushel, R. and Stern, R. (1996). Knots, links, and 4-manifolds, AMS-preprint, dg-ga 9612014.
Fintushel, R. and Stern, R. (1998a). Constructions of smooth 4-manifolds, in

Proceedings of the 1998 ICM Doc.Math.J.DMV Extra Volume ICM, Vol. **II**, pp. 443–452, aMS-preprint, alg-geom 9505018.

Fintushel, R. and Stern, R. (1998b). Knots, links, and 4-manifolds, *Inv. Math* **134**, pp. 363–400, (dg-ga/9612014).

Freed, D. and Uhlenbeck, K. (1990). *Instantons and 4-manifolds*, second edition edn., Math. Sci. Res. Inst. Publ. (Springer Verlag, New York).

Freedman, M. (1982). The topology of four-dimensional manifolds, *J. Diff. Geom.* **17**, pp. 357 – 454.

Freedman, M. (1983). The disk problem for four-dimensional manifolds, in *Proc. Internat. Cong. Math. Warzawa*, Vol. **17**, pp. 647 – 663.

Freedman, M. (1984). There is no room to spare in four-dimensional space, *Noticess Amer. Math. Soc.* **31**, pp. 3–6.

Freedman, M. and Luo, F. (1989). *Slected Applications of Geometry to Low-Dimensional Topology*, ULS (AMS).

Freedman, M. and Quinn, F. (1990). *Topology of 4-Manifolds*, Princeton Mathematical Series (Princeton University Press, Princeton).

Freedman, M. and Taylor, L. (1977). λ splitting 4-manifolds, *Topology* **16**, pp. 181–184.

Freedman, M. and Taylor, L. (1986). A universal smoothing of four-space, *J. Diff. Geom.* **24**, pp. 69–78.

Freedman, M. and Teichner, P. (1995). 4-Manifold topology i:subexponential groups, *Inv. Math.* **122**, pp. 509–529.

Friedman, R. and Morgan, J. (1994). *Smooth Four-Manifolds and Complex Surfaces*, Ergebnisse der Mathematik und ihrer Grenzgebiete.3.Folge/A Series of Modern Surveys in Mathematics vol. 27 (Springer Verlag, Berlin-Heidelberg-New York).

Furata, M. (1998). Stable homotopy version of seiberg-witten invariant, Preprint, now incorporated in [Bauer and Furuta (2002)].

Furata, M. (2001). Monopole equation and the $\frac{11}{8}$-conjecture, *Math. Res. Lett.* , pp. 279–291.

Gallot, S., Hulin, D. and Lafontaine, J. (1980). *Riemannian Geometry* (Springer-Verlag, Berlin).

Geroch, R. (1985). *Mathematical Physics* (University of Chicago, Chicago).

Glimm, J. (1960). Two cartesian products which are euclidean spaces, *Bull. Soc. Math. France* **88**, pp. 131–135.

Golubitsky, M. and Guillemin, V. (1973). *Stable Mappings and their Singularities*, Graduate Texts in Mathematics 14 (Springer Verlag, New York-Heidelberg-Berlin).

Gompf, R. (1984). Infinite families of casson handles and topological disks, *Topology* **23**, pp. 395–400.

Gompf, R. (1985). An infinite set of exotic $\mathbb{R}^4$'s, *J. Diff. Geom.* **21**, pp. 283–300.

Gompf, R. (1988). On sums of algebraic surfaces, *Inv. Math.* **94**, pp. 171–174.

Gompf, R. (1989a). A moduli space of exotic $\mathbb{R}^4$'s, *Proc. Edinburgh Math. Soc.* **32**, pp. 285–289.

Gompf, R. (1989b). Periodic ends and knot concordance, *Top. Appl.* **32**, pp. 141–148.

Gompf, R. (1991). Nuclei of elliptic surfaces, *Topology* **30**, pp. 479–511.
Gompf, R. and Stipsicz, A. (1999). *4-manifolds and Kirby Calculus* (American Mathematical Society).
Goresky, M. and MacPherson, R. (1988). *Stratified Morse Theory* (Springer).
Goto, T. (1971). Relativistic quantum mechanics of one-dimensional mechanical continuum and subsidiary condition of dual resonancs model, *Prog. Theor. Phys.* **46**, p. 1560.
Greenberg, M. J. and Harper, J. R. (1981). *Algebraic Topology* (Addison-Wesley, New York).
Griffiths, P. and Harris, J. (1994). *Principles of Algebraic Geometry* (John Wiley and Sons, Inc., New York).
Gromoll, D. and Meyer, W. (1974). An exotic sphere with nonnegative sectional curvature, *Ann, Math.* **100**, pp. 401–406.
Grove, K. and Ziller, W. (1999). Curvature and symmetry of Milnor spheres, Preprint.
Haag, R., Łopuszański, J. and Sohnius, M. (1975). All possible generators of supersymmetries of the S-matrix, *Nucl. Phys. B* **88**, p. 257.
Harvey, F. and Lawson, H. (1993). *A theory of characteristic currents associated to a singular connection*, astérisque 213 edn. (Société Mathématique De France).
Hatfield, D. (1992). *Quantum Field Theory of Point Particles and Strings* (Addison-Wesley, New York).
Hehl, F., McCrea, J., Mielke, E. and Ne'eman, y. (1995). Metric affine gauge theory of gravity, *Phys. Rept.* **258**, pp. 1–171.
Hironaka, H. (1964). Resolution of singularities of an algebraic variety over a field of characteristic zero, *Ann. of Math.* **79**, pp. 109–326.
Hirsch, M. W. (1976). *Differential Topology* (Springer-Verlag, New York).
Hirzebruch, F. (1973). *Topological methods in algebraic geometry* (Springer Verlag, Berlin-Heidelberg-New York).
Hirzebruch, F. and Hopf, H. (1958). Felder von Flächenelementen in 4-dimensionalen Mannigfaltigkeiten, *Math. Annalen* **136**, p. 156.
Hitchin (1974). Harmonic spinors, *Adv. Math.* **14**, pp. 1 – 55.
Howard, E. ., D. and Stachel, E. ., J. (1986). *Einstein and the history of general relativity. Proceedings of Osgood Hill Conference* (Birkhaeuser), einstein Studies, vol. 1.
Husemoller, D. (1966). *Fiber Bundles* (MacGraw-Hill Book Co., New York-London-Sydney).
Husemoller, D. (1994). *Fiber Bundles*, 3rd edn. (Springer-Verlag, New York).
Husemoller, D. and Milnor, J. (1973). *Symmetric Bilinear Forms* (Springer Verlag, Berlin).
Jammer, M. (1960). *Concepts of Space* (Harper).
Jones, J. and Westbury, B. (1995). Algebraic K-theory, homology spheres, and the η-invariant, *Topology* **34**, pp. 929–957.
Jones, V. (1983). Index of subfactors, *Invent. Math.* **72**, pp. 1–25.
Kaku, M. (1993). *Quantum Field Theory - A Modern Introduction* (Oxford University Press, Oxford).

Karoubi, M. (1978). *K-Theorie - An Introduction*, Vol. 226, grundlehren der mathematischen wissenschaften edn. (Springer Verlag, Berlin Heidelberg New York).
Kervaire, M. (1965). Le théorème de Barden-Mazur-Stallings, *Comm. Math. Helv.* **40**, pp. 31–42.
Kervaire, M. and Milnor, J. (1958). Bernoulli numbers, homotopy groups and a theorem of Rohlin, in *Proc. Int. Congress of Math.*, Vol. **77** (Edinburgh), pp. 454– 458.
Kervaire, M. and Milnor, J. (1961). On 2-spheres in 4-manifolds, *Proc. Nat. Acad. Science USA* **47**, pp. 1651–1657.
Kervaire, M. and Milnor, J. (1963). Groups of homotopy spheres: I, *Ann. Math.* **77**, pp. 504 – 537.
Kirby, R. (1978). A claculus for framed links in s^3, *Inv. Math.* **45**, pp. 36–56.
Kirby, R. (1989). *The Topology of 4-Manifolds*, Lecture Notes in Mathematics (Springer Verlag, Berlin-New York).
Kirby, R. (1997). Problems in low-dimensional topology, in W. Kazez (ed.), *Geometric Topology (1993 Georgia International Topology Conference)*, Lecture Notes in Mathematics, Vol. **2** (AMS/IP), pp. 35–473.
Kirby, R. and Siebenmann, L. (1977). *Foundational essays on topological manifolds, smoothings, and triangulations*, Ann. Math. Studies (Princeton University Press, Princeton).
Kirby, R. and Tylor, L. (1998). A survey of 4-manifolds through the eyes of surgery, Tech. rep., math.GT/9803101.
Kobayashi, S. and Nomizu, K. (1963). *Foundations of differential geometry, vol. 1* (Interscience, New York).
Kreck, M. (1999). Surgery and duality, *Ann. of Math.* **149**, pp. 707–754.
Kreck, M. and Stolz, S. (1988). A diffeomorphism classification of 7-dimensional homogeneous einstein manifolds with $su(3) \times su(2) \times u(1)$-symmetry, *Ann. Math.* **127**, pp. 373–388.
Kronheimer, P. and Mrowka, T. (1994). Recurrence relations and asymptotics for four-manifold invariant, *Bull. AMS* **30**, pp. 215–221.
Labastida, J. and Pernici, M. (1988). A gauge invariant action in topological quantum field theory, *Phys. Lett. B* **212**, p. 56.
Lance, T. (2000). Differential structures on manifolds, in J. R. S. Cappel, A. Ranicki (ed.), *Surveys on Surgery Theory*, Annals of Mathematics Studies, Vol. **1** (Princeton University Press), pp. 73–104.
Lawson, H. B. (1985). *Theory of Gauge Fields in Four Dimensions* (American Mathematical Society, Providence).
Lawson, T. (1998). On the minimal genus problem, Preprint.
Lawson, T. (2003). *Topology: a geometric approach* (Oxford Science Publications).
Lee, R. and Wilczynski, D. (1993). Representing homology classes by locally flat 2-spheres, *K Theory* **7**, pp. 333–367.
Lickorish, W. (1962). A representation of orientable combinatorical 3-manifolds, *Ann. Math.* **76**, pp. 531–540.
Lu, N. (1992). A simple proof of the fundamental theorem of kirby calculus on

links, *Trans. Amer. Math. Soc.* **331**, pp. 143–156.

Manolescu, C. (2003a). A gluing theorem for the relative Bauer-Furuta invariants, Math/0311342.

Manolescu, C. (2003b). Seiberg-Witten-Floer stable homotopy type of three-manifolds with $b_1 = 0$, *Geom. Top.* **7**, pp. 889–932.

Markov, A. (1958). Insolubility of the problem of homeomorphy, in *Proceeding Intern. Congress Math.*, pp. 300–306.

Massey, W. S. (1967). *Algebraic Topology: An Introduction* (Harcourt, Brace and World, New York).

Mather, J. (1971). Stability of C^∞ Mappings. VI: The nice dimensions, in *Proccedings of the Liverpool Singularities Symposium* (Springer Lecture Notes in Mathematyics, Vol. 192), pp. 207 – 253.

Matsumoto, Y. (1982). On the bounding genus of homology 3-spheres, *J. Fac. Sci. Univer. Tokyo Sect. IA Math.* **29**, pp. 287–318.

Matveev, S. and Polyak, M. (1994). A geometrical presentation of the surface mapping class group and surgery, *Comm. Math. Phys.* **160**, pp. 537–556.

McMillan, D. (1961). Cartesian products of contractable open manifolds, *Bull. Amer. Math. Soc.* **67**, pp. 510–514.

Milnor, J. (1956a). Construction of universal bundles I, in [Milnor (1956b)], pp. 272 – 284.

Milnor, J. (1956b). Construction of universal bundles II, *Ann. Math.* **63**, pp. 430 – 436.

Milnor, J. (1956c). On manifolds homeomorphic to the 7-sphere, *Ann. Math.* **64**, pp. 399 – 405.

Milnor, J. (1958). On simply connected 4-manifolds, in *"Symposium International de Topologia Algebraira"* (Univ. Nac. Autonoma de Mexico and UNESCO, Mexico City), pp. 122–128.

Milnor, J. (1959). Differentiable manifolds which are homotopy spheres, Note, Princeton University.

Milnor, J. (1963). *Morse theory*, Ann. of Mathematical Study **51** (Princeton Univ. Press, Princeton).

Milnor, J. (1964). Microbundle I, *Topology* **3(Suppl. 1)**, pp. 53 – 80.

Milnor, J. (1965a). *Lectures on the h-cobordism theorem* (Princeton Univ. Press, Princeton).

Milnor, J. (1968). A note on curvature and fundamental group, *J. Diff. Geom.* **2**, pp. 1–7.

Milnor, J. (2000). The discovery of exotic spheres, in J. R. S. Cappel, A. Ranicki (ed.), *Surveys on Surgery Theory*, Annals of Mathematics Studies (Princeton University Press), p. 17.

Milnor, J. (2003). Towards the poincaré conjecture and the classification of 3-manifolds, *Notices of the AMS* **50**, pp. 1226–1233.

Milnor, J. and Stasheff, J. (1974). *Characteristic Classes*, Ann. Math. Studies,76 (Princeton Univ. Press, Princeton, N.J.).

Milnor, J. W. (1965b). *Topology from the Differentiable Viewpoint* (University Press of Virginia).

Misner, C., Thorne, K. and Wheeler, J. (1973). *Gravitation* (Freeman, San Fran-

cisco).
Moise, E. (1952). Affine structures on 3-manifolds, *Ann. Math.* **56**, pp. 96–114.
Moore, G. and Witten, E. (1998). Integration over the u-plane in donaldson theory, *Adv. Theor. Math. Phys.* **1**, pp. 298–387, hep-th/9709193.
Moore, J. D. (2001). *Lectures on Seiberg-Witten Invariants* (Springer-Verlag, New York).
Morgan, J. (1996). *The Seiberg-Witten Equations and applications to the topology of smooth four-manifolds* (Princeton University Press, Princeton).
Morgan, J., Szabo, Z. and Taubes, C. (1996). A product formula for the seiberg-witten invariants and the generalized thom conjecture, *J. Diff. Geom.* **44**, pp. 706–788.
Morgan, J., Szabo, Z. and Taubes, C. (1997). Product formulas along t^3 for seiberg-witten invariants, *Mathematical Research Letters* **4**, pp. 915–929.
Mosher, R. E. and Tangora, M. C. (1968). *Cohomology Operations and Applications in Homotopy Theory* (Harper and Row, New York).
Munkres, J. (1960). Obstructions to the smoothing of pieceswise-differential homeomeomorphisms, *Ann. Math* **72**, pp. 621–554.
Munkres, J. R. (1975). *Topology: A first course* (Prentice-Hall, New Jersey).
Nakahara, M. (1989). *Geometry, Topology and Physics* (Adam Hilger, Bristol and New York).
Nambo, Y. (1970). Duality and hydrodynamics, Lectures at the Copenhagen Summer Symposium.
Nash, C. (1991). *Differential Topology and Quantum Field Theory* (Academic Press, London-San Diego-New York-Boston-Sydney-Tokyo-Toronto).
Norton, J. D. (1984). How einstein found his field equations: 1912–1915, *historical Studies in the Physical Sciences* **14**, pp. 253–315.
Norton, J. D. (1992). Einstein, nordström and the early demise of scalar, lorentz-covariant theories of gravitation, *Archive for History of Exact Science* **45**, p. 17.
Ocneanu, A. (1988). Quantized groups, string algebras and Galois theory for algebras, in Evans and Takesaki (eds.), *Operator Algebras and Applications*, pp. 119–172.
Papakyriakopoulos, C. (1943). New proof of the invariance of the homology groupsof a complex (article in greek), *Bull. Math. Soc. Greece* **22**, pp. 481–488.
Perelman, G. (2002). The entropy formula for the ricci flow and its geometric applications, ArXiv:math.DG/0211159.
Perelman, G. (2003a). Finite extinction time for the solutions to the ricci flow on certain three-manifods, ArXiv:math.DG/0307245.
Perelman, G. (2003b). Ricci flow with surgery on three-manifolds, ArXiv:math.DG/0303109.
Rado, T. (1925). Über den Begriff der Riemannschen Fläche, *Acta Litt. Scient. Univ. Szegd* **2**, pp. 101–121.
Ramond, P. (1990). *Field Theory: A Modern Primer*, 2nd edn. (Addison-Wesley, New York).
Reeb, G. (1952). Sur certain propriétés topologiques des variétś feuilletées, *Actual. sci. industr.* **1183**, pp. 91–154.

Reshetikhin, N. and Turaev, V. (1990). Ribbon graphs and their invariants derived from quntum groups, *Comm. Math. Phys.* **127**, pp. 1–26.
Reshetikhin, N. and Turaev, V. (1991). Invariants of three-manifolds via link polynomials and quantum groups, *Inv. Math.* **103**, pp. 547–597.
Rohlin, V. (1952). New results in the theory of 4-dimensional manifolds, *Dokl. Akad. Nauk. SSSR* **84**, pp. 221–224, (in Russian).
Rolfson, D. (1976). *Knots and Links* (Publish or Prish, Berkeley).
Rosenberg, J. (1994). *Algebraic K-theory and its application* (Springer).
Rourke, C. and B.J., S. (1972). *Introduction to Piecewise-linear Topology*, Vol. 69, ergebnisse der mathematik edn. (Springer Verlag).
Salamon, D. (1995). Spin Geometry and Seiberg-Witten Invariants, University of Warwick, manuscript.
Seiberg, N. (1988). Supersymmetry and non-perturbative beta functions, *Phys. Lett. B* **206**, pp. 75–80.
Seiberg, N. and Witten, E. (1994a). Electric-magnetic duality, monopole condensation, and confinement in N=2 supersymmetric Yang-Mills theory, *Nucl. Phys.* **B 426**, pp. 19–52.
Seiberg, N. and Witten, E. (1994b). Monopoles, duality and chiral symmetry breaking in N=2 supersymmetric QCD, *Nucl. Phys.* **B 431**, pp. 581–640.
Serre, J.-P. (1951). Homologie singulière des espace fibrés. Applications, *Ann. Math.* **54**, pp. 425–505.
Serre, J.-P. (1953). Groupes d'homotopue et classes de groupes abéliens, *Ann. Math.* **58**, pp. 258–294.
Sładkowski, J. (1999). Strongly gravitating empty spaces, Preprint.
Smale, S. (1961). Generalized poincaré's conjecture in dimensions greater than four, *Annals of Mathematics* **74**, p. 391.
Smale, S. (1962). On the structure of manifolds, *Amer. J. of Math.* **84**, pp. 387–399.
Spanier, E. (1966). *Algebraic topology* (McGraw-Hill).
Stallings, J. (1962). Piecewise-linear structure of euclidean space, *Proc. Cambridge Phil. Soc.* **58**, p. 481.
Steenrod, N. (1951). *The Topology of Fiber Bundles* (Princeton University Press, Princeton, N.J.).
Steenrod, N. (1999). *Topology of Fibre Bundles* (Princeton University Press, Princeton).
Stingley, R. (1995). *Singularities of maps between 4-manifolds*, Ph.D. thesis, State University of New York at Stony Brook.
Stromberg, K. R. (1981). *Introduction to Classical Real Analysis* (Wadsworth, Belmont, California).
Taubes, C. (1982). Self-dual Yang-Mills connections on non-self-dual 4-manifolds, *J. Diff. Geom.* **17**, pp. 139–170.
Taubes, C. (1987). Gauge theory on asymptotically periodic 4-manifolds, *J. Diff. Geom.* **25**, pp. 363–430.
Taylor, L. (1997). An invariant of smooth 4-manifolds, *Geom. Top.* **1**, pp. 71–89.
Taylor, L. (1998). Smooth Euclidean 4-space with few symmetries, Published in math.GT/9807143.

Thom, R. (1954). Quelques propertiétés global des variétés différentiables, *Comment. Math. Helv.* **28**, pp. 17 – 86.
Thurston, W. (1997). *Three-Dimensional Geometry and Topology*, 1st edn. (Princeton University Press, Princeton).
tom Dieck, T. (1991). *Topologie* (Walter de Gruyter, Berlin-New York).
Trace, B. (1982). On attaching 3-handles to a 1-connected 4-manifold, *Pacific J. Math.* **99**, pp. 175–181.
Trautman, A. (1984). *Differential Geometry for Physicists* (Bibliopolis, Napoli).
Veneziano, G. (1968). Construction of a crossing-symmetric Regge-behaved amplitude for linearly rising Regge trajectories, *Nuovo Cim.* **57A**, p. 190.
Vick, J. W. (1994). *Homology Theory* (Springer-Verlag, New York).
Visser, M. (1996). *Lorentzian Wormholes* (American Institute of Physics, New York).
von Neumann, J. (1955). *Mathematical Foundations of Quantum Mechanics* (Pinceton University press, Princeton).
Wall, C. (1964a). Diffeomorphisms of 4-manifolds, *J. London Math. Soc.* **39**, pp. 131–140.
Wall, C. (1964b). On simple-connected 4-manifolds, *J. London Math. Soc.* **39**, pp. 141–149.
Wall, C. (1970). *Surgery on compact manifolds* (Academic Press).
Weinberg, S. (1972). *Gravitation and Cosmology* (Wiley, New York).
Weinberger, S. (1994). *The topological classification of stratified spaces*, Chicago Lect. in Math. (Univ. of Chicago Press).
Wess, J. and Bagger, J. (1992). *Supersymmetry and Supergravity*, second edition edn. (Princeton University Press, Princeton).
West, P. (1990). *Introduction to Supersymmetry and Supergravity*, second edition edn. (World Scientific, Singapore).
Wheeler, J. A. (1962). *Geometrodynamics* (Academic Press, New York).
Whitehead, J. (1935). A certain open manifold whose group is unity, *Quart. J. Math. (Oxford)* **6**, pp. 268–279.
Whitehead, J. (1940). On c^1 complexes, *Ann. Math.* **41**, pp. 809–824.
Witten, E. (1982). Supersymmetry and morse theory, *J. Diff. Geom.* **17**, pp. 661–692.
Witten, E. (1985). Global gravitational anomalies, *Comm. Math. Phys.* **100**, pp. 197–229.
Witten, E. (1988). Topological quantum field theory, *Comm. Math. Phys.* **117**, pp. 353–386.
Witten, E. (1994). Monopoles and four-manifolds, *Math. Research Letters* **1**, pp. 769–796.
Witten, E. (1998). D-branes and k-theory, *JHEP* **9812**, p. 019, hep-th/9810188.
Wraith, D. (1998). Exotic spheres with positive ricci curvature, Preprint.

Index

Akbulut cork, 193, 284
Alexander's trick, 212
algebra
   Clifford, 79
   exterior, 39
attaching
   map, 162
      framing of, 165
   region, 162
   sphere, 162

band-sum, 188
base space, 55
belt sphere), 162
Betti number, 31
biholomorphic, 40
boundary, 26
   operator, 27
branched covering, 284
Brieskorn sphere, 211
bundle, 55
   fiber, 57
   group, 56, 57
   homotopy theory of, 88–89
   Hopf, 208
   micro, 223
   microtangent, 223
   non-trivial, 58–59
   of frames, 66
   projection, 56
   repere, 66
   section of, 56

   sphere, 207
   tangent vector, 56
   transition function, 57
   trivial, 56
   universal, 226

canceling pair, 167
Casson handle
   definition, 186
   skeleton, 183
CAT structure, 224
   concordance, 225
category
   functional, 16
category theory, 6
chain group
   simplicial, 26
   singular, 26
characteristic class
   Chern class, 102–106
   Euler class, 111
   Pontrjagin class, 106–107
   universal, 102
   Weil homomorphism, 107–109
characteristic classes, 101–114
   as obstructions, 101
   Chern class, 102
   Pontrjagin class, 102
   Stiefel-Whitney class, 102
characteristic element, 198
Chern character, 105
Chern class, 102–106

definition, 103
differential form description, 110
total, 105
Christoffel symbol, 66
classification
  1-manifolds, 169
  2-manifolds, 169–172
  definite intersection form, 200
  indefinite forms, 198, 199
  topological 4-manifolds, 204
  topological manifolds, 236
classification theorem
  Freedman, 204
  Kirby-Siebenmann, 228
classifying space, 97–101
  definition, 100
  homotopy properties of, 101
cobordant, 159
  oriented, 166
cobordism, 159
coboundary
  operator, 27
cochain group, 27
cocore, 162
coefficient
  group, 26
  universal theorem, 30
cohomology
  de Rham, 41–43
  deRham, 21
cohomotopy invariant, 265
compact, 20
compactness, 18
complete intersection, 202
complex structure, 11
connected sum, 203
connection, 64–67
  form, 52, 65
  singular, 285–286
  torsion free,metric, 52
coordinate patch, 36
coordinates
  affine, 201
  homogeneous, 201
core, 162
curvature

form, 53, 69
CW complex, 33

deformation retract, 31
Dehn
  surgery, 174
  coefficient, 174
  slope, 174
derivative
  covariant, 51, 62, 67–69
  exterior, 39, 42
diagram
  commutative, 30
diffeomorphic, 8
diffeomorphism, 36
differentiable structure, 5, 36
  standard, 37
differential p-form, 42
disk, 18
Donaldson polynomials, 133–135
  definition, 134
Donaldson theory
  counterexample of smooth
    h-cobordism theorem, 135
  Donaldson polynomials, 134
  equation, 127
  irreducible connection, 129, 131
  main theorem, 132
  moduli space, 129, 132
  reducible connection, 129, 130
Donaldson's theorem, 133

Eilenberg-Steenrod axioms for
  homology, 34
electromagnetic field, 71
eta invariant, 304
Euler characteristics, 31
Euler class
  definition, 111
excision, 30
exotic
  $\mathbb{R}^4$, 249
expectation value, 143
exponential map, 40

fiber, 55

fiber bundle
  bundle coordinates, 60
  reduction of bundle group, 61
fiber bundles
  associated, 61
fiber bundle, 59
  base space, 60
  bundle equivalent, 60
  bundle space, 60
  coordinate, 59
  fiber, 60
  section of, 60
  structure group, 60
  transition function, 60
form
  intersection, 193
  symplectic, 259
  unimodular, 194
frame
  reference, 36, 70
framing, 164–165
functor, 29
functorial property, 29
fundamental group, 22

gauge potential, 73
gauge theory, 120
gauge transformation, 73
Gauss-Bonnet theorem, 112
genus, 20
Grassmannian, 98
Grothendieck construction, 93

h-cobordant, 178
h-cobordism, 178
  theorem, 178
handle
  n-dimensional, 162
  addition, 188
  attaching
    map, 162
    region, 162
    sphere, 162
  belt sphere, 162
  canceling pair, 167
  cancellation, 187

cocore, 162
core, 162
creation, 187
index, 162
kinky, 184
moves, 187
slide, 167
subtraction, 188
handlebody, 163
  relative, 163
handlebody decomposition, 163
Hausdorff separable, 19
Higgs potential, 148
Hodge $*_g$-operator, 256
Hodge star, 47
homeomorphic, 20
homogeneous polynomial, 201
homology
  relative, 29
  simplicial, 25
  singular, 21
homology group
  definition, 27–29
homology three-spheres, 33
homotopic, 22
homotopy, 21, 22
homotopy sphere
  abelian group of, 217
  S-parallelizable, 218
homotopy-equivalent, 22
horizontal lift, 65
hypersurface, 201

Instanton, 127
instanton, 83
intersection form
  parity
    even, 237
intersection form, 194
  E_8, 199, 202, 237
  H, 195, 198, 199
  2-manifolds, 171–172
  4-manifolds, 193–197
  direct sum, 198
  parity, 197
    even, 197

odd, 197
positive (negative) definite, 198
rank, 197
signature, 197
unimodular, 197
intersection number
  algebraic, 179
  geometric, 179

join of topological spaces, 100

K-theory, 92–97
  algebraic, 302
  as generalized cohomology theory, 96
  as semi-group completion, 94
  Bott periodicity, 97
  definition, 94
  examples, 95
  of spheres, 95–96
K3-surface, 202
kinky handle, 184
Kirby Calculus, 187
Kirby diagram, 187

Lie algebra, 40, 62, 63
  graduate, 139
  super, 139
Lie bracket of vector fields, 39
Lie group, 40
lifting, 25
line bundle, 61
link
  band-sum, 188
logarithmic transformation, 287
Lorentz group, 70

manifold
  complex, 40
  complex algebraic, 201
  connected sum of, 203
  construction of, 200–203
  Grassmannian, 98
  piecewise-linear (PL), 16
  smooth (DIFF), 16, 36
  smooth with boundary, 36

topological, 5, 19
topological (TOP), 16
map
  attaching, 162
metric, 18
  tensor, 50
micro bundle, 223
  micro-isomorphism, 223
microtangent bundle, 223
model geometry, 176
moduli space, 124–126
  general definition, 124
  homotopy properties, 126
  of $U(1)$ connections, 125
monopole equation, 257
monopole map, 264
Morse
  theory, 154
Morse function, 154
Morse theory
  Witten's supersymmetric approach to, 142

null-cobordant, 160

obstruction, 59
obstruction theory, 17, 227
one-form, 39
operator
  annihilation/creation, 136
orbit, 39
oriented cobordant, 166

path integral, 143
Poincaré Conjecture, 173
  Generalized, 180, 217
Poincaré conjecture, 31
Pontrjagin class, 102, 106–107
  definition, 106
  differential form description, 111
Pontrjagin-Thom construction, 219
primitive element, 198
principal (fiber) bundle, 61
Product structure theorem, 225
projective
  line (complex), 201

Witten type, 144
Witten's conjecture, 147
topology, 2, 4
  algebraic, 20
  differential, 2
  product, 18
  quotient, 19

universal bundle, 97–101
  definition, 100
  Milnor construction, 100–101

vacuum expectation value, 143
vector bundle, 61
  bundle classification problem, 88
  bundle isomorphism, 87
  bundle map, 87
  characteristic classes, 101
  characteristic classes, 90, 114
  classifying space, 99
  definition, 87
  isomorphism classes, 88

K-theory of, 94
operation on, 92
pullback, 88
reduction of the bundle group, 91
stable, 93
stable equivalence, 93
transition function, 86–87
universal bundle of, 99

wedge product, 39
Weierstraß function, 10, 17
Weil homomorphism, 107–109
  definition, 108
Whitehead continua, 9
Whitehead torsion, 181–182
Whitney
  disk, 190
  trick, 179, 190

Yang-Mills theory, 120–123

Index

plane (complex), 201
space (complex), 200

rational surgery, 174
relativity
 general, 71
 special, 69

s-cobordism theorem, 180
Seiberg-Witten
 basic class, 258
 function, 258
 gauge group, 256
 invariant, 258
 main theorem, 257
Seiberg-Witten equations, 149
Seiberg-Witten invariant
 knot surgery, 261
 logarithmic transform, 260
 non-vanishing, 259
 vanishing, 258
Seiberg-Witten theory
 configuration space, 256
 equations, 149, 256
 moduli space, 256
sequence
 exact, 29
 Mayer-Vietoris, 30
simplex, 26
 singular, 26
singular support, 293
singularity
 double point, 183
 stable, 282–283
smoothness, 2
space
 classifying, 97–101
 contractible, 32
 covering space, 25
 Euclidean, 18
 horizontal, 64
 metric, 18
 simply connected, 25
 triangulable, 25
sphere, 18
sphere bundle, 207

Spin bundle, 117
Spin structure, 117
Spin structure
 pseudo-Riemannian case, 118
 existence, 117
$Spin_C$ connection, 255
$Spin_C$ group, 118–119
$Spin_C$ structure, 119
 construction, 255
 determinant line bundle, 254
 Dirac operator, 255
Spinor, 117
Stiefel-Whitney class, 102, 113–114
string algebra, 299–300
structure
 CAT, 224
 stable CAT, 225
super group, 139
supersymmetry, 139, 142
surgery, 161, 165–166
 Dehn, 174
 integer, 175
 rational, 174
 theory, 165

tangent vector, 38
tangent vector field, 38
Temperley-Lieb algebra, 296
theorem
 de Rham, 46
 Gauss-Bonnet, 112
 Hodge, 48
 Hurewicz, 31
 Poincaré duality, 48
 Stokes, 46
Thurston's Geometrization
 Conjecture, 177
topological charge
 topological quantum number, 127
Topological QFT, 143–146
 action, 145
 Donaldson-Witten theory, 144–146
 Observables, 146
 relation to Donaldson polynomials, 147
 Schwarz type, 144